ECOLOGY OF CENTER CITY, PHILADELPHIA

ECOLOGY OF CENTER CITY, PHILADELPHIA

KENNETH D. FRANK

Published by Fitler Square Press
2015

Published by Fitler Square Press,
2508 Pine Street, Philadelphia, PA 19103
fitlersquarepress@gmail.com

ISBN: 978-0-692-50648-6

Library of Congress Control Number: 2014919692

Designed by Meena Mehta | www.TwoMs.com

Contents

Spotlights

PREFACE

During the latter half of the eighteenth century and the beginning of the nineteenth, Philadelphia established itself as the country's center for the study of natural history. The city's contribution to the natural sciences has been beautifully presented in two recent books: *A Glorious Enterprise,* by Robert McCracken Peck and Patricia Tyson Stroud;[1] and *Knowing Nature: Art and Science in Philadelphia, 1740–1840,* by Amy R. W. Meyers.[2]

This scientific industry produced a trove of historical records of Philadelphia's flora and fauna. For some species, the accounts are the first recorded observations and took place within city limits. The records exist not only in the form of publications but also in specimens in museums, especially the Academy of Natural Sciences of Drexel University (formerly the Academy of Natural Sciences of Philadelphia).[3]

Coexisting with this scientific legacy is a brick-and-mortar heritage. The streetscape of residential neighborhoods in downtown Philadelphia looks today much as it did in the nineteenth century. Municipal landmarks cited over a century ago in accounts of plants and animals remain in place. Environmental change has been recorded in historical photographic collections.[4] Municipal records document the evolution of pollution of air and water in nineteenth-century industrial Philadelphia and the city's largely successful efforts at control.

This book examines the flora and fauna of the city's downtown district known as Center City. Each of the book's first twenty-seven chapters focuses on a different species, starting with earliest accounts I could find in the vicinity of Philadelphia. The book highlights additional species in one-page "spotlights." Most of the species are common here, but some are common just outside downtown, and others were once common but are now locally extinct. The chapters explore how they succeeded or failed to establish local populations. They look at pollution—light, sound, water, air, and thermal. A recurrent topic is the effect of prejudice, both positive and negative, on the fate of species downtown.

If this book has a unifying theme, it is the many ways people have shaped communities of plants and animals that inhabit downtown, and the ways these communities have defied human control and survived in spite of, or because of, dense urban development. The iconic landmarks included in many of my photographs convey this theme's immediacy. The ecology of Center City has been dynamic and resilient—qualities that I expect will endure.

One final reason for choosing downtown Philadelphia: My wife and I have lived here for almost four decades. What a pleasure it has been to observe natural history just beyond our front stoop. I hope this book will entice people living downtown in other cities to explore ecology close to home.

ACKNOWLEDGEMENTS

Many staff members and resources at the Academy of Natural Sciences of Drexel University contributed to this book:

David Hewitt, research associate in botany at the Academy, mentored me in urban botany and Center City's ecological history. Alfred Ernest Schuyler, curator emeritus in the Academy's Botany Department, provided a list of Center City trees classified according to whether they were present before European settlement. Alina Freire-Fierro, collection manager of the Academy's Botany Department, retrieved from the Academy's herbarium nineteenth-century specimens of mugwort collected on ballast dumps near the port of Philadelphia. Tatyana Livshultz, assistant professor in the Department of Biodiversity, Earth and Environmental Science of Drexel University, introduced me to her basic research on the evolutionary biology of milkweeds. Jason D. Weintraub, lepidopterist and entomological collection manager at the Academy, showed me the historic cynthia moths in the Academy's collections, and included me on an expedition searching for the cynthia moth in Philadelphia. He also collaborated with me in photographing the polyphemus moths in the historic Titian Ramsay Peale Butterfly and Moth Collection, which he curated. Greg Cowper, in the Entomology Department, retrieved for my examination the Academy's historic Chinese mantids, collected around the time of the species' initial North American introduction in Philadelphia. Jon K. Gelhaus, professor in the Department of Biodiversity, Earth and Environmental Science of Drexel University and curator in the Entomology Department at the Academy, called my attention to the colony of sand wasps (stink bug hunters) nesting in Logan Circle, across the street from the Academy. Richard J. Horwitz, professor in the Department of Biodiversity, Earth and Environmental Sciences of Drexel University and fisheries section leader in the Academy's Patrick Center for Environmental Research, identified the bryozoan I found, *Cristatella (Pectinatella) magnifica*. Nathan H. Rice, ornithology collection manager at the Academy, identified the northern parula I found dead on a sidewalk after a storm. Cathy Buckwalter, reference librarian at the Academy, and Jennifer Vess, the Academy's Brooke Dolan Archivist, retrieved from the library's archives Thomas Say's copy of Frederick Valentine Melsheimer's *A Catalogue of Insects of Pennsylvania*. Robert McCracken Peck, senior fellow of the Academy, provided helpful consultation about publication of this book.

People and resources in other institutions in Philadelphia made important contributions:

Joel T. Fry, curator at Bartram's Garden, called my attention to many of John Bartram's observations, especially his written description of the black and yellow mud dauber. Susan Glassman, director of The Wagner Free Institute of Science of Philadelphia, and Lynn Dorwaldt, librarian at the Wagner, made available for my inspection artifacts and documentation from the excavation of the Subway Tree, on display at the Institute. Joseph A. Perillo aquatic biologist for the Philadelphia Water Department, told me about his observations confirming establishment of northern snakeheads in the tidal Schuylkill River. A volunteer librarian at the McLean Library of the Pennsylvania Horticultural Society discovered in the library's collection and made available to me Thomas Meehan's nursery catalog from 1858. Joseph K. Sheldon, professor of biology at Messiah College, gave me a firsthand account of his discovery of the flatworm *Bipalium pennsylvanicum* near Philadelphia. Keith Russell, of the National Audubon Society, advised me on interpretation of his online postings of winter bird counts in Philadelphia. Brenda Malinics told me about her rehabilitation of local silver-haired bats, including the one I brought to the Wildlife Rehabilitation Clinic of the Schuylkill Valley Center for Environmental Education. Leo Sheng reported observations based on his fishing in the Schuylkill River; he reported the behavior and local distribution of brown bullheads and took the photograph that I included in this book. Bradley Maule, coeditor of *Hidden City Daily,* generously allowed me to use his photograph of a red-tailed hawk eating a squirrel in Fairmount Park. Martin F. Heyworth, physician at the Philadelphia Veterans Affairs Medical Center, identified beetles that we encountered in Center City. Many neighbors graciously endured my poking around their properties and taking photographs.

Susan Alix Williams in Rowe, Massachusetts, identified the tree mosses, *Orthotrichum pumilum* and *Syntrichia papillosa,* I found in Center City. Through email correspondence, Susan Munch, associate professor emerita of biology at Albright College, guided me toward final identification of the liverwort *Reboulia hemisphaerica* in Center City. Sara M. Lewis, professor of evolutionary and behavioral ecology at Tufts University, hypothesized in correspondence with me that light pollution might be the reason that *Photuris* fireflies are absent in Center City. In correspondence, Peter K. Ducey, professor of biological sciences, State University of New York at Cortland, counseled me on how to search for predatory terrestrial flatworms.

Joel Katz read the first manuscript for this book and provided encouragement and guidance; Joel Katz Design Associates produced the map of Center City. Heather Diacont Rinehart drew a diagnostic picture of *Bipalium pennsylvanicum* based on photos and descriptions in the species' original description. Edward S. Barnard mentored me during my completion of the manuscript and my preparing it for submission for publication. Diane Fredrick, my editor, has improved this book in countless ways.

Finally, my wife, Susan, has been my partner in every stage of my studies of natural history.

I have used captions to credit photographs taken by people other than me. Although many people helped me produce this book, I take responsibility for any errors.

Philadelphia fleabane (*Erigeron philadelphicus*) named after this city by Carl Linnæus in 1753.[1] It is native to North America and grows wild in Center City.

CENTER CITY, PHILADELPHIA

Chapter 1	The Subway Tree	Eighth and Locust Sts
Chapter 2	Red tailed Hawk's nest	Franklin Institute's north facade
Chapter 3	House sparrow	Independence Mall
Chapter 4	Bagworm	23rd and Delancey Sts
Chapter 5	Old ailanthus tree	Lemon Hill, Fairmount Park
Chapter 6	Woodland Cemetery	West Philadelphia
Chapter 7	Northern parula	23rd and Walnut Sts
Chapter 8	Polyphemus moth	Inside Academy of Natural Sciences
Chapter 9	Bridge Spiders	Schuylkill River Trail at Chestnut St
Chapter 10	John Bartram's House	Bartram's Garden
Chapter 11	European paperwasp	Fairmount Water Works
Chapter 12	Firefly habitat	Garden of the College of Physicians
Chapter 13	Site of Joseph Leidy's Home	Philadelphia Criminal Justice Center
Chapter 14	Cowbird egg	2400 block of Waverly St
Chapter 15	CSX Railroad Tracks	Parallel to Schuylkill Trail
Chapter 16	Pillbug habitat	Courtyard garden, 25th & Pine Sts
Chapter 17	Milkweed plant	22nd and Sansom Sts
Chapter 18	Purple-stemmed cliffbrake	Eastern State Penitentiary
Chapter 19	Mugwort	Broad and Bainbridge Sts
Chapter 20	Site of fresh water sponge	Fairmount dam
Chapter 21	Power plants	Christian St and Schuylkill Ave
Chapter 22	Redback salamander	Fitler Square
Chapter 23	Alley with liverwort	25th and Delancey Sts
Chapter 24	Silver-haired bat	2500 block of Pine St
Chapter 25	Canada Geese	Schuylkill River Trail by Spring Garden St
Chapter 26	Stinkbug hunter wasp	Logan Square
Chapter 27	Reading Viaduct	9th and Green Sts

INTRODUCTION

Center City, Philadelphia, along the Schuylkill River, 2007, with University City in the lower left quadrant of the photo. (Aerial photo credit: Bill Cobb at SkylineScenes.com)

Downtown Philadelphia is commonly called Center City. Interstate highways surround Center City on three sides, but destinations within Center City are accessible on foot. Within this urban core are residential streets lined with trees and row houses located a few blocks from bustling commercial districts with vehicular congestion and tall buildings.

Service industries such as health care, finance, education, and tourism have replaced heavy industry and manufacturing, which once dominated Philadelphia's economy. People are wealthier, more educated, and increasing in number in Center City compared to the rest of Philadelphia,[1] which ranks as the nation's fifth most populous city.[2]

For Philadelphia as a whole, the proportion of the population below the poverty level is 26 percent, which exceeds that in Boston, New York, Baltimore, and Washington, DC, and is twice that for Pennsylvania.[3] Philadelphia shares with all these cities a geographic location along the East Coast and a heritage of nineteenth-century neighborhoods.

Two rivers border Center City, the Schuylkill on the west, and the Delaware on the east. Center City has green space in the form of public squares, playgrounds, ball fields, community gardens, pocket parks, and a trail along the Schuylkill River, but these places are manicured; Center City has essentially no urban forest or naturalized areas. Bulkheads, highways, and development along both rivers have eliminated all traces of riparian habitat. On residential streets, most green space is private and hidden behind row houses, which commonly have rear courtyards and gardens. Vacant lots are rare.

Wild plants sprout in almost any sliver of soil, including cracks in pavement and strips between sidewalk and street. Street trees are abundant. Despite heat and desiccation typical of cities, Philadelphia's temperate climate is favorable to growth of plants. The city's annual precipitation of 104 cm is more than that of London (59 cm), Beijing (62 cm), Moscow (63 cm), Rome (65 cm), Chicago (84 cm), and Portland (101 cm), but less than that of Atlanta (120 cm), Seoul (134 cm), and Bangkok (140 cm).[4]

In a comparison of eight American cities, Philadelphia's flora most closely resembled that of New York and Washington, and to a lesser degree Boston; and least that of Detroit, Chicago, Minneapolis, and St. Louis. One third of the flora in these cities was classified as nonnative. The fraction of species of plants common to all eight cities was just over 10 percent for native species and just under 10 percent for nonnative.[5]

Even though it is the core of a large metropolitan area, Center City supports populations of wild plants and animals. Within its dense matrix of streets and buildings are fragments of habitat varied in composition and size. This book explores how these habitats and their wild inhabitants have fared over time.

1

THE SUBWAY TREE

(Bald cypress, *Taxodium distichum*)

A tree stump over 36,000 years old unearthed in downtown Philadelphia provided clues to the city's geologic past and future.

Figure 1.1 Fragment of the Subway Tree, unearthed in 1931 at 8th and Locust Streets. Carbon-14 dating puts its age at greater than 36,600 years. This specimen is on exhibit at the Wagner Free Institute of Science of Philadelphia. (Courtesy of the Wagner Free Institute of Science of Philadelphia. Specimen accession number 15868.)

In 1931 construction workers in Philadelphia discovered well-preserved tree stumps underground at a depth of 12 meters—3 meters below sea level. They were digging a subway tunnel at 8th and Locust Streets, one block west of the Tomb of the Unknown Soldier in Independence National Historical Park. The stumps were positioned upright in swamp sod buried in clay. One stump, referred to as "the Subway Tree," had a circumference of over 5 meters.[1]

Investigation of the Subway Tree

The consulting engineer of the project immediately notified the Academy of Natural Sciences, and a team of six scientists investigated. They identified the Subway Tree as bald cypress, *Taxodium distichum,* a southern wetland species whose current natural range begins almost two hundred kilometers to the south of Philadelphia and extends to Florida and the Gulf Coast. They examined the clay around the stump for diatoms; that they found none was evidence that the sediment that buried the swamp was not from the sea.[2]

Horace G. Richards, a member of the team, had earlier discovered marine fossils indicating that climate in this region before the last glaciation had been warmer than it is currently. He had also found fossil evidence that the sea level off the New Jersey coast had once been 91 meters (300 feet) lower than it is today. Richards hypothesized that the Subway Tree grew in a warm interglacial period before the last (Wisconsin) glaciation and that meltwater flowing down the Delaware River from distant glacial ice had inundated the swamp, buried it in sediment, and contributed to a rise in sea level. Richards acknowledged that against his hypothesis was the well-preserved state of the wood of the Subway Tree. In this location, preservation of wood older than the last ice age seemed implausible but possible.[3]

In 1960, almost three decades after Richards published his hypothesis, carbon-14 dating of the Subway Tree confirmed its antiquity—older than 36,600 years.[4] The Subway Tree was indeed older than Pennsylvania's last glacial ice, which existed north of Philadelphia 17,000 to 22,000 years ago.[5] Richards wrote, "After twenty-eight years, I have achieved vindication!"[6]

Pieces of the Subway Tree are on display at the Wagner Free Institute of Science of Philadelphia. Accompanying the wood is a chunk of clay from the excavation. Lynn Dorwaldt, librarian at the Wagner, told me that the Wagner accessioned these objects at the time of the excavation. The wood feels light in weight, not stony like petrified wood and not blackened like coal. Its grain is visible and smooth, with splintered ends in cross section. Nothing about the wood's general appearance gives a clue to its age, or to the glacial sediment that buried it.

Floods

Glacial meltwater had long been recognized as the source of extensive sedimentary deposits in the region of what is now Center City, Philadelphia. In two papers published in the early 1880s Henry Carvill Lewis, geologist at the Academy, described how receding glaciers produced these deposits:

During the melting of the great Northern Glacier, whose southern terminus crossed the river probably near Belvidere, the flooded Delaware, then a great torrent five or ten miles wide and at least 150 feet deeper than it is now, deposited at first gravels and afterwards, when quieter, clays; while floating ice carried down already rounded boulders and dropped them upon its bed....It thus appears that during the Glacial epoch the waters of the Schuylkill emptied into those of the Delaware at Falls of Schuylkill, the city proper being entirely submerged.[7]

The great glacier which covered the whole northeastern portion of our continent, and which, as a great sea of ice, flowed in a continuous stream across Labrador, the Laurentian highlands of Canada, the Adirondacks, the Catskills and the Alleghenies, was proved to have finally stopped within sixty miles of our city. At the extreme edge of the glacier it heaped up a terminal moraine, composed of rock fragments brought from more northern regions, which moraine was shown to stretch in a continuous line completely across our State.[8]

Figure 1.2 Rock fragments mark the surface of the terminal moraine of the Wisconsin glaciation at its southern limit in the Delaware River basin. Photographed near Belvidere, New Jersey, 100 kilometers north of Philadelphia in 1916. (U.S. Geological Survey photo # 871-awc00871, by W. C. Alden)

Philadelphia brick clay

A recent geologic map shows glacial sediments distributed as Lewis described them in Philadelphia, although some of these deposits preceded the last glaciation.[9] The glacial deposits left bountiful brick clay close to the surface. Called Philadelphia brick clay, it was still in place when Lewis explored Center City's geology:

THE PHILADELPHIA BRICK CLAY. The built-up portion of the city stands upon an extensive deposit of brick clay and gravel, sections of which are exposed in every cutting. The brick clay invariably overlies the gravel.[10]

Philadelphia brick clay supported the city's early building boom. During the eighteenth and nineteenth centuries, it was quarried in brickyards within the city, and Philadelphia produced more bricks than did any other city in America.[11] A map of the city in 1794 identifies fourteen brick kilns, including one in the area that currently is Rittenhouse Square.[12] By 1857, the city had fifty brickyards, each of which produced on average a million bricks per year and employed about thirty men and boys. By the end of the nineteenth century, Philadelphia's brickyards were producing more than 200 million bricks per year, mostly by hand.[13]

Philadelphia's sprawling inventory of eighteenth- and nineteenth-century red brick buildings showcases its geologic and industrial past. Once substrate for the region's plants and animals, Philadelphia brick clay became the principal raw material for kilometers of row houses. James Stoops, whose brickyard and kiln were between 9th and 10th and Race and Vine Streets, produced bricks for the construction of the Pennsylvania State House, now called Independence Hall.[14] He molded and fired them from Philadelphia brick clay, a legacy of glacial meltwater and climate warming.

Figure 1.3 Independence Hall, constructed with bricks made from clay deposited a few blocks away by glacial meltwater.

Rising rivers

Glaciers no longer threaten to inundate the Delaware Valley with torrents of meltwater, but melting ice threatens to raise sea levels and flood low-lying areas within Philadelphia.[15] Center City and neighborhoods east and south are surrounded on three sides by tidal water that rises and falls with lunar cycles and also with interglacial cycles.[16] Climate warming will continue to produce meltwater, raising the level of Philadelphia's tidal rivers, creeks and marshes, and submerging wetland trees,[17] reminiscent of what happened to the Subway Tree.

Ann Fowler Rhoads and Timothy A. Block, in *Trees of Pennsylvania,* describe a kind of tree peculiar to cities. It grows on urban riverbanks and floodplains. It is a backcross between American sycamores (*Platanus occidentalis*) and London plane trees (*Platanus × acerifolia*), which themselves are sycamore hybrids.[18] These trees have the bark of the stately London plane trees lining old streets downtown, but they grow in wet habitats typical of American sycamores. Today these backcrossed hybrids thrive along a narrow intertidal zone on the east bank of the Schuylkill River, just downstream from the Philadelphia Museum of Art. The rising sea is slowly submerging them.

Figure 1.4 Plane trees (*Platanus* hybrids) at high tide on the east bank of the Schuylkill River downstream from the Philadelphia Museum of Art, looking south toward Center City, October 27, 2012. Rising sea level is submerging the bases of these trees.

SPOTLIGHT

HUMAN BODY LOUSE

Human body louse (*Pediculosis humanus humanus*) that I collected from the clothing of a homeless man in a primary care clinic in Philadelphia.

In 1837 W.W. Gerhard, a physician at the Philadelphia Alms House Infirmary, described an epidemic of typhus in the city the year before. It afflicted 230 to 250 people, especially in a neighborhood bounded by Lombard and Shippen (Bainbridge) Streets, and Fifth and Eighth Streets. He wrote:

> The origin of the disease is unknown...It attacked those who were sunk in poverty and intemperance, and huddled together in confined apartments.[1]

The human body louse (*Pediculosis humanus humanus*) was later shown to transmit the pathogen that causes epidemic typhus (not to be confused with murine typhus, transmitted by fleas, as mentioned in Chapter 2). Unlike head lice and pubic lice, the human body louse resides in its host's clothing. Only while feeding does it move onto its host. Human body lice are closely related to head lice and likely evolved from them. In contrast to infestations of body lice, infestations of head lice in Center City are common, especially on children, and are not associated with transmission of disease.[2]

I diagnosed no case of typhus and encountered only one patient with human body lice in Philadelphia during my practice of primary care medicine, which spanned almost four decades. This patient changed his clothes only when they wore out, and he never washed them—even though he understood that his clothes harbored lice.

2

EASTERN GRAY SQUIRREL

(*Sciurus carolinensis*)

Eastern gray squirrels endure in Center City despite a history of persecution and new exposure to predators and poisons.

Figure 2.1 An eastern gray squirrel hesitates before leaving the safety of a garden fence for an offering of a peanut in Rittenhouse Square.

In 1748 the Swedish naturalist Peter Kalm came to Philadelphia to document the region's plants and animals. After naming eleven species of nut-bearing trees in Pennsylvania, he described how gray squirrels had shifted their food preference to corn, the cultivation of which had increased "infinitely." Squirrels devastated corn crops both in fields and in storage.[1]

Bounties for squirrels

Governments in Pennsylvania posted squirrel bounties of three pence per head. In 1749 they paid out £8,000—equivalent in bounties to 640,000 dead squirrels. Bounty hunting became so lucrative that young men abandoned employment to shoot squirrels. After payouts exhausted local treasuries, governments in Pennsylvania reduced the bounty by half. In other colonies, the squirrel bounty was two pence. In Maryland, mandates required every citizen to present to colonial officials four squirrel heads annually.[2]

Kalm described how adept the squirrels were at evading shooters:

> Though a grey squirrel does not seem to be very shy, yet is very difficult to kill; for when it perceives a man, it climbs upon a tree, and commonly chooses the highest about it. It then tries to hide itself behind the trunk, so that the shooter may not see it, and though he goes ever so fast around the tree, yet the squirrel changes its place as quickly, if not quicker; if two boughs bend towards each other, the squirrel lies in the middle of them, and presses itself so close that it is hardly visible. You may then shake the tree, throw sticks and stones to the place where it lies, or shoot at it, yet it will never stir. If three branches join, it takes refuge between them, and lies as close to them as possible, and then it is sufficiently safe. Sometimes it escapes on a tree where there are old nests of squirrels, or of large birds; it slips into such, and cannot be got out, either by shooting, throwing or any thing else; for the grey squirrels seldom leap from one tree to another, except when extreme danger compels them.[3]

Affection for squirrels

Despite persecution, squirrels were occasionally kept as pets. Kalm wrote:

> Of all the wild animals in this country, squirrels are some of the easiest to tame, especially when they are taken young for that purpose. I have seen them tamed so far that they would follow the boys into the woods, and run about everywhere, and when tired would sit on their shoulders. Sometimes they only ran a little way into the woods, and then returned home again to the little hole that had been fitted up for them. When they eat, they sit almost upright, hold their food between their fore feet and their tail bent upward. When the tame ones got more than they could eat at a time, they carried the remainder to their habitations, and hid it amongst the wool that they lay upon. Such tame squirrels showed no fear of strangers, and would suffer themselves to be touched by everybody, without offering to bite. They sometimes would leap upon strangers' clothes, and lie still on them in order to sleep. In the farmhouses, where they were kept, they played with cats and dogs.[4]

In the mid-eighteenth century squirrels were abundant outside the city, but within Philadelphia, wild squirrels disappeared, casualties of both hunting and deforestation. Etienne Benson at the University of Pennsylvania recently reconstructed the historic ebb and flow of populations of squirrels in downtown Philadelphia in the nineteenth century. He found that beginning in the late 1840s, the city introduced squirrels into public squares for the amusement of visitors. In 1864 the Committee on Entomology of the Pennsylvania Horticultural Society blamed squirrels for adverse effects on populations of birds and insect pests; in response, the city captured or killed the squirrels and removed squirrel nest boxes from the squares. Starting in the 1870s, landscaped urban parks and renewed public support for squirrels improved conditions for squirrels, which dispersed widely in big East Coast cities, including Philadelphia.[5]

Safe haven from predators

The squares in the city gave squirrels safe haven from large raptors such as red-tailed hawks. In the early nineteenth century Alexander Wilson, whose interest in ornithology began in Philadelphia, had trouble getting close enough to red-tailed hawks to describe them. In his *American Ornithology,* he begins his account of them with a disclaimer:

> Birds naturally thinly dispersed over a vast extent of country; retiring during summer to the depth of the forests to breed; approaching the habitations of man, like other thieves and plunderers, with shy and cautious jealousy; seldom permitting a near advance; subject to great changes of plumage; and, since the decline of falconry, seldom or never domesticated—offer to those who wish eagerly to investigate their history, and to delineate their particular character and manners, great and insurmountable difficulties.[6]

In 1885, the legislature of the Commonwealth of Pennsylvania passed the "Scalp Act" establishing a fifty-cent bounty for a slain hawk or owl.[7] In the following two years, counties in Pennsylvania paid $90,000 in bounties for killing raptors.[8] Near the end of the nineteenth century, red-tailed hawks were still elusive, as reported by Benjamin Harry Warren, ornithologist for the Pennsylvania State Board of Agriculture:

> This hawk—the most abundant of our raptorial birds—is the detested "Hen Hawk" of the farmer. The Red-tailed Hawk is exceedingly shy and wary, and is taken with difficulty, unless approached on horseback or in a sleigh or wagon.[9]

In 1944, John A. Gillespie, a local birder, published an account of the birds of Rittenhouse Square, based on sixteen years of observation. The number of species he and his friends observed totaled ninety-four, including five species of raptors, but no red-tailed hawks.[10] In 1975 numbers of red-tailed hawks nesting in suburban Philadelphia were declining.[11]

In the second half of the twentieth century, a series of events coalesced to benefit red-tailed hawks. These included publication of Rachel Carson's *Silent Spring;*[12] banning of DDT;[13] passage of protective legislation and implementation of enforcement;[14] development of Hawk Mountain Sanctuary in Berks County, Pennsylvania;[15] and promotion of recreational birding.[16]

End of the safe haven

In the last four decades, numbers of red-tailed hawks in Pennsylvania have quadrupled, according to the Breeding Bird Survey.[17] Parks and campuses in the Philadelphia metropolitan area, including Rittenhouse Square, have become their hunting grounds. In a news story with the title "City's New Pastime: Talon Shows," Inga Saffron, writer for the *Philadelphia Inquirer,* described the transformation:

So many YouTube videos document hawk kills in the city that they practically constitute a genre. Besides recording the mayhem on Market Street, humans have filmed hawks in mid-bite in Rittenhouse Square, on the University of Pennsylvania campus, in the Philadelphia Museum of Art's sculpture garden, and in the yards of Bella Vista row houses. One local bystander narrowly missed becoming collateral damage when a large redtail dived for a squirrel outside the museum. The squirrel got away.[18]

Photos accompanying her story show a crowd of bystanders photographing a young red-tailed hawk devouring a pigeon on the roof of a car parked at 8th and Market Streets.

In 1998, Marie Winn's *Red-Tails in Love* described red-tailed hawks returning over a succession of years to a nest on the façade of a building on Fifth Avenue across the street from Central Park, Manhattan.[19] In 2009 a pair of red-tailed hawks began nesting on a window ledge of the Franklin Institute, overlooking the Benjamin Franklin Parkway.[20] The Institute constructed a supporting structure for the nest and installed a video camera linked to the Internet for live monitoring on the Web. In 2012, it began providing dead rats for the mother of newly hatched chicks after her mate died in a collision with a truck on Interstate 76 outside 30th Street Station. Within days after his death, another male appeared, bonded with the female, and helped raise her chicks.[21] As of the fall of 2012, red-tailed hawks in this nest have raised nine chicks.

Figure 2.2 A pair of red-tailed hawks tends their brood on the Franklin Institute, May 26, 2012. The adult male has replaced the biological father, who died in a collision with a truck on the Schuylkill Expressway.

Figure 2.3 Fledgling red-tailed hawk stretches its wings.

Figure 2.4 Red-tailed hawk at Memorial Hall, Fairmount Park, Philadelphia. (Photo by Bradley Maule, Phillyskyline.com)

Resilience of populations of gray squirrels

Unlike urban deer and geese, urban red-tailed hawks continue to fit Alexander Wilson's description as "thinly dispersed over a vast extent of territory." Even though scarce compared to other urban birds, red-tailed hawks have the potential to deplete localized, vulnerable populations of prey. In one instance, red-tailed hawks reduced an adult ground squirrel population by over 90 percent.[22] Fear of hawks, independent of actual predation, has suppressed reproduction in sparrows.[23]

Populations of gray squirrels have proved resilient despite losses. Hunting killed 38 percent of gray squirrels in Virginia woodlots, but had no measurable impact on mortality rates.[24] In Ireland, where gray squirrels introduced from North America have proliferated, intensive trapping reduced populations of gray squirrels only temporarily; within ten weeks, young squirrels immigrating from neighboring areas restored these populations.[25]

The reproductive life of a wild female gray squirrel may last as long as 12½ years.[26] Females have produced two litters annually with around three offspring per litter.[27] High mortality offsets this high reproductive potential. In a North Carolina woodland, 75 percent of squirrels died in their first year, and mean life expectancy at birth was only one year.[28] Evidently the gray squirrel's mobility and reproductive potential can maintain populations despite high mortality, including that from red-tailed hawks.

Parks in downtown Philadelphia no longer endow squirrels with safe havens from red-tailed hawks, but they do provide them with refuge from hunters and other predators, including foxes, coyotes, bobcats, weasels, owls, snakes, and other raptors.[29] I have seen feral cats stalking gray squirrels in our backyard, but not in public squares.

In the past decade in Rittenhouse Square, traffic of people and their dogs has increased, preempting space on the ground where squirrels forage and bury nuts. In Independence National Historic Park, the crowds occupying squirrels' home ground are even bigger.

Figure 2.5 Rittenhouse Square, Sunday, April 15, 2012. In recent years, crowds have increased, preempting territory where squirrels forage.

The gray squirrel's natural rhythm of activity separates it from these crowds. In summer, its peak activity occurs shortly after sunrise and before sunset, circumventing the times when crowds peak. In winter, when crowds are small, its activity is distributed evenly during daylight hours.[30]

Figure 2.6 Gray squirrel behind Independence Hall. Squirrels forage early in the morning, before the arrival of crowds of tourists.

Threats from Norway rats

Squirrels in Rittenhouse Square must contend not only with people, dogs, and hawks, but also with Norway rats (*Rattus norvegicus*), which are conspicuous near park benches in the evening. In 1831 John Davidson Godman, professor of natural history at the Franklin Institute in Philadelphia, described their depredations:

> The common, brown, or Norway rat, now so extensively diffused over this country, is not indigenous to our soil, but introduced from Europe, which received it from Asia in the eighteenth century, as late as the year 1750. There are few parts of the world now visited by navigators where this animal has not been introduced, and the immediate consequence of its introduction has been that all the native rats have been destroyed, or obliged to withdraw beyond the reach of this subtle and implacable enemy...It was brought to this country in European ships, and has been gradually propagated from seaports over the greater part of the continent.
>
> He is one of the most impudent, troublesome, mischievous, wicked wretches that ever infested the habitations of man. To the most wily cunning he adds a fierceness and malignancy of disposition that frequently renders him a dangerous enemy, and a destroyer of every living creature he can master. He is a pure thief, stealing not merely articles of food, for which his hunger would be sufficient justification, but substances which can be of no possible utility to him.
>
> The brown rat takes up its residence about wharves, storehouses, cellars, granaries &c. and destroys the common black rat and mouse, or entirely expels them from the vicinities it frequents. To chickens, rabbits, young pigeons, ducks and various other domestic animals, it is equally destructive when urged by hunger and opportunity. Eggs are also a very favorite article of food with this species, and are sought with great avidity; in fact, everything that

> is edible falls prey to their voracity, and can scarcely be secured from their persevering and audacious inroads.
>
> When attacked and not allowed an opportunity of escaping, he becomes a dangerous antagonist, leaping at his enemy and inflicting severe and dangerous wounds with his teeth. The most eager cat becomes immediately intimidated in the presence of one of these rats thus penned up, and is very willing to escape the dangers of an encounter.
>
> The cunning of these rats is not less than their impudence; it is almost impossible to take them in traps after one or two have been thus caught, as the rest appear perfectly to understand the object of the machine, and afterwards avoid it with scrupulous care, however tempting may be the bait it contains. The surest way to remove them is by poison, which, however, they frequently detect and avoid.[31]

Gray squirrels distance themselves from Norway rats. Gray squirrels nest in trees, whereas Norway rats nest in holes in the ground. Gray squirrels forage on the ground during the day, whereas Norway rats forage mostly at night.

Although they avoid contact with rats, they are vulnerable to infestations of rats depleting their food supply. Norway rats are omnivorous and have been reported to consume even acorns.[32] Both species scavenge leftovers from people eating in the square. Competition for food may take a toll on gray squirrels. In a public park on the campus of the University of Kansas, scarcity of food in the spring contributed to deaths of young gray squirrels.[33]

Danger from rat poison

Gray squirrels are vulnerable to poisons used to control rats. In New York, postmortem examinations showed that rat poisons killed many kinds of wild animals, including red-tailed hawks and gray squirrels.[34] In April 2004 rat poison was blamed for the disappearance of squirrels in Rittenhouse Square. News reports attributed deaths of squirrels here to bromethalin, a neurotoxin the city's Vector Control team used against rats in Rittenhouse Square.[35]

Danger of rats to people in Philadelphia

The City of Philadelphia's Department of Health has long struggled with rats and their control. In 1891 the city's coroner listed rat poison (arsenic) sold under the name "Rough on Rats" as the most common poison used in suicides; eight cases were reported that year.[36] In June of 1912, epidemics of bubonic plague broke out in Cuba and Puerto Rico, threatening port cities such as Philadelphia. Rats are a reservoir for plague bacteria (*Yersinia pestis*). That year the city's Bureau of Health, offering bounties for rats dead or alive, examined 2,510 rats and found no evidence of plague.[37] In 1932 another survey in the city examined rats,[38] this time for a particular species of rat flea (*Xenopsylla cheopis*) that had been shown just the year before to transmit the pathogen responsible for a distinctive type of typhus, today called endemic, or murine, typhus.[39] Sixty percent of 4,629 fleas taken from 2,765 rats in the survey turned out to be this species.[40] This discovery occurred in the wake of past epidemics of typhus in Philadelphia.[41]

In 1967 a tugboat engineer who fell into the Schuylkill River was hospitalized with leptospirosis,[42] a potentially fatal disease caused by *Leptospira icterohaemorrhagiae,* a bacteria found in 11 percent of rats (*R. norvegicus*) sampled in Philadelphia; the rats' urine contained the pathogen.[43] From 1974 to 1996 the city received reports of over 600 rat bites, primarily involving children under age five bitten between midnight and 8 a.m. in poor neighborhoods.[44] The list of human pathogens potentially transmitted from rats to people is long.[45]

Figure 2.7 Philadelphia's Bureau of Health Rat Receiving Station at a wharf along the Delaware River near Pine Street, 1914. On the right is a baffle designed to prevent rats from using ropes to crawl from ship to shore. The city offered bounties for rats, dead or alive, that it received here. (Courtesy City of Philadelphia photo archives)

Norway rats breed all year long, and on average a female produces more than thirty-five offspring a year. In one month a population of Norway rats can increase in size by 50 percent, making up for losses due to predation or poison.[46] Red-tailed hawks are known to prey on Norway rats,[47] but in Rittenhouse Square they have not prevented outbreaks.

Barriers to rat control

Even if a predator or a poison eliminated every rat in the square, rats from surrounding areas would soon recolonize it. Philadelphia's nineteenth-century sewer system harbors rats. Instead of separate systems for storm and sewer drainage, one system serves both.[48] Infestation of rats in sewers is positively correlated with sewers' concentration of suspended solids[49] and with sewers' age; most occur in sewers over thirty years old.[50] Drain grates by the curb give sewer rats access to the street.

After poison kills rats, their numbers quickly rebound to levels set by availability of food and nesting sites.[51] The same principle applies to rat control by other methods, such as contraceptives, trapping, and fumigation of burrows.[52] In Rittenhouse Square, sustained reduction in populations of rats through poisoning requires ongoing application of poison. This is the strategy of the city's Vector Control unit. In theory, the ideal strategy would be to effect complete removal of the food that people who use the park leave behind, but twice-daily removal of trash in Rittenhouse Square has not eliminated rats.

Protection of squirrels from rat poison

After the accidental poisoning of squirrels in 2004, populations of squirrels in Rittenhouse Square recovered. Workers from the city's Vector Control unit locate the rats' holes, pour poison down the holes, and then cover up the holes. They monitor the holes, and if a hole they have covered opens, they repeat the process.[53] The poisoning of rats theoretically benefits squirrels by reducing competition. In 2004, the city's Vector Control unit reported that its rat poison in Rittenhouse Square reduced the number of rat burrows from fifty to six.[54]

How populations of squirrels in Rittenhouse Square endure

1. Adaptation

The population of gray squirrels in Rittenhouse Square survives in part because it is well adapted. It nests in trees, safe from people and dogs. It supplements its diet of acorns and nuts with handouts and leftovers. Its daily rhythm keeps it away from rats and crowds. It habituates to the bustle of the city. It tolerates infection by West Nile virus[55] and squirrel pox virus.[56] It does not defend territory;[57] city parks have supported population densities as high as fifty individuals per hectare.[58] Its reproductive power can buffer its population from losses, such as from red-tailed hawks and rat poison.

2. Luck

The gray squirrel survives here also because of good fortune. Rittenhouse Square is endowed with an abundance of nut-bearing trees, including oaks (*Quercus*) and horse chestnuts (*Aesculus sp.*). It has no understory brush to support ticks and chiggers, which infest squirrels in woodlands.[59] Unlike rats, the gray squirrel does not endanger public health. Exterminators spare it. Rittenhouse Square offers squirrels no downspouts to clog, attics to invade, birdfeeders to rob, or crops to ravage. In this setting, people and squirrels can coexist with impunity.

3. Charisma

Neither luck nor adaptation alone is sufficient to explain the success of gray squirrels in populating downtown. In contrast to Norway rats, gray squirrels have charisma. The affection that people reserve for squirrels dates back to our earliest records, when farmers in Pennsylvania kept them as pets despite the existence of bounties for killing them. People take pleasure in the anthropomorphic way gray squirrels sit upright, holding nuts between their two front paws. They enjoy their antics and the look of their white chests. On the other hand, appreciation of squirrels is far from universal. In Philadelphia, charisma has brought less attention to squirrels than to red-tailed hawks, which attract paparazzi and webcams.

4. Personality

The gray squirrel may misjudge its welcome and crawl too close to a visitor in Rittenhouse Square, but it can sense hostility and withdraw. It is skilled at calibrating an optimal stance, be it hiding behind a tree or soliciting a handout. Its treatment of people as patrons or predators matches people's treatment of it. Downtown, its way with people may be the gray squirrel's greatest strength.

Recently my wife and I observed squirrels in Rittenhouse Square, where feeding animals is forbidden but tolerated if the offering consists of only a single peanut. As we approached squirrels, they ignored us or scampered off. Extending her hand holding a peanut, my wife caught the attention of one squirrel foraging beneath shrubbery behind a low garden fence. Through the bars of the fence it eyed the nut, but did not budge. She backed off, holding the nut toward it, but the squirrel stayed behind the bars. She then placed the nut on the sidewalk, but kept a finger on the nut; the squirrel crept toward the nut, but stopped a meter shy. Finally she stepped back and the squirrel inched forward. It hesitated, reversed course, then continued, paused again, and eventually crawled just within reach of the nut. With all four feet on the pavement, it craned its neck forward, grabbed the nut in its teeth, and scampered off.

Figure 2.8 Squirrel watching until the hand by the peanut is withdrawn.

Squirrels in Rittenhouse Square are part of a community of people, dogs, rats, hawks, pigeons, and sparrows. The community endures even though relationships among some members are antagonistic. Historically, the relationship of the gray squirrel to people has been ambiguous, a mixture of hostility and affection, but always, at least for the gray squirrel, fraught with danger.

Gray squirrels in Rittenhouse Square have recently declined in number. During the day when I strolled through the square I used to see them consistently; now I see them only rarely.

3

HOUSE SPARROW

(English sparrow; *Passer domesticus*)

House sparrows, introduced into Philadelphia in 1869 to control insect pests, are declining in numbers.

Figure 3.1 House sparrow (*Passer domesticus*) attacking winged insect in Independence Mall, December 13, 2012. At this time of year, flying insects are rare.

Figure 3.2 Male house sparrow, showing winter plumage and insect prey in beak. It is the same individual as in figure 3.1.

In his report to City Councils in 1862 about infestations of insects in shade trees, Joseph Leidy noted that Philadelphia failed to attract insect-eating birds. He advised introducing turkeys, guinea fowl, and chickens into the public squares, and diverting water from fire hydrants to pools to attract wild insect-eating species. He recommended that the city avoid planting silver maples, and instead plant pest-resistant species, such as ailanthus; and that stiff brushes be used to sweep insects off trunks and larger branches.[1]

Introduction of house sparrows into Philadelphia

Over the next five years, infestations of insects in Philadelphia increased, as did petitions demanding that City Councils import English sparrows from Europe to consume them. (The term "house sparrow" has replaced "English sparrow," a former common name for *Passer domesticus,* a species whose ancestry is broadly distributed in Europe, Asia, and North Africa.[3]) Since 1851 these birds had been introduced for pest control in cities in New York, Maine, Rhode Island, Massachusetts, Connecticut, and Texas.[4] In 1868 John W. Bardsley of Germantown decided to take matters into his own hands and set sail for England, where he planned to collect house sparrows for introduction into Philadelphia. While he was in England, City Councils in Philadelphia officially designated him as its authorized agent for importing the birds. He brought back more than a thousand and surrendered them to city authorities, who released them in 1869.[5] On May 18, 1869, *The Evening Telegraph* reported that Philadelphia's mayor, Daniel M. Fox, signed an ordinance appropriating "the sum of one hundred ($100) dollars…to pay John W. Bardsley for services rendered in procuring sparrows lately imported by this City."[6]

Figure 3.3 Plate VII from Thaddeus William Harris's *Treatise on Some of the Insects Injurious to Vegetation,* published in 1862.[2] The colorful caterpillar on the upper left is a pest that Joseph Leidy reported to City Councils in Philadelphia in 1862. It is the larva of the white-marked tussock moth (*Orgyia leucostigma*), a male of which is illustrated just below it. Above it are two wingless females of this species. The other insects are all moths in various stages of development, and all were regarded as pests.

Exponential proliferation

Twenty years later, populations of house sparrows were multiplying so fast that they prompted a federal investigation that produced a 440-page report, *The English Sparrow* (Passer domesticus) *in North America: Especially in Its Relations to Agriculture.*[7] The author, William Bradford Barrows, concluded that populations of house sparrows had increased geometrically and now covered over a million square miles of North America. Barrows's report did not put to rest a bitter controversy over whether house sparrows were on balance helpful or harmful.

The sparrow wars

In Philadelphia, the chief protagonist in the "sparrow wars" was Thomas Gentry, who deplored introduction of the house sparrow.[8] His main adversary was Boston's Thomas M. Brewer, who defended the species.[9] Allied with Gentry against Brewer was Elliot Coues,[10] an accomplished ornithologist based in Washington, DC.

Gentry complained that Philadelphians, by feeding sparrows, spoiled them as agents of biological control:

> Charities poured in upon them from every source, and the gullible Philadelphian soon commenced to lavish more than usual attention upon these creatures of foreign extraction. The birds often fared much better than their poor human brethren. These fancied "saviors of vegetation" finally became well housed and well fed. Their good qualities were loudly applauded, and the law was constrained to throw around them its ægis of protection.
>
> But a change soon came over the aspect of affairs. Too much pampering had engendered a spirit of laziness. Accustomed to an easy life, the birds assembled three times a day to receive their allowances of food. The results of such folly soon began to be apparent. The squares became alive with caterpillars. The rusty vaporer crawled everywhere. Sparrows were never more plentiful. They abandoned their carnivorous propensities, in a great measure, and took to vegetable diet with a cheerful chirp.[11]

Gentry also held Bostonians culpable:

> The sparrow is rapidly exterminating the native songsters and insect-eating birds from our cities and large towns…It was only the other day that the shrikes *(Collurio borealis)* made their appearance upon Boston Common and began to decimate the ranks of the sparrows a little, when a crusade was instituted against them, by some person or persons who had the affair at his or their whimsical command. This was undoubtedly the first indication of a natural healthy reaction against the sparrows which has occurred, but it was most fatuitously nipped in the bud.[12]

Gentry condemned the house sparrow on grounds that were economic, ecological, aesthetic, moral, racist, and chauvinistic. The sparrow wars spread to the popular press, captured a wide audience, and engaged the American people in the first great national conversation on biological control. The controversy itself is controversial. One historian has argued that anti-immigration sentiment drove it.[13] Another has contended that anti-immigration sentiment had nothing to do with it; he views it as a scientific milestone: one of the first major debates among professional scientists in America, and one of this country's earliest ecological battles.[14] Sociologists have interpreted it as a metaphor for diverse social concerns of the day.[15]

Figure 3.4 Engraving published in 1889. The caption is: "OUR IMPORTED PROTECTORS, MUTUAL DISGUST. English Sparrow to Irish Guardian of American Peace—'Do your own nahasty work, sir: W'english sparrows, sir, didn't come 'ere to eat hup your nahasty H'american worms."[16]

Methods to control population explosion

Barrows's report, published by the U.S. Department of Agriculture, came down decisively on the side of Gentry, despite Gentry's lack of standing as an ornithologist.[17] It concluded that "the English sparrow is a curse of such virulence that it ought to be systematically attacked and destroyed before it becomes necessary to deplete the public treasury for that purpose."[18] It found the species to be harmful to agriculture, horticulture, and native birds. The report adopted Gentry's proposal to repeal laws protecting the English sparrow. It recommended enactment of laws legalizing the killing of the English sparrow and the destruction of its nests, eggs, and young, and making it a misdemeanor to give the English sparrow food and shelter. It called for enactment of laws protecting its predators: the great northern shrike, sparrow hawk, and screech owl. Finally, it proposed that every town and village appoint an official whose duty it would be to "bring about the destruction of English sparrows in the streets, parks and other places where the use of fire-arms is not permitted."[19]

Decline in abundance

At the beginning of the twentieth century, house sparrow populations declined—but not because of Barrows's recommendations. When automobiles replaced horses, sparrows in cities lost an abundant supply of food in the form of spillage of oats in horse feed and undigested seeds in horse droppings.[20] The decline in sparrow populations at the beginning of the twentieth century plateaued until about fifty years ago. Since 1966 numbers of house sparrows have dropped by 85 percent in the United States and by 62 percent in Pennsylvania.[21] In Canada they have similarly declined, the species becoming rare to absent in much of the Maritime Provinces.[22] Similar trends have occurred in Western Europe, particularly in cities, where in some cases the species has disappeared.[23] In India, declines have prompted calls for protection.[24]

Recent declines in populations of house sparrows have been attributed to many causes, none of which alone is sufficient to account for geographic differences in rates of decline. Purported reasons for the decline in cities include predation and fear of predators, particularly cats and raptors; shortages of food, including seeds and insects; competition, such as from house finches; and loss of nesting sites, especially eaves of roofs. Other putative causes include herbicides, pesticides, pollution, pathogens, parasites, vehicular traffic, and even exposure to microwaves and radio waves. Evidence in all cases is inconclusive. Decline in populations of house sparrows is probably multifactorial.[25]

House sparrows today are plentiful in Center City, but their numbers may be dropping, given recent declines in Pennsylvania and the United States. A decline in abundance of these birds downtown may not become obvious until their populations drop to levels that make the birds scarce.

Figure 3.5 Flock of house sparrows eating bread in Independence Mall. House sparrows may appear plentiful even as their numbers plummet.

Figure 3.6 European starling (*Sturnus vulgaris*), iridescent in direct sunlight in Independence Mall. Like the house sparrow, it is a common urban species that is declining in Pennsylvania and also generally in North America, including Canada, and in Europe.

The decline in populations of urban birds in Pennsylvania includes the house sparrow but also the European starling (*Sturnus vulgaris*), an introduced species whose omnivorous diet and disturbed habitats resemble those of the house sparrow.[26] The decline of both species in Europe[27] and North America,[28] including Canada,[29] suggests a common cause.

Decline in supply of wild seeds

The house sparrow's dietary supply of seeds from wild herbaceous plants in downtown Philadelphia is thin. Gentry reported that house sparrows in winter eat seeds of ragweed (*Ambrosia*), goosefoot (*Chenopodium*), pigweed (*Amaranthus*), dock (*Rumex*), goldenrod (*Solidago*), and asters.[30] Ragweed is now absent from Center City except for rare plants near railroad tracks along the Schuylkill River. Goldenrod is rare or absent, and, outside of gardens, asters are absent except for heath aster (*Symphyotrichum pilosum*). Small patches of goosefoot, pigweed, and dock are scattered about, but vacant lots filled with wild plants have practically disappeared.

Homeowners and maintenance crews apply herbicides to vegetation in pavement cracks, the last refuge for wild seed-bearing plants in commercial and residential areas, outside of gardens. Green space in Center City, such as parks, is manicured. Neighborhoods with the most wild flora (i.e., "weeds") lie outside prosperous districts downtown.

Pollen counts provide a quantitative measure of changes in the regional abundance of weeds. The best published data on long-term pollen counts for this region cover the northern New Jersey–New York City metropolitan area from 1993 to 2002. They show total pollen counts decreasing by over half, particularly for herbaceous weeds, including ragweed, goosefoot, pigweed, and dock.[31]

Figure 3.7 The old South Street Bridge, facing Center City, August 2007. Wild vegetation here disappeared when the bridge was torn down and replaced.

Declines in numbers of insects downtown

House sparrows are omnivorous, but must eat at least some insects to realize their full reproductive potential.[32] They eat insects in all common orders, plus spiders and earthworms.[33] The scarcity of weeds in Center City deprives insects of food and habitat. Grasshoppers common only a decade ago are now uncommon or absent because the patches of wild plants that supported them are gone.

Figure 3.8 Pearl crescent (*Phyciodes tharos*) on ornamental ironwork along the sidewalk of the old South Street Bridge, October 2007. (The ironwork is also shown in figure 3.7.) Heath aster (*Symphyotrichum pilosum*), one of its larval food plants, grew in cracks on the bridge. The insect, common in Center City a decade ago, is now rare here.

Figure 3.9 Male herringbone grasshopper (*Melanoplus differentialis*) off Martin Luther King Drive in Fairmount Park, Philadelphia. A decade ago in Center City it was common in a field of grasses and forbs extending the length of Center City along the east bank of the Schuylkill River. Construction of a recreational park with a paved path eliminated this habitat.

To control injurious insects, Joseph Leidy's report to City Councils recommended that the city of Philadelphia plant insect-resistant species such as ailanthus, an exotic tree. Douglas Tallamy, chairman of the department of entomology at the University of Delaware, has concluded that introduction of exotic ornamentals harms populations of birds that depend on insects for food.[34]

Exotic plants

In Philadelphia, the number of exotic species of plants that have naturalized is 627—more than in any other county in Pennsylvania,[35] even though Philadelphia is geographically the second smallest county in the state. This number is over half the total for all of Pennsylvania.[36] Alfred Ernest Schuyler, botanist at the Academy of Natural Sciences, has classified Center City trees according to whether the trees were present before European settlement. Of the 130 species on his list, only 43 met this definition of native.[37]

Ginkgo biloba, native to China, exemplifies a common exotic street tree in Center City. Around 1784 William Hamilton imported this species from England to his Philadelphia estate at Woodlands—the first introduction of ginkgo into North America.[38] Ginkgo is entomologically unusual, in that no species of insect specializes in eating it—even in China. Although insects that are generalized consumers of plants occasionally eat it, the species is remarkably pest-free.[39]

Figure 3.10 Eleven ginkgo trees line both sides of the 2200 block of Delancey Street. No insects specialize in eating ginkgoes.

In Center City, about a mile away from Woodlands, eleven stately ginkgoes line the 2200 block of Delancey Street. By contrast, a mix of tree species native to Pennsylvania grows in Fitler Square half a block away. On an evening in late August, I strolled down Delancey Street toward Fitler Square and listened for the songs of tree crickets (*Oecanthus* sp.). I heard no crickets singing on Delancey Street, but in Fitler Square their chorus filled the air.

Figure 3.11 Fitler Square. Native trees include sugar maple, northern red oak, eastern redbud, American elm, and flowering dogwood. On late summer nights, tree crickets sing here.

Figure 3.12 Four-spotted tree cricket (*Oecanthus quadripunctatus*) on screen of author's house on Pine Street, two blocks from Fitler Square.

Despite the rarity of insects on ginkgoes, exotic plants do support populations of insects in cities. Arthur M. Shapiro at the University of California, Davis, found that native butterflies in local urban-suburban gardens bred mostly on alien plants, especially naturalized weeds. Almost half of these native butterflies had no known native host plants in the vicinity.[40] Others have found that adding native plants to community gardens in New York City did not increase diversity of butterflies, bees, and wasps.[41] From the perspective of house sparrows in Center City, the concern about exotic plants is less important than the pervasive loss of wild herbaceous vegetation ("weeds"), both native and exotic.

A century and a half ago, City Councils approved the importation of house sparrows to control infestations of caterpillars defoliating municipal shade trees. Such outbreaks are now rare, and when they do occur, they are typically self-limiting. Center City and its ecosystems have aged. Enemies of insect pests have had time to move into the city and establish populations sufficient to suppress such plagues, as discussed in the next chapter.

The controversy over whether house sparrows are helpful or harmful has lost relevance. The increasing scarcity of house sparrows has reduced the competitive pressure they exert on other birds. The ecological significance of house sparrows has shifted from the birds themselves to the environmental changes responsible for depleting their numbers.

SPOTLIGHT

BUCKEYE

Buckeye (*Junonia coenia*) sunning itself near Martin Luther King Drive, Philadelphia.

Buckeyes in the southern United States are resident all year around, while those in the north are migratory, recolonizing habitat every summer. Half a century ago populations in the Delaware Valley were found to be both residential and migratory, the proportions varying from year to year depending on temperature and location. In theory, global warming and Philadelphia's heat island could increase the proportion that overwinter here. In Center City host plants for buckeye caterpillars are common; they are English plantain (*Plantago lanceolata*) and blackseed plantain (*Plantago rugelii*).[1] Buckeyes are rare in Center City but occasionally appear during the fall migration.

4

BAG WORM

(*Thyridopteryx ephemeraeformis*)

Infestations of bagworms ravaged shade trees in nineteenth-century Philadelphia.

Figure 4.1 Bagworm bag in winter. It is on a recently planted street tree. It might contain bagworm eggs or parasites, or it might be empty. In September, male bagworm moths emerge from the bottom of their bags and fly to females, which mature, mate, and lay eggs inside their bags. In the spring the eggs hatch and larvae crawl out the bottom and make new bags, in which they develop.

Joseph Leidy's report in 1862 to City Councils about insects injurious to shade trees describes five species, all Lepidoptera (moths) except one, a scale insect. The species that Philadelphians are most likely to see today, although infrequently and only in small numbers, is the bagworm, *Thyridopteryx ephemeraeformis,* which, according to Leidy:

> ...is among the most curious of insects. It is common on our shade trees, but especially infests the maples, larches, and arborvitae. Just at this period, July, the writer observes a large number on the cypress trees in front of the United States Mint, on Chestnut Street. The worms, after escaping from the eggs, immediately compose for themselves cases composed of silk, interwoven with fragments of their food...As the worms grow, they enlarge their silken and leafy habitations, until they reach an inch or two in length. In the latter part of summer, these insects are often noticed dangling from the trees of our sidewalks, suspended from the boughs by a silken thread, and enclosed in a dark, rough, spindle-shaped sack. They never leave the latter, but when they have reached their full growth, they fasten their silken case securely to a branch of the tree, and within it undergo transformation into a pupa. From the latter is produced the moth, the male of which awaits the night to leave his habitation in search of a mate. The female never leaves her silken dwelling, nor does she even throw aside her pupa garment; it is her nuptial dress and her shroud. Within it she deposits her eggs, enveloped in the down stripped from her body. The eggs, thus protected and enclosed within the mother's habitation, remain suspended from the branches of the tree, secure from storms and the cold of winter, until the following season.
>
> They are easily destroyed. All that is required to get rid of them, is to remove their silken cases when the trees are trimmed in the spring. With the cases, the accumulations of eggs are destroyed, which otherwise would give origin to new colonies of worms.[1]

Figure 4.2 Group of three bagworm bags, including the one in figure 4.1.

Protection from the bag

The tough silken sacks that protect bagworms are covered with twigs and leaves that camouflage them and function like armor. They also shield them from solar radiation, wind, and rain. Although silken cocoons are commonplace during the immobile pupal stage of the life cycle of moths, the bagworm family, Psychidae, is unique for the portable sack its caterpillars carry.[2] The family includes 1,000 species worldwide,[3] but *Thyridopteryx ephemeraeformis* is the most conspicuous species in the northeastern United States,[4] where the term "bagworm" refers typically to just this species.

Leidy did not offer a clue to the mystery of how male bagworm moths manage to mate with flightless females enclosed within two defensive layers—pupal cases inside silk bags.

Mating through two defensive layers

In 1927, Frank Morton Jones, like Leidy a member of the Academy of Natural Sciences of Philadelphia, published a paper revealing how male bagworm moths penetrate the two enclosures that wall females off from the outside world. Mating takes place in September or October. Inside her pupal case, which in turn is inside her silk bag, the female is positioned head down, her genitalia facing the bag's top. At the bottom of the bag is a hidden opening through which the male will insert his genitalia at the tip of his abdomen. When ready to mate, the female—a maggot-like creature without wings, legs, antennae, or functional eyes—splits her pupal case open a crack near her head and emits a pheromone. Navigating using olfactory and visual cues, the male moth flies to the bag, grabs onto it, and probes its bottom with the tip of his abdomen, searching for the hidden opening. While the moth clings onto the outside of the bag, he inserts his genitalia, which occupy the tip of his abdomen, through the opening and then through the crack in the pupal case near the female's head. At this point, the male's genitalia are still far from the female's genitalia at the opposite end of the bag. The moth generates pressure that telescopes his abdomen and propels his genitalia past the female's head. The abdomen continues to elongate though the space between the female's body and the inside wall of the pupal case, finally apposing the genitalia of both sexes.[5]

After copulation, the moth's abdomen retracts to its normal length, and the moth flies away, capable of mating again. Almost immediately after mating, the female fills her pupal shell with eggs. Now in a shriveled, weak state, she exits her pupal case and seals the eggs inside. Contrary to Leidy's account, she then maneuvers herself through the hole in the bottom of the bag and drops to the ground to die. In the spring when the eggs hatch, the minute caterpillars emerge through the hole, crawling away onto nearby branches or floating away on strands of silk blown by the wind, to construct new bags.[6]

Figure 4.3 Moth (*Thyridopteryx ephemeraeformis*) reared from a bagworm feeding on eastern red cedar *(Juniperus virginiana)* planted along the Schuylkill River Trail. The moth is male and does not feed. To mate, he flies to a female confined inside her bag; her pheromone guides him. After landing on her bag, he inserts his genitalia (the orange structures at the tip of his abdomen) through an opening in the bottom of the bag and telescopes his abdomen inside the bag toward her genitalia at the top of the bag.

Thousands of eggs concentrated at one point

The capacity of this species to defoliate a tree is apparent from the size of a single brood: up to 1,200 eggs per bag.[7] Because the entire lot of eggs is stored in one bag, the release of bagworms is concentrated on a single point. One tree may harbor dozens of bagworm bags that collectively have the potential to release tens of thousands of bagworms. The larvae of this species can completely denude its host, although the host usually recovers.[8] Newly hatched caterpillars dangling on silken threads disperse by ballooning, blown by the wind to new host plants.[9] Bagworms have been recorded feeding on more than 128 species of plants in 45 families.[10]

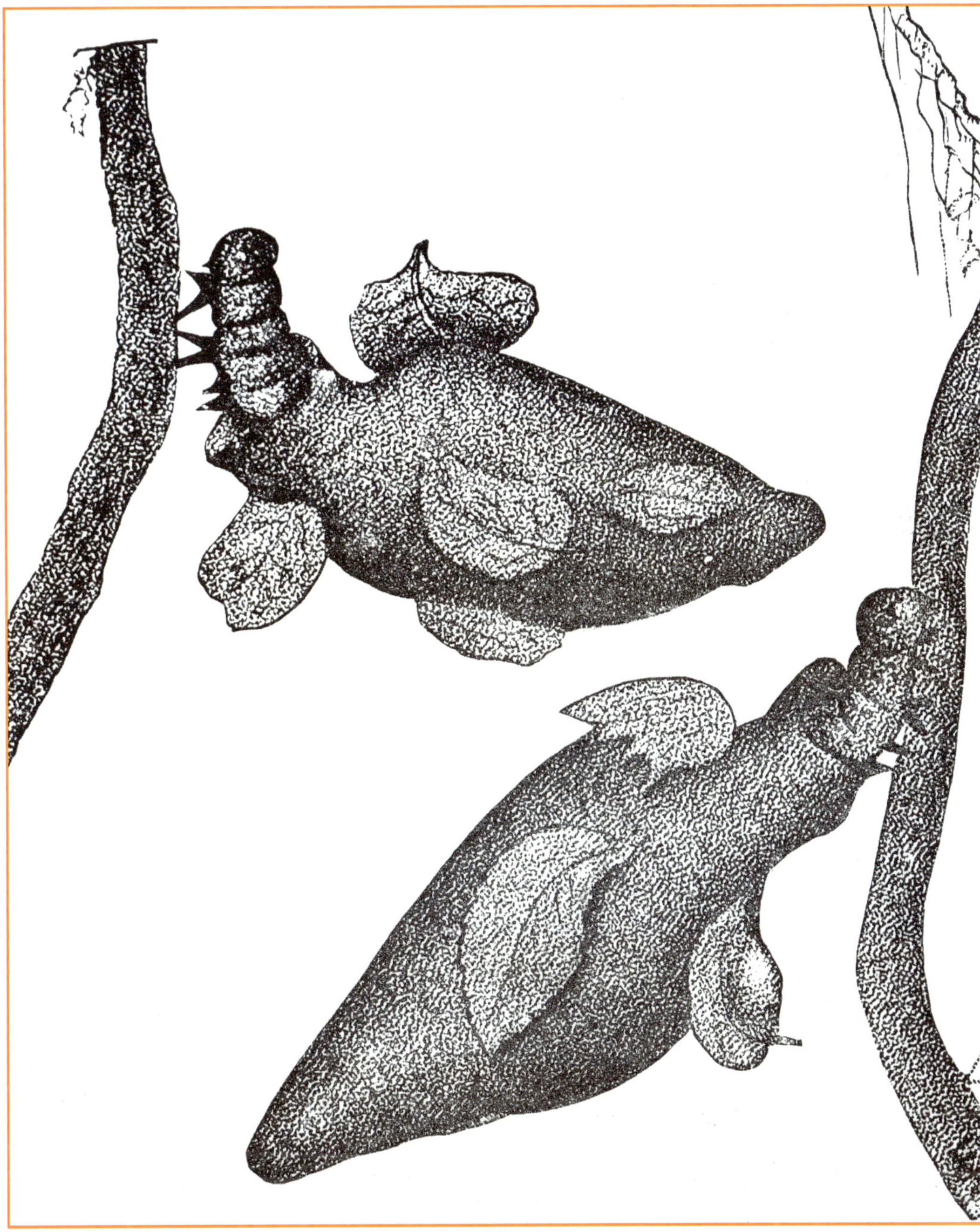

Figure 4.4 "How the bag-worms walk and climb." (Caption and cartoon from Henry McCook, *Tenants of an Old Farm. Leaves from the Note Book of a Naturalist. Illustrations from Nature.* [Fords, Howard and Hulbert, New York, 1889], 395.)

Infestations of bagworms in downtown Philadelphia are currently common only on recent plantings from nurseries. Why have bagworm infestations in Philadelphia declined since 1862, when Philadelphia City Councils enlisted Leidy's help in controlling them? In 1831 a horticultural report on Philadelphia's public squares indicated that horticultural development of Philadelphia's southwest square (now Rittenhouse Square) was a plan yet to be realized.[11] One hypothesis to explain the high prevalence of outbreaks of bagworms in nineteenth-century Philadelphia is that horticultural expansion required stock from nurseries, which then, like today, introduced bagworms that caused outbreaks.

More hypotheses to explain nineteenth-century outbreaks of bagworms

A second hypothesis is that enemies of bagworms in nineteenth-century Philadelphia had yet to establish populations sufficiently diverse and large to prevent outbreaks. Bagworms have been found to escape parasites by colonizing new areas and new host plants.[12] New plantings of municipal trees along streets and in other public places in the nineteenth century would have provided bagworms with host plants distant from established populations of specialized enemies, such as parasitic wasps. Joseph Leidy's report to City Councils in 1862 referred to a bagworm outbreak he had just witnessed on cypress trees in front of the United States Mint on Chestnut Street.[13] According to the escape-from-enemies hypothesis, this outbreak occurred because the bagworm's enemies had yet to colonize this area in numbers sufficient to keep the bagworm population under control.

Figure 4.5 Adult virgin female *Thyridopteryx ephemeraeformis* resembling a maggot, photographed in September. She developed in the bag of a bagworm I collected along Schuylkill River Trail in in August. Ordinarily, females mate and deposit their eggs inside their bags, and then, in a shrivelled state free of eggs, they emerge from their bags, drop to the ground, and die. Kept indoors and isolated from males, this female emerged as a virgin still laden with eggs. Whether she could have mated outside her bag is unknown. Her head and three pairs of diminutive legs are on the right; her genitalia are on the left.

Figure 4.6 Unidentified species of male ichneumon wasp attracted at night to black light in our backyard.

Figure 4.7 Unidentified species of female ichneumon wasp with long ovipositor, which she uses to lay eggs on (or in) her host. She was attracted at night to a pillowcase illuminated by black light in the rear of our home in Center City. Six species of ichneumon wasps are known to parasitize the bagworm (*Thyridopteryx ephemeraeformis*).

Figure 4.8 Plundered bagworm bag with interior silk exposed, September 16, 2013, along the Schuylkill River Trail, Center City. The predator had attacked most of the bagworm cases on this cultivated eastern red cedar, which hosted more than twenty. I did not witness this action. The ripping apart of bagworm bags (which are strong) attached to the tips of fine branches is consistent with depredation by white-footed mice (*Peromyscus leucopus*), nocturnal arboreal insectivores that prey on bagworms and inhabit Philadelphia.

A third hypothesis is that methods used to control bagworms in the nineteenth century were counterproductive. Leidy's recommendation to pick and destroy egg-laden bags in the winter had the potential for destroying parasitized bags by mistake, since the bags containing parasites or bagworm eggs look the same. Unintended destruction of parasites could be high if rates of parasitism were also high. Entomologists from the U. S. Department of Agriculture in the late nineteenth and early twentieth century recommended spraying arsenic-based insecticides on trees infested with bagworm bags that could not be picked off by hand. This, too, may have killed parasites and interfered with biological control. Charles Valentine Riley, chief entomologist for the U.S. Department of Agriculture, recommended a simple modification of Leidy's approach: Instead of destroying bagworm bags, store them in a container that would allow parasites to fly out, but would deny newly hatched bagworms access to food and sites for ballooning.[14] I do not know whether Philadelphians tried this parasite-conserving method.

A fourth hypothesis is that shady promenades popular in nineteenth-century Philadelphia favored bagworm outbreaks. A recent horticultural restoration of the south garden of the Fairmount Water Works exemplifies such a promenade, planted with sweet gum cultivars and London plane trees emulating the design of Frederick Graff, Jr., one of the garden's nineteenth-century landscape architects.[15] The striking horticultural feature of this garden is the absence of flowerbeds. In 1874 Fanny Kemble, who lived on Rittenhouse Square, commented on the square's absence of flowerbeds and flowering shrubs.[16] Flowers provide food for ichneumon wasps, whose larvae are the primary parasites of bagworms. In a study designed to test the utility of flowers for control of bagworm infestations, it was found that bagworms feeding on shrubs surrounded by flowering forbs were parasitized at rates 71 percent higher than were bagworms feeding on shrubs not surrounded by flowers.[17]

A fifth hypothesis is that abundance of native host plants in nineteenth-century Philadelphia promoted infestations of bagworms. Populations of bagworms, which are native to North America, may be more likely to proliferate on native plant species compared to exotics, which took time to spread and naturalize in the Philadelphia area over the past two hundred years. Douglas Tallamy and his colleagues at the University of Delaware attempted to rear bagworms on sixteen species of exotic ornamentals currently naturalized in the mid-Atlantic area. Bagworms on thirteen of the species starved; those on the remaining three species grew at unsustainably low rates.[18] In another study, female bagworms feeding on Japanese maple *(Acer palmatum),* an exotic ornamental common in Center City, grew and matured normally, but did not produce eggs.[19]

A sixth hypothesis is that nocturnal darkness undisturbed by electric lighting promoted reproductive success of bagworms in nineteenth-century Philadelphia. Male bagworm moths fly to electric light, as demonstrated by their capture in electric light traps.[20] Bagworm moths in the laboratory live for only a day,[21] so moths diverted from mating to lamps might not have a second chance to mate. Electric lighting has been invoked to explain decreases in populations of moths in habitats already compromised by other disturbances.[22]

Despite the potential for artificial lighting to disrupt reproduction, its impact on populations of bagworms is doubtful. The moth's mating flights peak from 3 to 6 p.m. and are usually over by dusk.[23] *T. ephemeraeformis* may tolerate artificial lighting better than do most other moths; lamps cannot attract its wingless females, which mate and lay eggs in their bags. Flight to lamps may disturb bagworm moths less than their enemies, such as parasitic flies and wasps.[24]

Perpetual influx of new enemies

The decrease of outbreaks of bagworms in Philadelphia today compared to a century and a half ago is likely multifactorial in origin. Causes may have changed over time; in particular, bagworms in the city likely encountered a steady increase in enemies over the last century and a half. In 1986 *T. ephemeraeformis* was found for the first time to be parasitized by *Coccygomimus disparis,* an ichneumon wasp repeatedly introduced into North America from Asia from 1972 to 1984 to control the gypsy moth, *Lymantria dispar.*[25] In one survey published in 2005, *C. disparis* was the most common parasitoid found on *T. ephemeraeformis.*[26] (A parasitoid is a parasite that kills its host.) Enemies of *T. ephemeraeformis* include at least sixteen parasitoids, four predators, ten pathogenic fungi, a polyhedrosis virus, and bacilliform bacteria. These enemies include four orders and nine families of insects. Predators of bagworms include house sparrows and white-footed mice *(Peromyscus leucopus),* which inhabit Philadelphia.[27]

How long might it take for all potential enemies of *T. ephemeraeformis* to disperse into Philadelphia and establish populations here? In principle, evolution and importation of new enemies could continue indefinitely, a process exemplified by the annual influx of new viral strains that cause human influenza. The status of Center City as a nexus of commerce and transportation promotes this process.

Termination of recent outbreaks

During winter about a decade ago, I collected ten bagworm bags from saplings planted in Schuylkill River Park. I stored them in a breadbox to observe young bagworm larvae emerging from their mothers' bags. Bagworms had been partially defoliating trees in the park for several years. Four of the bags turned out to be empty; I had mistaken empty bags for egg-laden bags. Some of the empty bags may have been from males that had hatched in the fall, but others could have been leftovers from previous years. By leaving empty bags dangling conspicuously from branches, *T. ephemeraeformis* has established a system of decoys capable of fooling predators, including me.

The bags I collected yielded parasitic wasps and flies, but no bagworms. Leidy had observed wasps emerge from bags, but the attack rate in my small sample was 100 percent. I had hoped my removing bagworm bags from the trees in winter would reduce the infestation of bagworms, but I had succeeded only in removing the bagworm's natural enemies. I did not think to release the wasps back into the park. That summer I found no bagworms in Schuylkill River Park. The parasites had ended the outbreak of bagworms—despite my meddling.

Since then, I found new bagworm bags on newly planted river birches 100 meters away along the Schuylkill River Trail. A cluster of these trees had fifteen bagworm bags hanging from their branches, which showed moderate defoliation. This time I left the bags alone. After two years, I found no defoliation and no new bags.

Figure 4.9 Bag of the bagworm, *Thyridopteryx ephemeraeformis,* in winter on a river birch recently planted along the Schuylkill River Trail in Center City. New bags stopped appearing on these trees two years after appearance of the first bags.

SPOTLIGHT

ORIENTAL COCKROACH

Female oriental cockroach (*Blatta orientalis*), also known as water bug, on the floor of a row house in Center City.

Wings of female oriental cockroaches are vestigial, incapable of flight. How have these insects, which are up to 3 cm (one inch) long, dispersed so successfully in Center City?

Storm drains in Center City connect to sewers that offer oriental cockroaches food, water, shelter, and safe passage underground. They also provide these cockroaches subterranean access into buildings. Crevices in masonry walls in nineteenth century buildings offer cockroaches crawl spaces into row houses and apartments.

Nocturnal activity and black profiles help keep oriental cockroaches hidden. Outdoors, nocturnal travel protects them from solar radiation and desiccation. Female oriental cockroaches that disperse into territory without males can reproduce asexually.

The oriental cockroach has had a long time to adapt to human habitation. In Britain, archeological excavation turned up remains of this species in a Roman town from the fourth century. The species originated in Africa, despite its name.[1]

5

CYNTHIA MOTH

(Ailanthus silkmoth; *Samia cynthia*)

In Philadelphia in 1861, the cynthia moth was introduced into North America for the purpose of manufacturing silk, but the industry never developed, and the moth, after thriving in the wild, became extinct.

Figure 5.1 Cynthia moths, larvae, cocoons, and their parasites in a museum drawer of the Academy of Natural Sciences of Drexel University. All specimens are from the late nineteenth to early twentieth century. The moth is now extinct in Philadelphia. (Courtesy of the Academy of Natural Sciences of Drexel University)

In 1862 Joseph Leidy recommended planting ailanthus trees because of their resistance to insect pests.[1] Fifteen years earlier Andrew Jackson Downing, horticultural expert, praised ailanthus trees for the same reason:

> The variety of trees for cities—densely crowded cities—is but small; and this, chiefly, because the warm brick walls are such hiding places and nurseries for insects, that many fine trees—fine for the country and for rural towns—become absolute pests in cities. Thus, in Philadelphia, we have seen, with regret, whole rows of the European Linden cut down within the last ten years, because this tree, in cities, is so infested with odious worms that it often becomes unendurable. On this account that foreign tree, the Ailanthus, the strong scented foliage of which no insect will attack, is every day becoming a greater metropolitan favorite.[2]

Figure 5.2 Old ailanthus tree at the historic Lemon Hill mansion in Fairmount Park, Philadelphia.

Ailanthus trees become disreputable

Leidy's advice, however, might have been controversial. By 1852 Downing had reversed his endorsement of ailanthus:

> The vices of the Ailanthus—the incurable vices of the bygone favorite—then, are two-fold. In the first place it smells horribly, both in leaf and flower—and instead of sweetening and purifying the air, fills it with a heavy, sickening odor; in the second place it suckers abominably, and thereby over runs, appropriates and reduces to beggary, all the soil of every open piece of ground where it is planted. These are the mortifications which everybody feels sooner or later, who has been seduced by the luxuriant outstretched welcome of its smooth round arms.[3]

Figure 5.3 Colony of ailanthus saplings outside O'Connor Swimming Pool on Lombard Street. They are sprouting from roots tracking along the crack at the base of the wall.

Joseph Leidy may have favored this disreputable species out of desperation. Alternative methods for controlling infestations of insects defoliating shade trees in Philadelphia had been disappointing.

Introduction of the ailanthus silkmoth

Leidy also may have considered a presentation made the year before at the Academy of Natural Sciences by fellow member Thomas Stewardson. In 1861 Stewardson had reported that a silkworm he had just imported into America could establish a prosperous silk industry here. After procuring eggs of this species from Paris, Stewardson succeeded in rearing the silkworms in a private garden in the city and producing eighty cocoons. At a meeting in the Academy, Stewardson exhibited a sample of cloth fabricated in France from silk of this species. He also exhibited the moth,[4] which today is called the cynthia moth (*Samia cynthia*), a strikingly colored giant silkmoth with a wingspan up to 14 centimeters (5.5 inches).[5] He displayed a live caterpillar feeding on a leaf of its food plant—*Ailanthus altissima.*[6]

Cynthia moths are native to China, where they have a long history of cultivation for commercial production of silk.[7] In China cynthia moths thrive in the wild, unlike the domesticated species of silkworm, *Bombyx mori,* which feeds on mulberry and had already been part of an ill-fated sericulture industry in Philadelphia. In 1769 Benjamin Franklin, then serving as agent of the colonies in England, had sent the American Philosophical Society a letter recommending allocation of public funds for construction of a filature (factory for producing silk thread from cocoons) to promote sericulture based on mulberry.[8] In 1771 a filature located on 7th Street between Market and Arch received over a ton of *B. mori* cocoons for processing.[9] But in 1840 the silk industry in Philadelphia collapsed due to a speculative bubble that lead to the deliberate destruction of 90 percent of the mulberry plants cultivated around Philadelphia.[10]

Aspirations for the silk industry based on ailanthus silkmoths

News of Stewardson's presentation at the academy spread quickly and sparked unbridled enthusiasm for the future of cynthia sericulture:

> The cultivation of this worm is an employment well adapted to the poor, or the aged, or the very young who are not capable of performing any severe labor. As the worm, from the time of its exclusion from the egg to the spinning of the cocoon requires only about forty days at the furthest, an occasional supervision during eighty days (two broods are reared) of the most pleasant season of the year, is all that is required for the production of millions of cocoons, and all this can be done by a smart child of ten years of age, or an infirm or aged person. Wherever the ailanthus can grow the worm can be reared, and even in the extreme northern States [where] only one brood a year can be raised, still the profits will be large enough to justify the enterprise. With no labor worth mentioning, and with no outlay or money, a textile material, holding a middle place between the silk of the mulberry worm and other materials, as wool, hemp and cotton, can easily be raised, which will prove richly remunerative in furnishing a cheap, substantial and lasting material for apparel. The material would be cheap, and thus favorable to the poor. The coarser sorts could be manufactured into various articles of underclothing at a much less price than is now paid for them; they are tough and strong, and will wear longer than any textile material now used. It is said that garments made of it by the Chinese last through several generations of constant wearing.
>
> Reliable estimates of the cost of raising a pound of this silk can only be proximately made, but under any circumstances it could not amount to one-fourth the cost of raising a pound of mulberry silk. The fact is that it would cost nothing but a little care, and as the worm is so hardy it can be left to do its work without any particular oversight. The unwinding of the

> cocoons would cost a little, but this could be done by young or aged people at very little expense.[11]

Naturalization of cynthia moths in the eastern United States

Despite the ease of rearing cynthia larvae, Stewardson's dream of establishing a silk industry based on this insect never materialized. In the United States no practical method was found for reeling the silk off the cocoons.[12] Stewardson delegated the task of rearing the silkworms to the Academy's assistant librarian, Edward J. Nolan, who in 1863 released 200 on a large ailanthus tree growing in the yard of a laboratory of the University of Pennsylvania, then located on 9th Street just above Chestnut Street. Nolan forgot about them until the winter of 1864, when he discovered 40 cocoons on the tree. He left these undisturbed, effectively releasing them to propagate in the wild.[13]

The introduction of cynthia moths into North America bore similarities to the introduction of its host plant. In the 1780s William Hamilton imported *Ailanthus altissima* from England to his estate, Woodlands, in Philadelphia, the site of the species' first cultivation in North America.[14] Like the moth, ailanthus is native to China and had only recently been imported to Europe before its introduction here. The tree and moth escaped cultivation, naturalized, and disseminated by repeated introductions elsewhere in the United States.[15]

Initially populations of cynthia moths expanded rapidly to other urban areas. The species became established in cities in New York, New Jersey, Connecticut, and Virginia, and in Washington, DC, and later in Georgia, west to Indiana.[16] In 1881 cynthia caterpillars were observed feeding on nearly all the trees and shrubs in New York City's Central Park, but only the caterpillars feeding on ailanthus developed normally; nearly all the others died before completing their life cycles. In 1880, ichneumon parasitoids had been noted for the first time emerging from cynthia cocoons in Central Park.[17] By 1900 the population explosion of cynthia moths had abated:

> It became so common in Philadelphia and Washington, D.C. at one time as to be a pest, and threatened the destruction of the trees; but the parasites and birds seem now able to cope with it and hold it in check.[18]

Parasites attack Samia cynthia

By the second half of the twentieth century in Philadelphia, parasitism rates were high and the moth had become uncommon, as reported by the lepidopterist Arthur Shapiro, who grew up in Philadelphia:

> The moth was apparently quite common early in the century. I heard this from old-timers when I was a kid...By the late 50's – early 60's the cocoons were not at all easy to find, but tended to be highly clumped. I found them near the Frankford Arsenal, in South Philadelphia, and along Passyunk Avenue, and occasionally at the foot of Arch Street near the river and sometimes rather commonly in the old RR yard in South Camden, behind the J. B. Van Sciver Co. warehouse. They would not be in all those places in the same year, as a rule. The tree of course is nearly ubiquitous in the city. The parasitization rate was incredible. I believe the parasite was *Spilochalcis mariae*—check on this, as I am retrieving stuff through a lot of memory!—and some whole batches were bad—certainly the average was at least 85% parasitized. I caught single adults once at International Airport, while waiting for a bus; once on

> the windows of the main Lit Brothers store; and once at a shopping center near Norristown, Montgomery County, the only one I ever saw in the "country" (but downtown Norristown was pretty seedy!).[19]

Others have reported high rates of parasitism of cynthia pupae by *S. mariae.*[20] Populations of cynthia moths became limited to ailanthus's harshest urban habitats, such as railroad yards[21] and verges along major highways.[22] The species was not found on ailanthus growing in rural or suburban areas.[23] When cynthia larvae were experimentally placed on food plants in a rural area, predatory wasps destroyed them all within ten days.[24]

Extinction of the cynthia moth in Philadelphia

The last report of the species in the wild in Philadelphia was in 1992: Christopher Clark, a moth collector, recalls finding half a dozen cynthia cocoons in southwest Philadelphia near the Eastwick SEPTA train stop. The most recent sighting before that was in 1970, when about fifty cynthia cocoons were found on a small ailanthus tree growing on the property of an American Legion Post then located at 34th and Market Streets. Cynthia moths—big, showy popular insects—have spawned a cottage industry cultivating and selling cynthia cocoons to hobbyists; accidental or intentional reintroductions could account for occasional sightings of this insect in "the wild."

A century and a half after its importation from France and its naturalization in Philadelphia, the cynthia moth is locally extinct here, despite the abundance of ailanthus. Had it been introduced as a biological control agent against ailanthus trees, it would have been deemed a failure. Twenty-five years ago, I theorized that the enemies of the moth could not tolerate gritty nineteenth-century industrial Philadelphia, which afforded it safe haven. According to this theory, as Philadelphia became less polluted and greener, predators and parasites moved into the city, which no longer served as a refuge for cynthia moths. The cynthia moth became a fugitive species with nowhere to go, a vestige of a bygone era.[25]

Establishment of populations of parasites in Philadelphia

How did populations of parasites move into the moth's urban refuges? The cynthia parasite that Arthur Shapiro remembered, a tiny 4 millimeter wasp called the golden-yellow chalcid (*Spilochalcis mariae* [*Conura maria*]), was first found parasitizing cynthia pupae in 1881 in New York City. Museum specimens of this parasite date back to 1869, when the species was isolated from cocoons of the bagworm, the same species (*Thyridopteryx ephemeraeformis*) that plagued shade trees in Philadelphia.[26] The parasite has also been isolated from three native giant silkmoth species—*Hyalophora cecropia, Antheraea polyphemus,* and *Callosamia promethea*[27] —that I have found in Philadelphia. These hosts of the parasite could have maintained parasite populations even when cynthia populations were low or absent. In the list of species that have been identified as hosts of the chalcid parasite, the most common in downtown Philadelphia is the bagworm—which could have indirectly contributed to the cynthia moth's extirpation.

A tachinid fly, *Lespesia frenchii,* parasitized over ten times more *S. cynthia* than did the chalcid wasp in one survey.[28] A catalog of hosts of this fly lists many species common in Philadelphia. They include the tent caterpillar *(Malacosoma americana),* tiger swallowtail *(Papilio glaucus),* black swallowtail *(Papilio polyxenes),* spicebush swallowtail *(Papilio troilus),* red admiral *(Vanessa atalanta),* painted lady *(Vanessa cardui),* and cabbage white *(Pieris rapae).*[29]

The most abundant of these species in Philadelphia is the cabbage white, which in this region feeds on sixteen species of plants in the mustard family (Brassicaceae), including common weeds.[30] Like the cynthia moth, the cabbage white was introduced into North America from Europe. It was first recorded in Quebec in 1860 and spread south and west.[31] By 1908 individuals of this species were numerous in suburban Philadelphia.[32] It is possible that introduction of the cabbage white into North America provided an alternate urban host for cynthia's tachinid parasite. In Philadelphia the cabbage white, like the bagworm, may have indirectly contributed to the ailanthus silkmoth's extirpation.

Figure 5.4 Cabbage white (*Pieris rapae*) in the community garden at 25th and Spruce Streets in Center City. It is an alternate host of a fly (*Lespesia frenchii*) that is a parasite of *Samia cynthia.* It was introduced into North America from Europe.

After the cabbage white, the most common species on the list of hosts of the tachinid fly is the red admiral, at least in Center City. Its abundance here coincides with the abundance of one of its food plants. The plant is Pennsylvania pellitory *(Parietaria pensylvanica),*[33] an inconspicuous native herbaceous weed that grows in cracks in pavement at the base of buildings. By supporting red admirals, the establishment of Pennsylvania pellitory as a weed in downtown Philadelphia could have contributed to the cynthia moth's local extinction.

Figure 5.5 Red admiral (*Vanessa atalanta*) sunning itself on a sidewalk near 25th and Pine Streets in Center City. It is another alternate host of the cynthia parasite *Lespesia frenchii*.

Figure 5.6 Pennsylvania pellitory (*Parietaria pensylvanica*), with small green flowers along the stem, near 25th and Pine Streets in Center City. It is a food plant of the larvae of the red admiral butterfly *(Vanessa atalanta)*. In this neighborhood, pavement cracks at the base of buildings are its favorite habitat.

S. cynthia's tachinid parasite, *L. frenchii,* disappeared in New England after introduction of a competing tachinid parasite, *Compsilura concinnata,* to control gypsy moths.[34] *C. concinnata* parasitizes at least 180 species and has been blamed for reductions in New England's populations of native giant silkmoths.[35] In Virginia, *C. concinnata* is itself attacked by a parasite,[36] which may explain why populations of giant silkmoths there have not declined.[37] How such complex interactions might play out over time in Philadelphia remains to be seen, but recovery of populations of cynthia moths here after an absence lasting decades would seem improbable.

A fruitless search for cynthia cocoons on ailanthus trees

Recently Chris Clark escorted Jason Weintraub, lepidopterist at the Academy of Natural Sciences, and me on a tour of the site near the train stop in Eastwick where he had found cynthia cocoons two decades ago. The density and numbers of ailanthus trees here were greater than any I had seen elsewhere in Philadelphia. The site included highways and rail lines—ideal habitat for cynthia moths. Jason had picked the date—November 8—to maximize the likelihood of spotting any cocoons that might be present. By this date, most of the ailanthus leaflets have dropped, which would expose cocoons hanging from the main stems of the compound leaves. A few weeks later the stems and cocoons would have fallen to the ground, where they would have blended in with leaf litter. We spotted many curled up leaflets that resembled cocoons, but no cocoons.

Later Jason showed me cynthia moths in the collection in the Academy of Natural Sciences. He pulled out a glass-topped wooden drawer filled with rows of pinned specimens from the late nineteenth and early twentieth century. The drawer included specimens of the moths and their caterpillars, plus pinned specimens of their parasites. Today the last refuge of cynthia moths in Philadelphia is here, inside these museum drawers, at the same institution where in 1861 the moth, as the charismatic star of a scientific meeting, made its North American debut.

The disappearance of this moth paradoxically exemplifies increased biodiversity. As parasites and their alternate hosts populated downtown, the safe haven that protected the moth for a century in Philadelphia ended. The same forces that currently suppress outbreaks of bagworms contributed to the local extinction of the cynthia moth.

SPOTLIGHT

COMMON BLUE VIOLET

Two varieties of the common blue violet (*Viola sororia*) found growing wild in Center City.

Two centuries ago in Philadelphia the habitat of *Viola sororia* was reported to be dry woods along the Schuylkill River.[1] Nearly a hundred years later in this city, this violet was found to be hybridizing. This observation prompted speculation that hybridization could produce new forms on which natural selection could act, causing evolutionary change.[2] Today in Center City this violet thrives in lawns and pavement cracks, and its flowers vary in color and pattern. Perhaps adaptation of this species to Center City expresses, at least in part, evolution through hybridization, as hypothesized over a century ago.[3]

6

AILANTHUS WEBWORM MOTH

(*Atteva aurea*)

While cynthia moths in Philadelphia went extinct, another ailanthus moth thrived—the ailanthus webworm moth.

Figure 6.1 Ailanthus webworm moth taking nectar at white snakeroot (*Ageratina altissima*), a common wildflower in Center City.

In 1911 Carl Ilg, an entomological laboratory assistant, submitted a one-paragraph note to *Entomological News, and Proceedings of the Entomological Section of the Academy of Natural Sciences of Philadelphia:*

> It was at the later part of August when I was out collecting, that my attention was called to a web which looked to me like a spider's nest, on a small ailanthus bush. By investigating more closely, I saw a chrysalis suspended in the web. Not knowing what it was, I took it home, and several days after, a small moth emerged and proved to be *Atteva aurea.* As I knew the food plant now, I looked in the same neighborhood and found several similar webs containing newly hatched, as well as full grown, larvae and also chrysalids in them. The full grown larva is about 1¼ inches long, blackish, with a distinct brown stripe all along its back, while the sides are dotted with fine white spots...As far as I could find out, there is no record as to food plant or life history of this little moth, but should any other collectors have made any observations in this respect, I would like to hear from them.—Carl Ilg, 2728 Somerset St., Philadelphia.[1]

Figure 6.2 Pupa of ailanthus webworm moth in its web. The web is in an ailanthus sapling growing along the Schuylkill River Trail in Center City. Until Carl Ilg of Philadelphia discovered such a pupa and identified the moth that emerged from it, nobody knew that the species made webs or that it ate ailanthus.

A mystery solved and a mystery created

Ilg's discovery solved a mystery. In 1857 Asa Fitch, entomologist for New York state, described *Atteva aurea*—but only the moth. Fitch had never seen the moth's larva and did not know the identity of its food plant.[2] The fact that its life history had gone unrecognized for so long is remarkable given the moth's beauty: it has four metallic gold bands offset by brilliant white spots embedded in iridescent blue and black. It is 12 millimeters (half an inch) long. Its species name, *aurea,* is from *aureus,* meaning "golden" in Latin. Ilg's discovery led to the moth's common name, "ailanthus webworm moth."

Ilg solved one mystery but created another: where did *Atteva aurea* come from? It could not be native to Philadelphia, since ailanthus, its only host plant here, is not native. It is not known from Europe or Asia. The specimen Fitch described came from Savannah, but the same puzzle existed in both cities.

Daniel Janzen, ecologist at the University of Pennsylvania, has been conducting a long-term inventory of moths in a Costa Rican nature preserve, *Área de Conservación Guanacaste.* His inventory includes two confusing species of *Atteva* with wing patterns that look almost identical. Using DNA fingerprinting and other data, he and his colleagues compared these two species with *Atteva aurea* collected in North America, including the mid-Atlantic region and Canada. They concluded that one of the two species in Guanacaste is *Atteva aurea.* They also determined that a species of *Atteva* in southern Florida is also *Atteva aurea.* Thus Philadelphia's ailanthus webworm moth ranges from Costa Rica to Canada.[3] In Guanacaste[4] and southern Florida[5] it feeds on the paradise tree, *Simarouba glauca,* which, unlike ailanthus, is native to tropical and subtropical areas in North and Central America. *Simarouba glauca* and *Ailanthus altissima* belong to the same family, Simaroubaceae.[6]

These findings lead to a hypothetical scenario explaining the mystery of the origin of the ailanthus webworm moth in Philadelphia. The chain of events begins around 1784 when William Hamilton introduces *Ailanthus altissima* into North America by planting it in Woodlands, his estate in west Philadelphia. When the distribution of ailanthus trees extends around the country, it approaches populations of *Atteva aurea* feeding on the paradise tree in Florida. *A. aurea* then encounters ailanthus trees for the first time and begins to feed on this close relative of its native host plant. Thriving on ailanthus trees, it expands its range north, moving into ailanthus's new territory, including Savannah by 1857, Philadelphia by 1911, and later, urban and suburban areas throughout the eastern half of this country and southern Canada.[7] One variant of this scenario is possible: *A. aurea* may have switched to ailanthus in southern Texas, which like Florida has native plants in the family Simaroubaceae.[8]

Differences between the cynthia moth and ailanthus webworm moth

In Philadelphia the ailanthus webworm moth is the most common moth at flowers such as goldenrod during the day in late summer and fall. What might account for this species' survival, in contrast to the extinction of *S. cynthia,* in Philadelphia? One might suppose the two species would share the same fate, since they have so much in common: both are moths that arrived in Philadelphia over a century ago and are specialized feeders on the same plentiful host plant. *A. aurea,* however, is smaller— about one fifth as large by wingspan—requiring less food for development and offering potential predators fewer calories and a smaller target. During the day its caterpillars are protected inside a web, in contrast to *S. cynthia,* whose caterpillars are fully exposed.

The most obvious difference, however, is in behavior and coloration. *A. aurea* is brilliantly colored and visits flowers during the day, whereas *S. cynthia* flies at night, and in the adult stage does not feed.[9] While visiting flowers, *A. aurea* is indifferent to its surroundings, in the sense that it does not fly away when a person approaches it. This fearlessness makes it easy to photograph. The overall syndrome—daytime flight, bright colors, and insensitivity to danger—is common in bees and wasps, but rare in moths. The three traits suggest that *A. aurea* possesses some kind of protection. Since it cannot sting or bite and has no sharp spines or urticating hairs, one might suspect that *A. aurea*'s protection is chemical, and its bright colors *aposematic,* warning potential predators.

Aposematic coloration

In Philadelphia, the most familiar example of a chemically defended species is the monarch butterfly, whose conspicuous black and orange pattern distinguishes it from other species at a distance of a dozen yards or more, barring confusion with its mimic, the viceroy butterfly, which is rare here. In Philadelphia, monarch caterpillars feed mostly on common milkweed, *Asclepias syriaca,* which is widely scattered in old fields in Fairmount Park. Like the monarch butterfly, the monarch's caterpillars are distinctively marked. They have bold yellow, black, and white stripes along their entire length. Feeding on milkweed blossoms, the caterpillars contrast sharply against the pink flowers. Any milkweed patch of a dozen or more stalks is likely to host other conspicuously colored insects, including bright red and black beetles (*Tetraopes tetrophthalmus*) and bugs (*Lygaeus kalmii and Oncopeltus fasciatus*).

Figure 6.3 Monarch butterfly (*Danaus plexippus*) taking nectar in the community garden at 25th and Spruce Streets in Center City. It is poisonous and aposematic (warningly colored).

Figure 6.4 Monarch caterpillars on tropical milkweed (*Asclepias curassavica*) in a garden in Center City. Like monarch butterflies, they are aposematic. They obtain their protective poisons from milkweed and retain them after they undergo metamorphosis into butterflies.

Figure 6.5 Large milkweed bug (*Lygaeus kalmii*) on common milkweed (*Asclepias syriaca*) off Martin Luther King Drive in Fairmount Park. Like the monarch butterfly, it is aposematic.

Figure 6.6 Red milkweed beetle (*Tetraopes tetrophthalmus*), another aposematic species, on common milkweed in Fairmount Park.

The showy species noted above all contain cardenolide poisons that match those present in their milkweed (*Asclepias*) food plants.[10] One might hypothesize that these insects gain protection by sequestering noxious chemicals synthesized by milkweed. Lincoln Brower, who took me on as an assistant in college, tested this hypothesis. He succeeded in breeding a strain of monarch caterpillars that ate cabbage. I recall his frustration when he tried to select such a strain and the elation when caterpillars finally started eating cabbage—and then pupated and hatched into monarch butterflies. Caged blue jays initially rejected all monarch butterflies offered on sight, but they were eventually conditioned to eat cabbage-reared monarchs, which they consumed without ill effects. When Brower substituted monarchs reared on milkweed *(Asclepias curassavica)* for those reared on cabbage, the jays vomited within fifteen minutes of eating even one. As described by Brower and colleagues:

> In great contrast to the cabbage-fed monarchs, those reared on *Asclepias curassavica* caused all eight birds to become sick. Ingestion of these was followed uniformly by violent retching and vomiting of the partially digested insects and fluid…Other less objective indications of unpalatability included excessive billwiping, crouching, alternate fluffing and flattening of the feathers, erratic movements about the cage, jerky movements of head, wings, and thoracic regions, partial closure of the eyes, eating of sand, twitching, and a generally sick appearance.[11]

One might suspect that *Atteva aurea,* like the monarch, is unpalatable due to poisons it sequesters from its host plant. Its host plant belongs to a family that makes bitter compounds known as quassinoids. Leaves from *Ailanthus altissima* have yielded forty-nine volatile compounds with diverse biological activity: cytotoxic, phytotoxic, antiproliferative, antifeedant, insecticidal, and insect growth regulating.[12] Two investigators reported that birds find *A. aurea* unpalatable, but the number of observations was small.[13] No studies have investigated the chemical composition of *A. aurea*.

Richard Peigler, an expert on the ailanthus silkmoth, *Samia cynthia,* wrote:

> I agree that *Atteva* is aposematic, but I do not have any evidence that *Samia* moths are also toxic. Blue jays did swoop down and catch and eat flying *cynthias* that I released into my back yard in South Carolina.[14]

Other protective traits

The dramatic coloration of *Atteva aurea* might have functions unrelated to poisons. Its uniqueness could discourage predation by birds that avoid novelty.[15] Birds avoid attacking prey they perceive as unfamiliar. Ray Coppinger, working with Lincoln Brower, found that hand-raised blue jays and red-winged blackbirds in cages tended to reject novel-appearing insects offered as food. He demonstrated that rejection of novel insects was due to novelty per se and not experience or innate preference.[16] Sexual selection may also favor evolution of bright colors.

Poisons may protect *A. aurea* from birds, but not from other predators, such as insects. Ants, tachinid flies, and *Polistes* wasps attack monarch eggs and larvae, which in one study had survival rates of less than 12 percent.[17]

A. aurea's web provides barriers against invertebrate attack. The caterpillars stay motionless inside their web during the day; they leave it to feed only at night. Diurnal parasitic wasps would have to penetrate a hatchwork of threads to reach larvae in the web. A unique feature of *A. aurea's* web is that all life stages except adults occupy it simultaneously. Moths lay eggs in the web, and caterpillars develop and pupate in the web.[18] A bird that attacked a noxious caterpillar in the web would presumably learn to avoid other caterpillars in the web. Since the web hosts more than one generation of *A. aurea,* this protective benefit could span generations.

Fungal threat to ailanthus trees

A. aurea is the most common moth attracted to outdoor lighting in our backyard in Center City. Despite its abundance, the moth's host plant in Pennsylvania is susceptible to an emerging lethal contagion. In 2003 a verticillium wilt was discovered to be killing ailanthus trees in the Tuscarora State Forest about 210 kilometers (130 miles) west of Philadelphia. Mark Schall and his colleagues at Pennsylvania State University have been investigating this outbreak. By 2008, Schall estimated that the fungal pathogen, probably a strain of *Verticillium albo-atrum,* had killed 10,000 ailanthus trees.[19]

Schall reported that the fungus spreads rapidly from tree to tree. It can overwinter in infected ailanthus trees or on fallen leaves. The primary infection begins in the spring and spreads circumferentially around the tree and up and down the trunk until the tree dies. Trees experimentally inoculated with the fungus died within one season. In severely affected parts of the forest, the fungus wiped out the entire ailanthus canopy and half of ailanthus seedlings and sprouts. Seedlings of red maple, striped maple, and sweet birch began to fill in forest gaps caused by deaths of ailanthus.[20] Schall and his colleagues are investigating the application of *Verticillium albo-atrum* as a biocontrol agent against ailanthus, which is classified as an invasive species in Pennsylvania.[21]

To what extent verticillium wilt will reduce the distribution of ailanthus over time is hard to predict. Its hyphal resting structures do not tolerate acidic soils.[22] It may have difficulty propagating in urban leaf litter, which tends to get discarded. *Verticillium* is a fungal genus with ten recognized species.[23] *V. albo-atrum* is highly adaptable, with strains differing in virulence and host specificity.[24] In the Tuscarora State Forest, the strain's lethality appears specific to ailanthus, but worldwide *V. albo-atrum* and other members of the genus *Verticillium* have infected over 200 species of plants[25] and have been blamed for billions of dollars in annual crop damage.[26]

Figure 6.7 Woodlands cemetery, formerly William Hamilton's estate, site of the introduction of *Ailanthus altissima* into North America around 1784. It is located in west Philadelphia, a short walk from Center City. A fungal contagion has been discovered to be killing stands of ailanthus trees 210 kilometers to the west of here. The lethal infection is a kind of verticillium wilt.

Although the recent establishment of ailanthus in North America began when William Hamilton imported it to his estate in Philadelphia,[27] ailanthus fossils in North America span approximately 40 million years, from the early Eocene to the middle Miocene. These fossils accompany fossils from temperate plant genera that, unlike ailanthus, never died out.[28] Why ailanthus disappeared in North America while so many other members of its fossil temperate plant community survived is unknown, but one possibility is an ailanthus-specific pathogen like the fungus currently attacking it in Pennsylvania.

The rapid colonization of *A. aurea* in Philadelphia and farther north is remarkable for an insect originating in subtropical and tropical habitats. Conceivably, the recent spread of *A. aurea* into North America represents repopulation of ancestral territory. Whether a progenitor of *A. aurea* was present in temperate North America in the Eocene when ailanthus grew here is unknown. The genus *Ailanthus* and its family Simaroubaceae are believed to have originated in North America,[29] so the moth and its host plant could have evolved here together. On the other hand, *A. aurea* belongs to a pantropical genus (*Atteva*) of fifty-three species,[30] pointing to a tropical, not temperate, origin. An unanswered question is whether *A. aurea* overwinters in Philadelphia or whether it annually recolonizes the region by migration from the south.

Survival of populations of ailanthus webworm in Philadelphia

The survival of the ailanthus webworm moth but not the ailanthus silkmoth in Philadelphia is a mystery. Adaptive traits that favor the ailanthus webworm moth include webs, small size, aposematic coloration, and probably poisons. These advantages alone do not resolve the paradox; the ailanthus silkmoth flourished in Philadelphia in the nineteenth and early twentieth century despite its lack of such traits.

When populations of the ailanthus webworm moth expanded north from their home ranges in subtropical and tropical America, they likely left behind native enemies such as parasites that were intolerant of cold or otherwise maladapted to temperate North America. Perhaps Philadelphia endowed the ailanthus webworm moth with a refuge from its tropical enemies.

The theory that the ailanthus webworm moth in Philadelphia escaped tropical enemies does not explain the ailanthus webworm moth's survival in Philadelphia. While the moth in late summer and early fall is abundant, the damage its larvae inflict on ailanthus trees is minor. Some forces are reining in populations of the ailanthus webworm moth while simultaneously allowing them to propagate.

The difference in the fate of the two exotic ailanthus moths defies easy explanation. Perhaps parasites of the ailanthus webworm moth, in contrast to those of the cynthia moth, do not have alternate hosts; or perhaps lowering the population density of *A. aurea* lowers its vulnerability to enemies, be they pathogens, parasites, or predators.

COMMON BLUE VIOLET – *continued*

Top: Common blue violet (*Viola sororia*) on sidewalk on Naudain Street.
Bottom: Closed flower, hidden on the ground under the leaves.

On sidewalks in Center City, pollinators are scarce, especially in the spring when violets bloom. The "cleistogamous" flower in the bottom photo will never open but can self-pollinate while closed. The violet's two flowering types—open and closed—allow cross-pollination when pollinators are present, and self-pollination when pollinators are absent. In addition, the common blue violet is able to propagate asexually in pavement cracks by stolons and rhizomes. On Naudain Street deep crevices between brick pavers protect these vegetative structures from trampling.[1]

7

NORTHERN PARULA

(Parula warbler; *Setophaga [Parula] americana*)

Northern parulas were among the most common victims of migratory bird collisions first noted at City Hall Tower at the start of the twentieth century.

Figure 7.1 Northern parula, a nocturnal migrant I found dead on the sidewalk at 23rd and Walnut Streets on October 4, 2010, after a storm the night before.

In 1916 the Delaware Valley Ornithological Club published a report by club member Delos E. Culver:

> ABOUT 10 AM, May 22d, 1915, there was received, at the Academy of Natural Sciences, Philadelphia, a call from the "Evening Bulletin" of that city for aid in the identification of a small "yellow and green" bird which had been picked up in the court yard of the City Hall. From this it was learned that hundreds of birds were lying about on the ledges surrounding the Public Buildings and City Hall Tower. Immediately upon receiving this information, I, accompanied by a "Bulletin" photographer, hurried to the scene, and the mortality, when ascertained, was really appalling.
>
> Upon reaching the courtyard, the areaways were first examined. Looking down into them, we found that although very few dead specimens were visible (most having been gathered by employees), there were many living birds continually flying up and down the full length of the areaways, apparently having lost all sense of direction. Maryland Yellow-throats were in evidence everywhere. Every areaway was full of fluttering birds of this species, and it was among them that the greatest mortality occurred. Upon entering the areaways from below, the following species were identified: Maryland Yellow-throat, Parula Warbler, Redstart, Red-eyed Vireo, Chewink, Long-billed Marsh Wren, Water-Thrush, Black-throated Blue, and Black-poll Warblers. Of the Vireo, Chewink and Wren but single specimens were observed. The former was caught alive and later liberated in the country, making little or no effort to escape when approached. The Wren was the most active of the three, while the Chewink, apparently hungry, was continually picking at dirt particles and other minute objects in search of something to eat.
>
> After making the above notes, we proceeded to the roofs for further examinations, and here the conditions proved even more pitiful than those below. Dead birds lay everywhere, while others, seemingly bewildered, flitted about on the ledges of the building, apparently too weak to resume their weary journey, or, as before stated, had lost all sense of direction. If such was not the case, the birds were certainly on the point of exhaustion, otherwise one cannot conceive anything to prevent them from resuming their northward journey from these upper ledges, high above the city, its noise and confusion.
>
> The birds in the areaways acted in the same way. When we entered from below they immediately flew to the top and alighted on the surrounding railings; but when we withdrew, the birds, instead of flying up to the roof and continuing their journey, immediately flew back down into the pits, which were sooner or later to be their tombs, apparently frightened by the crowds and continuous bustle. Most of these birds seemed very much exhausted, but were quite able to fly continually back and forth the full length of the areaways.
>
> Although many of the birds became exhausted from continuous fluttering about the lights and later succumbed to exposure, the greater number of the hundreds of lives lost were caused by coming in contact with hard structures, as the fractured limbs, bruised bodies, indented and blood-clotted skulls proved, when examinations were made after skinning the specimens…
>
> And now let us consider some of the most interesting points in the case; i.e. the cause of such an appalling destruction. Following an unusual cool period of weather for the month of May, on the 21st considerable moderation took place, and about 10 p.m. rain began falling. Prior to the rain quite a heavy mist hung about the city, but was later cleared away by the falling rain. By midnight and in the early morning hours the rain had turned to a thunderstorm with a terrific downpour, which continued well into the morning.
>
> As before stated the greater number of birds were killed by striking hard structures, and it is the writer's opinion that the birds, being forced to migrate low on account of the storm, were attracted by the bright lights, and apparently misconceiving them to be suspended in midair, attempted to fly past just above or below the center of illumination, and therefore struck the darker portions of the tower, which were unilluminated.

We must however bear in mind that this was but one immediate locality, and when we consider the number of towers, and equally as tall buildings through the city, we realize that the loss of life must have been tremendous, and can certainly not help but have a noticeable effect upon bird-life.[1]

Figure 7.2 City Hall Tower, with its "corona" of arc lamps near the top, 1917. On the evening of May 21, 1915, hundreds of migrating birds, including northern parulas, died in collisions with the tower. (Photo courtesy of PhillyHistory.org, a project of the Department of Records of the City of Philadelphia)

The dead "yellow and green" bird that prompted the phone call to the Academy in May was probably a warbler, perhaps the northern parula. A century earlier Alexander Wilson reported that the northern parula arrives in Pennsylvania from the south in May.[2]

The clock in City Hall Tower

In December 1898 the City of Philadelphia had installed a gigantic illuminated clock in City Hall Tower, then the world's tallest occupied building. The tower rose 167 meters above the street, and on each of its four sides it supported an illuminated dial 8 meters in diameter. Each minute hand was almost 5 meters long, including the counterweight, and weighed over 100 kilograms. The clock with all its parts weighed 50 tons. A hydraulic air compressor powered it, and 512 electric lamps, each with an output of 16 candlepower, illuminated the dials.[3]

To broadcast the time to the surrounding suburbs, the clock controlled a "corona" of arc lamps shining outward from the base of the statue of William Penn, the highest point of the tower. They were visible "twenty-five or thirty miles from the city, appearing like a delicate silver crescent suspended low against the horizon."[4] Every night the clock would turn the arc lamps off ten minutes before 9 p.m. and back on precisely at 9 p.m. *The Official Handbook of City Hall* instructed suburbanites on how to process the signals:

> Look towards the City Hall a few minutes before nine o'clock P.M. until the circle of light at the top of the tower disappears; then, when it reappears, set your watch or clock at the hour NINE, and, presto, you have secured correct time.[5]

Dead birds were first observed at City Hall Tower in 1899. William L. Baily, one of the founders of the Delaware Valley Ornithological Club, reported the phenomenon at a meeting of the American Ornithologist's Union in Philadelphia that year:

> In the centre of the city of Philadelphia, five hundred feet and more above the pavement, on top of the City Hall Tower, stands the colossal bronze figure of William Penn, encircled with a ring of arc lights which burn the night long. Unintentionally this beautiful circle, crowning the highest point for miles around, has been the destroyer of many birds during their nocturnal migrations between their winter and summer homes.[6]

Baily reported the dead birds collected from around the tower in 1899 consisted of 56 species and 452 individuals, including 67 northern parulas.[7] During the first decade of the twentieth century, *Cassinia,* the journal of the Delaware Valley Ornithological Club, published an annual tally of birds killed at City Hall Tower.

Lighthouses in the Gulf of Mexico

In 1904 Wells W. Cooke was able to pinpoint fall migration times based on deaths of migrant birds at powerful electric lights in the Gulf of Mexico.

> The largest single addition to the knowledge of movements of birds along the southern border of the United States is due to records of species striking the lighthouses off the south coast of Florida. Several thousands of these instances have been recorded. They furnish the best available data so far collected on the length of the migrating season, and afford also much-needed information concerning the time when many species of birds begin their migration in the fall. The keeper of the lighthouse at Sombrero Key, in particular, has taken

> much interest in the matter, and has spent many hours counting and identifying birds, either killed by flying against the glass protecting the light or resting bewildered on the balcony after striking. Eight hundred and sixteen records were received in five years from this one lighthouse. They comprise a total of 2,011 dead birds and 10,086 birds which struck the light with so little force that on the return of clear skies or daylight they were able to resume their flight. Warblers migrate chiefly by night and are so susceptible to the influence of a bright light that they constitute at least 80 percent of these thousands.[8]

Among warblers, the northern parula was the second most common to strike the lighthouse:

> The earliest fall movements of the parula warbler on land cannot be noted, for the migrants are not distinguishable from the breeding birds. When, however, the species begins to strike against the lighthouses of southern Florida, it is certainly migrating. It passes through Florida in countless thousands, being second only to the black-throated blue warbler in the frequency with which it strikes the lighthouses. Out of eighty-eight recorded dates of the striking of parulas in fall only eight are earlier than the second week in September.[9]

Disruption of visual cues used in navigation

The abundance of warblers (including northern parulas) migrating at night accounts at least in part for the large number of these birds colliding with City Hall Tower and Cooke's lighthouses. To navigate at night, warblers integrate many cues, including the pattern of stars in the sky, polarization of skylight, landmarks, and the earth's magnetic field.[10] Clouds and fog increase the risk of collisions with buildings, presumably by obscuring visual cues. Electric lighting at night exposes migrants' navigational systems to visual artifacts, compounding disorientation caused by overcast skies. Laboratory experiments and field trials suggest that artificial light disturbs magnetoreception, and that the least disruptive wavelengths lie in the green spectral region.[11]

Glass facades and windows

Eight skyscrapers, all constructed since 1987, now dwarf City Hall, which is dimly lit, its arc lamps long gone. The Comcast Center is taller than City Hall by 130 meters—greater than the length of a football field. The shortest of Philadelphia's top twenty tallest buildings is 23 meters higher than City Hall's big clock.[12] Illuminated windows highlight the sides of these buildings at night, and ornamental lighting decorates their tops and sometimes their facades. On overcast nights, when nocturnal migrant birds are most vulnerable to collisions with buildings, the upper stories disappear in a shroud of fog.

Figure 7.3 Comcast Center's glass façade blending in with the sky. One thousand birds per year have been estimated to die striking this and adjacent buildings.

The reflective blue-tinted glass of new skyscrapers compounds the danger their lights and height pose to migrating birds. Over the past two decades, Daniel Klem of Muhlenberg College in Allentown, Pennsylvania, has studied bird fatalities caused by collisions with glass. He has concluded that window glass kills more birds than does any other human disturbance except destruction of habitat,[13] and that the primary reason for collisions with glass is the failure of birds in flight to recognize clear or reflective glass as a barrier.[14]

The lethality of glass may appear obvious at ground level, where homeowners who have placed bird feeders near windows witness birds striking windowpanes.[15] On skyscrapers high above street level, the dangers are harder to observe. Klem and colleagues, collaborating with New York City Audubon, trained thirty volunteers to recover dead or injured birds from the base of seventy-three buildings in Manhattan during two migration periods. They recovered 549 birds—82 percent of which were dead—that included at least fifty species. The proportion of a facade that was glass correlated with the number of birds recovered from its base. Klem et al. concluded that glass, ranging from small windows to entire walls of buildings, is a lethal hazard for birds. Among the ten most common victims in this study were northern parulas.[16]

A study of birds killed striking buildings in Toronto compared the frequency of collisions to the proportion of windows that were illuminated at night. Based on recovery of 1,300 dead and injured birds from sixteen buildings, the investigators concluded that window lighting contributed to bird collisions.[17] A practical question is whether turning off lights in buildings at night reduces bird strikes at windows. Ornithologists at the Field Museum in Chicago had been acquiring new specimens by collecting dead birds at a large lakefront convention center called McCormick Place. To cut costs when the exhibition hall was not booked, the building manager began turning off the lights—and the number of dead birds per year plummeted by 80 percent.[18]

Figure 7.4 Illuminated facade of Cira Centre viewed across the Schuylkill River from Center City. The color and pattern of the ornamental lighting changes from one night to the next.

Removal of the arc lamps from the tower of Philadelphia's City Hall did not stop birds from colliding with other buildings in the city. In 2011, students monitoring bird kills on the campus of Temple University recovered 1,200 dead birds including thirty-one species, mostly around Paley Library, the student center, and Tyler School of Art.[19] Keith Russell from the National Audubon Society has monitored dead birds around the base of the Comcast Center. He has estimated that 1,000 birds per year fatally strike windows of buildings in this location.[20]

Figure 7.5 Temple University's Howard Gittis Student Center, with windows that reflect images of street trees, promoting collisions by birds. In 2011, 1,200 birds including thirty-one species were recovered around Temple's campus buildings, including this building.

Unilluminated structures

Even in the absence of windows and brightly illuminated facades, tall structures are a danger to birds migrating at night. Millions of birds have died striking power lines, guy wires, communication towers, and wind turbines.[21] Communication towers alone are estimated to kill 6.8 million birds per year.[22]

Artificial light aimed skyward

Artificial light can kill birds independent of collisions. In a single night in October 1954, an estimated 50,000 dead birds representing fifty-three species, including northern parulas, were strewn over the runways, taxi strips, and other surfaces of the Warner Robins Air Force Base in Macon, Georgia. They had flown or dropped downward in response to a ceilometer, a beam of light aimed up from the ground to measure height of cloud cover.[23] The birds may have circled around the light until they fell from exhaustion or disorientation.[24]

Polarized light reflected off buildings and pavement

Polarized light reflected off glass buildings on riverbanks attracts caddis flies (Trichoptera), which normally use polarized light reflected off water for navigation; glass, like water, polarizes light it reflects.[25] Bridges reflecting polarized light have disrupted flights of mayflies (Ephemeroptera), which normally breed in water.[26] Solar panels reflecting polarized light have attracted insects from three orders (Diptera, Trichoptera, and Ephemeroptera).[27] Roads reflecting polarized light have induced mayflies to lay eggs on asphalt,[28] and cars reflecting polarized light have attracted dragonflies, which laid eggs on their hoods.[29] Birds detect polarized light, which may explain why water birds at night become stranded on artificially illuminated asphalt parking lots.[30]

Birds and insects navigating downtown confront a cityscape filled with potentially disorienting visual artifacts. How they negotiate this visual noise is poorly understood. Dead birds and insects around buildings and lights demonstrate that visual noise can be lethal.

Impact of collisions on populations of the northern parula

The northern parula overwinters from the southern United States to Venezuela and Nicaragua. Records of this species colliding with towers and other man-made structures have been reported in Wisconsin, Missouri, Illinois, New York, Pennsylvania, Tennessee, and Florida.[31]

The species breeds in widely scattered locations throughout Pennsylvania, but not in Philadelphia,[32] where it stops to rest during its migratory journeys.[33] Breeding populations of northern parulas have recently been increasing in Pennsylvania,[34] but they have decreased or disappeared in southern New Hampshire, Massachusetts, Connecticut, Rhode Island, New Jersey, and Delaware.[35] Fatal collisions with man-made structures like City Hall Tower might have selectively killed northern parulas whose migration flyways exposed them to the densely urbanized Northeast corridor.

Many other conditions have been cited to explain declines in populations of neotropical migrants like the northern parula. Destruction of forests has reduced habitat in their northern breeding grounds or southern overwintering sites or both.[36] Acid rain has suppressed lichens (*Usnea sp.*) that northern parulas use for building nests.[37] Fragmentation of forests has stimulated growth in populations of cowbirds, which are nest parasites of neotropical migrants.[38] Declines in flying insects have reduced food supplies of neotropical migrants.[39] In Europe, global warming has desynchronized long-distance migrants from their insect prey, which now emerge earlier in the spring.[40]

None of these ecological problems is mutually exclusive, and together they reveal migrants' complex vulnerabilities, which defy easy analysis.[41] For example, some northern parulas build nests without lichens,[42] and compared to most other neotropical migrants, they are infrequent victims of nest parasitism by cowbirds.[43]

Controversy over effects of collisions on populations of birds

Collisions with buildings and communication towers have caused no measurable decline in bird populations, according to a recent statistical analysis of records of 240,000 collision fatalities involving more than 188 species; the analysis also included population records from the Breeding Bird Survey. The investigators collected data from records from New York, Chicago, and Toronto, plus records of collisions with communication towers outside of metropolitan areas. They pointed out that many of the species of birds killed are abundant and have high reproductive potential—sufficient to compensate for losses sustained in collisions with buildings.[44]

Partly in response to this paper, Scott R. Loss, at the Smithsonian Migratory Bird Center, and his colleagues have contended that data on causes of mortality of birds are currently insufficient for determining how any particular cause affects populations.[45]

Protective accommodations at city light show

During the peak of the fall migration over Philadelphia in 2012, the city hosted an outdoor light show at night on the Benjamin Franklin Parkway. Designed by Mexican-Canadian media artist Rafael Lozano-Hemmer, it included twenty-four moving searchlights aimed upward, somewhat reminiscent of the ceilometer that killed tens of thousands of birds in Georgia over half a century ago.

Alerted to the danger, Lozano-Hemmer collaborated with ornithologists to avoid injuring migrating birds. Bird watchers monitored the night sky for birds congregating around the light beams, which could be turned off in response. Blackout periods were programmed into the show to provide recovery time for the birds. The spotlights were kept in motion to avoid simulating the stationary behavior of lethal ceilometers. Beams were pointed away from buildings at risk of bird strikes. Filters reduced output of energy in the red and ultraviolet wavelengths within the light beams.[46] Watching the light show, I saw no birds concentrating around the lights.

Biophilia

On the morning of October 4, 2010, after a storm the night before, I found a dead warbler on the sidewalk along Walnut Street near 23rd Street. I submitted it to the Academy of Natural Sciences, which accessions such specimens. Nathan Rice, ornithological curator at the Academy, identified the warbler as a northern parula. The death of this iridescent yellow and green nocturnal migrant struck me as sorrowful. The most compelling reason for concern about lethal collisions of birds with buildings in downtown Philadelphia may be biophilia, the human bond with other species.[47]

8

POLYPHEMUS MOTH

(*Antheraea polyphemus*)

Polyphemus moths were collected in Philadelphia in the nineteenth century before the advent of electric lighting. Paradoxically, light pollution in Center City may protect them.

Figure 8.1 Display box from the Titian Ramsay Peale Butterfly and Moth Collection preserved at the Academy of Natural Sciences of Drexel University. The four largest moths are polyphemus moths. Records for those in the bottom left and top right specify Philadelphia, 1833. Historical notes, possibly Peale's, are partly visible in the background. (Curated by Jason D. Weintraub, entomological collection manager at the Academy of Natural Sciences of Drexel University. Photo courtesy of the Academy of Natural Sciences of Drexel University.)

In 1833 Titian Ramsay Peale collected a male and female polyphemus moth in Philadelphia.[1] Polyphemus moths are giant silkmoths (saturniids) with big three-dimensional-appearing eyespots on their hind wings. The species is native to most of the United States and southern Canada. Its larvae eat birch (*Betula*), willow (*Salix*), maple (*Acer*), and oak (*Quercus*). Although boldly patterned and with wingspans up to 15 cm, the moth is hard to find: it folds its wings and conceals its eyespots during the day; its larvae are cryptically colored; its cocoons usually drop and disappear into leaf litter; and the moth flies exclusively at night.[2]

Municipal lighting in Philadelphia

Conceivably, Peale collected his polyphemus moths at oil lamps. Municipal lighting in Philadelphia dates back at least to 1791, when an oil depot in Franklin Square supplied fuel to the city's streetlamps.[3] Records do not document whether Philadelphia's municipal oil lamps attracted giant silkmoths, but C. A. Frost reported that a kerosene lamp attracted polyphemus moths to a window in his house in Framingham, Massachusetts, in the beginning of the twentieth century.[4]

Figure 8.2 Lamplighter, Philadelphia. (Anonymous artist, from engraving in *History of Philadelphia...Containing a Correct Account of the City Improvements up to the Year 1839*[5])

By the time Peale collected these two moths, the city was planning municipal lighting powered by gas made from coal at the Philadelphia Gas Works, which was built in 1835 on the east bank of the Schuylkill River just north of Market Street.[6] Four years after its construction, the Philadelphia Gas Works was delivering gas through 23 miles of pipe to 11,802 burners, including 434 streetlamps.[7] Although Philadelphia's municipal gas lamps may have attracted polyphemus moths, conditions around the lamps may have deterred moth collectors. In a guide published in 1839, Daniel Bowen attempts to dispel fears about public safety downtown after dark:

> The City is well guarded at NIGHT by able bodied men of good character...Each of the Four Divisions has about 35 Watch-men whose duty it is to trim, light and extinguish the public lamps and gas-lights, to walk their rounds, and cry the hours while on duty and to secure the peace and quiet of the city.
>
> In addition to the Watch-men attached to particular stations, each Division has allotted to it 8 silent Watch-men whose duty it is to see that the stationed Watch-men attend to their prescribed duties and maintain watchfulness during the hours allotted them and to walk quietly through the Division, and to see that thieves &c. are not making inroads between the regular Watch-mens' rounds: they usually pursue their rounds, two in company.[8]

In 1881, four years before Peale died, the Brush Electric Company installed the city's first electric streetlights on Chestnut Street, where it erected Brush lamps, carbon arc lamps so bright that two of them in a hotel dining room replaced 144 gas burners.[9]

Moths once abundant at electric lights

In 1892 the Smithsonian Institution published *Directions for Collecting and Preserving Insects,* by Charles Valentine Riley, who advised where to search for moths:

> Collecting by the aid of strong light is a favorite means for moths as well as other insects, and nowadays the electric lights in all large cities furnish the best collecting places, and hundreds of species may be taken in almost any desired quantity.[10]

In 1900 Sherman F. Denton, author of a two-volume introduction to the butterflies and moths of the eastern United States, reported the rewards of collecting downtown at electric light:

> While employed in Washington, D.C., I made a splendid collection of the moths of that region simply by going the rounds of a number of electric lights every evening. The lamps about the Treasury Building were sometimes very productive of fine specimens and the broad stone steps and pillars were frequently littered with moths, May flies, beetles, etc., where one could stand and pick out his desiderata with little difficulty. I captured several of the Regal Walnut moths (Citheronia regalis) and a number of our largest and handsomest sphinxes. Besides making the acquaintance of a number of insects new to me, I met several entomologists who, like myself, had been attracted to the lights by the abundance of specimens.[11]

Light pollution

By 1988 Philadelphia had 100,000 high-pressure sodium streetlamps at an average density of almost 200 lamps per square kilometer. The radiant energy they emitted equaled more than 10 kilowatts per square kilometer, an order of magnitude greater than the energy of moonlight at full moon. Over the preceding four decades, the output (lumens) per lamp had increased sevenfold, while the number of lamps had tripled.[12] High-pressure sodium lamps emit minimal ultraviolet energy,[13] the spectral region most attractive to moths; but the energy they emit in the blue and green part of the spectrum does attract moths.[14]

Today light pollution in Center City is so diffuse that practically no outdoor locations are free of it, whether reflected from the sky or buildings, or transmitted directly from lamps. Even places that appear dark may be dark only relative to their artificially illuminated surroundings. Views of stars are washed out by electric light bouncing off the atmosphere. On clear nights that minimize atmospheric reflections, the Milky Way is visible, but its appearance is faint compared to that viewed in rural locations far from urban skyglow. Lamps in downtown Philadelphia attract few if any moths.

Harmful effects of light pollution on moths

Gerhard Eisenbeis, who monitored insects attracted to streetlamps in rural Germany, concluded that streetlamps deplete populations of nocturnal insects. He called the phenomenon a "vacuum cleaner effect."[15] Electric lamps can disrupt virtually every life function of moths that fly to them. These functions include feeding, mating, egg laying, dispersal, and migration. Electric lamps temporarily blind moths that approach them. The light probably resets their internal clocks. Predators such as birds and bats hunt insects attracted to light sources. By disturbing where moths land, the lamps spoil crypsis—the visual match between a moth and its background. Electric lamps desiccate or incinerate moths trapped inside their housings.[16] Based on Eisenbeis's conclusion, one might suspect that light pollution has depleted or extirpated Center City's populations of moths, such as the polyphemus moth.

Protective effects of light pollution on moths

A contrary view is that urban light pollution protects moths. By reducing background darkness, it suppresses the attraction of insects to lamps. Such attraction is the primary means by which artificial lighting harms moths. In 1997 José Luis Yela and Marcel Holyoak in Spain showed that moonlight reduced collections of insects attracted into light traps but not bait traps. Moonlight behaved like light pollution in the sense that it reduced background darkness and suppressed attraction of moths to artificial sources of light. Yela and Holyoak's findings showed that moonlight suppressed pathological behavior around artificial light, but allowed normal attraction to bait.[17]

Figure 8.3 Skyglow over Center City on a cloudy night. Background light from light pollution decreases attraction of moths to lamps. Cloud cover amplifies urban light pollution, which paradoxically protects moths from harm due to attraction to lamps.

Light pollution from one source of artificial light can protect moths from attraction to other sources. In 1950 H. S. Robinson and P. J. M. Robinson demonstrated how one lamp could reduce another's attractiveness to moths. If the two lamps are spaced so that they reduce the surrounding darkness that each requires for attracting moths, the two together will attract fewer moths in total than they would if operated separately.[18] Their experiments demonstrate how, in a densely illuminated area downtown, interactions among light sources clustered together suppress attraction of moths to lamps.

Cloud cover at night ordinarily screens out moonlight and starlight and increases darkness, boosting the attractiveness of lamps to moths. For example, Yela and Holyoak showed that cloud cover increased collections of moths in light traps.[19] Combined with light pollution, however, cloud cover may have the opposite effect: it reflects light pollution downward, magnifying it[20] and disrupting the darkness that flight-to-light behavior requires.

As a kid, I collected polyphemus moths at light in the George Washington National Forest in Virginia. Like any collector of moths, I soon discovered that the full moon was the worst time for collecting, and the new moon (i.e., no moon) with a cloud cover was the best. To maximize the attractiveness of my lamp, I took care to turn off all others nearby.

In Center City, light pollution may have contributed to the absence of polyphemus moths around city lights, but for reasons opposite to common wisdom. It's not that light pollution downtown harms moths; on the contrary, light pollution here protects them by suppressing flight to light. Light pollution may be construed as a double-edged sword, increasing or decreasing attraction to artificial light depending on circumstances.

Moths that fly to light in Center City

During the summer and fall of 2010, thirty-four species of moths flew to light in our backyard, which is shielded from streetlights. The number of individuals per night was small—usually none—and the moths were tiny compared to polyphemus moths. To attract them, I operated a 13-watt fluorescent blacklight (ultraviolet lamp) that illuminated a white pillowcase. The most common moth attracted to this light was the ailanthus webworm moth (*Atteva aurea*). Sometimes several of these colorful moths would arrive on a single night in the early fall; a mature ailanthus tree towers above the roof of a row house on our block. Larvae of most of the species that came to my blacklight are polyphagous: they feed on many kinds of local plants, both cultivated and wild. Figures 8.4–8.12 show some of the moths I photographed after they settled on the pillowcase or other surfaces near the lamp.

Figure 8.4 Green cloverworm moth (*Hypena scabra*). Its larvae feed on many kinds of plants, including clover (*Trifolium sp.*).[21]

Figure 8.5 Corn earworm moth (*Helicoverpa zea*). Larvae feed on corn and other crops.[22]

Figure 8.6 Boxwood leaftier moth (*Galasa nigrinodis*). Larvae feed on boxwood (*Buxus* sp.).[23]

Figure 8.7 Common looper moth (*Autographa precationis*). Larvae feed on many kinds of plants.[24]

Figure 8.8 Implicit arches moth (*Lacinipolia implicata*) on stucco wall of our house. Larvae feed on common dandelion (*Taraxacum officinale*) and many other plants.[25]

Figure 8.9 Suzuki's promalactis moth (*Promalactis suzukiella*). Larvae are found under bark of rotting logs.[26]

Figure 8.10 Morning glory plume moth (*Emmelina monodactyla*). Larvae feed on common morning glory (*Convolvulus sp.*), lambsquarters (*Chenopodium album*), and others.[27]

Figure 8.11 Little underwing moth (*Catocala minuta*). Larvae feed on honey locust (*Gleditsia triacanthos*),[28] a common street tree on our block.

Figure 8.12 Common tan wave (*Pleuroprucha insularia*). Larvae feed on many species, including goldenrod (*Solidago sp.*), bittersweet (*Celastrus scandens*), and oak (*Quercus sp.*).[29]

Why polyphemus moths did not come to my light

If so many kinds of moth in Center City flew to my lamp, why didn't the polyphemus moth? Giant silkmoths naturally live at low population densities, which protect them from parasitoids, microbes, and other enemies.[30] Polyphemus moths can find mates far away because males can detect minute concentrations of female sex pheromone[31] and efficiently home in on the source.[32] In one instance, a trap baited with pheromone captured a marked polyphemus male released the same evening at a distance of 7.5 kilometers.[33] Even if polyphemus moths did, like other moths, occasionally fly to lamps downtown, the large numbers of lamps and small number of moths would make the chance of finding a polyphemus moth at any particular lamp low.

A more compelling reason for the rarity of polyphemus moths at lamps in Center City is destruction of habitat. Daniel Janzen observed that declines of moths flying to electric light in Costa Rica coincided with agricultural destruction of their habitat.[34] From 1970 to 1990, urban sprawl in the 100 largest metropolitan areas in the United States increased by 37,671 square kilometers; Philadelphia's per capita increase in sprawl was 48 percent, the most of any metropolitan area.[35]

Unlike cocoons of giant silkmoths such as *C. promethea* and *H. cecropia,* cocoons of polyphemus moths usually fall to the ground rather than remaining suspended from tree branches.[36] In Center City, a cocoon on the ground is likely to be treated as litter, and trashed. In the suburbs, it is likely to be raked up with leaves, sent to a recycling center, and converted to mulch. Gray squirrels prey on polyphemus cocoons.[37]

A polyphemus moth shows up downtown

On the morning of July 24, 2011, looking out our third story window on Pine Street, I noticed a female polyphemus moth resting on a branch of a willow oak (*Quercus phellos*). This moth was just beyond arm's reach, and about 20 meters away from a high-pressure sodium streetlamp. Willow oak is a host plant of larvae of *Antheraea polyphemus.*[38] I contacted Jason Weintraub, a neighbor and lepidopterist at the Academy of Natural Sciences, who came over and, leaning out the window, used a long-handled net to capture the moth. She laid fertile eggs, and Jason reared her offspring.

Figure 8.13 Gravid female polyphemus moth on her host plant, willow oak (*Quercus phellos*), July 24, 2011, outside a third-story window of our home on Pine Street, a few doors down from a sodium streetlamp. Her wingspan when spread open was about 15 centimeters (6 inches).

In February several years later I found a polyphemus cocoon dangling from a low branch of a cultivated river birch tree (*Betula nigra*), another host plant of polyphemus caterpillars. It caught my attention after it cast a distinctive silhouette against a fresh layer of snow. This cocoon's failure to drop to the ground saved it from being trampled.

Figure 8.14 Cocoon of polyphemus moth highlighted against a fresh layer of snow in February. It is dangling from a cultivated river birch (*Betula nigra*) along the Schuylkill River Trail in Center City.

David L. Wagner recently reviewed the decline in abundance of moths in the northeastern United States, particularly in Connecticut. Among Connecticut's fifteen species of giant silkmoths (saturniids), he found an increase in abundance of only one: the polyphemus moth, which is now common. Most species of giant silkmoth in the state were either declining in numbers or extirpated. Why populations of polyphemus increased in Connecticut in the last ten years is a mystery.[39]

Whether populations of polyphemus moths are increasing in Center City is unknown. Declines in populations of house sparrows (*Passer domesticus*) could reduce predation on them. Common nighthawks (*Chordeiles minor*), nocturnal insectivores once common in Center City, are now rare. An epidemic fungal disease, white nose syndrome, is destroying local populations of little brown bats (*Myotis lucifugus*). The use of bug zappers outdoors to kill mosquitoes carrying West Nile virus may have helped polyphemus moths; in suburban Newark, Delaware, bug zappers killed fewer moths than parasites and predators of insects.[40]

Conceivably populations of polyphemus moths, like clothes moths, evolved resistance to attraction to artificial light; or they may have evolved increased fitness in some other way. Jason Weintraub has curated Titian Ramsey Peale's historic collection of moths preserved at the Academy of Natural Sciences of Drexel University. He has posted online images of nearly 100 boxes of specimens, including the two polyphemus moths Peale collected in Philadelphia in 1833. If polyphemus moths evolved traits that increased their fitness, changes in their DNA compared to that in specimens dating back to the nineteenth century might offer clues to the evolutionary steps.

Figure 8.15 Clothes moth (*Tineola bisselliella*) indoors in Center City. It is not attracted to light.

SPOTLIGHT

COMMON GROUNDSEL

Top: Common groundsel flowering at the base of a mailbox on a sidewalk on South Street, January 23rd. It is producing seeds.
Bottom: Close up of groundsel's flowers.

Common groundsel (*Senecio vulgaris*) in Center City blooms while rooted in cracks in concrete in the dead of winter. Such extreme conditions insulate it from competitors and enemies. Groundsel is endowed with toxins (pyrrolizidine alkaloids), toothed leaves, salt tolerance, lead tolerance, and herbicide resistance. It reproduces by self-pollination or cross-pollination, depending on season and conditions.[1] Its seeds disperse on silken threads blown by the wind. Its annual growth habit is adapted to ephemeral habitats, such as those in Center City. Introduced from Europe, it was established as a weed in the city at the time of publication of Philadelphia's first flora in 1818.[2]

9

BRIDGE SPIDER

(Gray cross spider; *Larinioides sclopetarius*)

Figure 9.1 Male bridge spider. The two hairy black spheres on arms in the lower center are palps, with which the male transfers sperm to the female. The dark structures between them are fangs, folded up.

In Center City, Philadelphia, the bridge spider snares insects exclusively at night. A Philadelphia arachnologist in the nineteenth century reported that this spider hunted primarily during the day.

On October 25, 1883, Henry C. McCook observed spiders dispersing by air and rail in downtown Philadelphia:

> At noon, while crossing the Chestnut Street Bridge, Philadelphia, I saw a great number of aeronautical threads floating in the air, streaming from the tips of the bridge balustrade and lodged upon the piers. One of the threads, a long filament, was sailing slowly toward the river as a Pennsylvania Railroad train dashed along the river track beneath the bridge. It was low enough to strike the cars as they rolled by, and so was carried on southward with its tiny voyager—another illustration of how artificial habits of man tend to the geographical distribution of life. The filaments were long, pure white, curled or wrinkled, about one millimetre wide or less, occasionally expanded into thicker wads.[1]

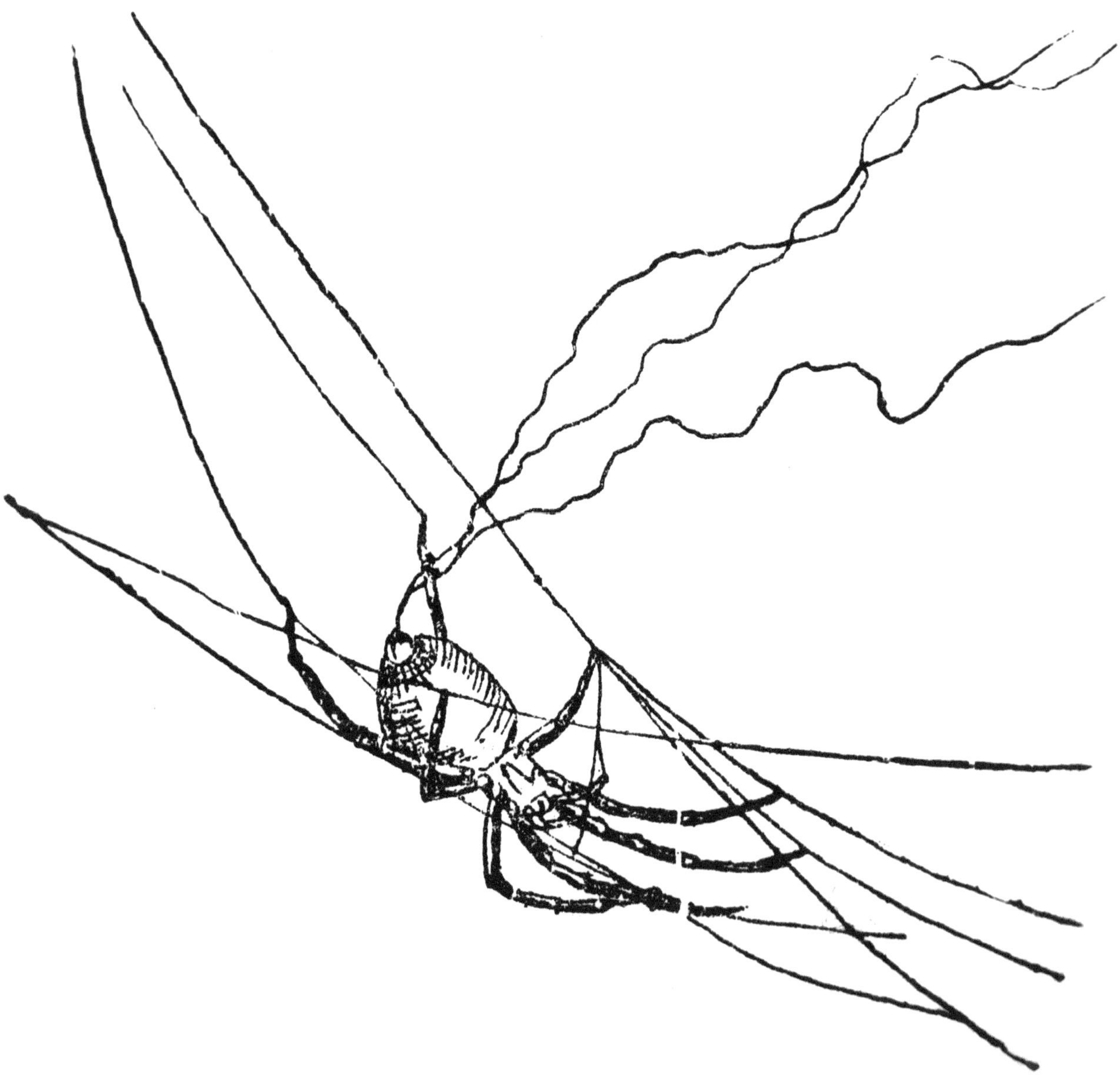

Figure 9.2 "Young spider sending out aeronautical threads." McCook observed spiders ballooning as he walked across the Chestnut Street Bridge. (Quote and illustration from H. C. McCook [1890] *American Spiders and Their Spinning Work*, vol. 2, Philadelphia, figure 276)

Attraction to artificial light

McCook brought an egg case of the bridge spider (*Lariniodes sclopetarius*) into his house, and noted that the newly hatched spiders resembled those he had observed floating off the Chestnut Street Bridge. He also noted that these spiders crawled toward the artificial light of his gas lamp.

> While reading on the evening of June 19th by the light of an argand burner, I glanced upward and observed that the lamp was covered with web lines that fringed the bottom of the porcelain shade and metal stand. Upon these lines forty or fifty spiderlings hung, in the full blaze of light. They had evidently just issued from the cocoon tent, and had been carried by the wind along a bookcase and across the desk to the lamp, a total distance of fourteen feet. A bridge line four feet long was strung from the bookcase to the lamp, along which the brood had clambered, attracted undoubtedly by the light. There was no reason why they should have sought that particular spot, and many reasons why they should have gone elsewhere, but the light dominated their action.[2]

Figure 9.3 Young bridge spiders attracted to gas lamp. (Illustration from H. C. McCook [1889] *American Spiders and Their Spinning Work,* vol. 1, Philadelphia, figure 141)

Capture of prey in daytime

In Atlantic City, New Jersey, McCook observed a large colony of bridge spiders preying on greenhead flies in webs constructed over water between pilings by boathouses. He watched them sitting on their webs continuously in broad daylight.[3] He commented that even though this spider is capable of capturing prey in the dark, it could not be classified as nocturnal. Infestations of greenhead flies *(Tabanus nigrovittatus)* still occur around buildings near salt marshes in New Jersey. Their flight activity peaks during daylight of moderate intensity (40,000 lux).[4]

Capture of prey during the night

McCook's observations about the daytime predatory behavior of the bridge spider contradict my observations of the same spider in Center City, but his findings of the attraction of the spider to light fit perfectly. In Center City I have observed bridge spiders catching prey exclusively at artificial light at night. They snare their prey in webs that they erect on municipal lamps. They construct them on lamp fixtures along the east bank of the Schuylkill River, where they prey on nocturnal insects such as midges (chironomids) that breed in the water and fly to lamps.

Bridge spiders are most common on lamps in sheltered locations, such as under bridges. They colonize lamps located in the open if the lamps have overhanging reflectors that can anchor and shelter webs. The bridge spider is the only kind of spider that I have observed on municipal lamps along the river. This spider is especially abundant on the illuminated walls of Lloyd Hall, also known as 1 Boathouse Row.

Figure 9.4 Webs of bridge spiders reflecting light from a lamp underneath Walnut Street Bridge. (See figure 9.12 for close-up.)

Figure 9.5 Lloyd Hall (1 Boathouse Row, on far right), whose light attracts large numbers of bridge spiders.

Figure 9.6 Lamp design attractive to bridge spiders. Overhanging reflectors shield webs from rain and anchor them. These lamps line the east bank of the Schuylkill River.

How might bridge spiders gain access to lamps on top of 5- or 6-meter-high metal lampposts? McCook's observations of spiderlings floating off the Chestnut Street Bridge show that bridge spiders could access municipal lampposts by air. His description of the spider's attraction to light suggests that a bridge spider ballooning through the air need not make a point landing precisely on a lamp fixture to gain access; it could float to a lamppost and then, by crawling or ballooning, follow the artificial light up the pole.

Figure 9.7 Orb web attached to municipal lamp fixture by the Schuylkill River, Center City. To reach this seemingly inaccessible location, the spider probably floated through the air on a strand of silk blown by the wind. After landing near the lamp, the spider could balloon or crawl the rest of the way, positioning its web according to the artificial light and structural support.

McCook's classification of the bridge spider (which he called the gray cross spider, or *Epeira sclopetaria*) as diurnal differs from the common view of this spider as nocturnal. Professionally, McCook was pastor of the Tabernacle Presbyterian Church (which still stands at 37th and Chestnut Streets), and technically he ranks as an amateur arachnologist, but his *American Spiders and Their Spinning Work*, published in three volumes of over 1,000 pages from 1889 to 1893, was at the time the most comprehensive work on the life history of North American spiders. Professional arachnologists continue to cite his work as authoritative. McCook's observations of the bridge spider capturing prey during the day were undoubtedly accurate.

Shift in activity from day to night

In 1999 Astrid M. Heiling at the University of Vienna confirmed McCook's finding that this species is attracted to artificial light. Heiling concluded that the attraction is innate because his laboratory-reared spiders sought light even though they had had no experience hunting at light. Heiling speculated that this behavioral trait evolved in response to concentrations of insects around moonlight reflected off water.[5] (He did not cite McCook's work, probably because McCook published it privately and copies were scarce until recently, when digital copies became freely available online.)

McCook's nineteenth-century report on the daytime activity of the bridge spider suggests that the bridge spider synchronizes its activity to that of its prey, day or night. Heiling and Marie E. Herberstein used sticky traps to monitor changes in prey availability over the course of the night while they observed bridge spiders on an illuminated footbridge. They found that the spider emerges from hiding and waits in its web when availability of prey is greatest.[6] This strategy of predator synchronizing with prey could explain why over a century ago the bridge spider preyed on greenhead flies in Atlantic City in broad daylight, while now in Center City it preys on midges at artificial light at night.

I have observed a shift from diurnal to nocturnal hunting in the case of the jumping spider, *Platycryptus undatus,* which ordinarily stalks and pounces on prey in bright sunlight. It relies on keen vision rather than a web to find prey. On Cape Cod after dark, I watched it appear at a porch lamp, where it seized insects attracted to the light. During the day it would hide in its silken retreat, and at night it would emerge to hunt at the lamp.[7]

Establishing a web on a lamp

Lamps by the river offer the bridge spider plentiful prey, but how do newly hatched spiderlings find sites suitable for anchoring their tiny webs on municipal lamp fixtures? McCook found that bridge spiderlings build their webs on abandoned webs of other spiders.[8] Abandoned webs festoon municipal lamps along the Schuylkill River.

Figure 9.8 Webs of juvenile bridge spiders in an abandoned web in a window. (Illustration from H. C. McCook [1890] *American Spiders and Their Spinning Work,* vol. 2, Philadelphia, figure 258)

Figure 9.9 Lamp under a bridge over the Schuylkill River Trail. Old abandoned webs support new webs of young bridge spiders.

Use of a visual lure to attract prey

In 2008 Chih-Yuan Chuang and colleagues at Tunghai University in Taichung, Taiwan, published a paper entitled "Deceptive color signaling in the night: A nocturnal predator attracts prey with visual lures." The investigators noted that many species of nocturnal orb weavers have bright markings on their otherwise dark ventral surface, which they display when they sit and wait in the center of their webs. They hypothesized that these bright markings lure prey into their webs. They showed that webs with spiders attracted more prey than did webs without spiders, and that webs with spiders attracted less prey when the bright markings were painted over. The spider they studied is in the same family as the bridge spider, and the reflective markings on the underside of the two species look similar.[9] Chuang demonstrated that a visual lure in another species of spider attracts prey during both day and night.[10]

Figure 9.10 Reflective lure on the underside of a bridge spider waiting in its web at a lamp along the Schuylkill River. View is from the lamp, looking out into the night.

The bridge spider's lure consists of yellow and orange reflective hairs and pigments that produce a bright pattern sharply demarcated against the dark background of the spider's thorax and abdomen. An insect attracted to the lure would fly directly into the spider's clutches. When the spider in its web faces a lamp, its lure faces the light, and viewed from the lamp, the light of the lure stands out against the blackness of the night.

Exploitation of a high concentration of prey

A characteristic of the bridge spider is its selection of habitats with prey in high concentration, as around lamps and boathouses. McCook reported that in Philadelphia the spider occurred around stables and outhouses.[11] Anja Kleinteich and Jutta M. Schneider at the University of Hamburg found that increases in food availability increase this spider's developmental growth rate. They hypothesized that such developmental agility enables the spider to exploit urban habitats.[12] The result can produce extraordinarily high concentrations of spiders. Bridge spiders attained densities of 100 individuals per square meter at artificial light at Cincinnati's Riverfront Coliseum Sports Arena.[13]

Figure 9.11 Bridge spider on pile of prey at lamp along the Schuylkill River.

Controlling aggression within a colony

Bridge spiders concentrated in dense colonies can consume massive numbers of insects attracted to a single lamp; but spiders are renowned for territoriality and cannibalism. These two traits may benefit a spider when prey is scarce and competition for food and space is high; but they may subject the spider to conflict that is counterproductive when prey is plentiful. How does a colony of bridge spiders regulate the aggressiveness of its members?

Figure 9.12 Colony of bridge spiders living in close proximity under a lamp (same lamp as in figure 9.4).

Additional studies at the University of Hamburg investigated this question. They measured "personalities" of bridge spiders. Individual spiders differed in the degree to which they expressed aggressiveness toward other members of the colony. As a trait, aggressiveness was in part inherited and, depending on conditions, linked to increased mortality. The investigators suggest that natural selection acting against this trait makes it self-regulating. They hypothesize that natural selection favors a mixture (polymorphism) of personalities within a population.[14]

Food security

Despite the efficiency with which a colony of bridge spiders is able to consume prey, these spiders exert a negligible threat to their own supply of food. Bridge spiders colonize only a minority of municipal lamps along the river, and the fraction of flying aquatic insects that fly to their lamps is small. Artificial light disrupts the mating and dispersal of insects attracted to it independent of predation by spiders.[15] The number of insects that bridge spiders consume is small compared to the total number of insects that breed in the Schuylkill River.

Mud daubers

The geographical range of the bridge spider spans much of temperate North America, Europe, and Asia, where it characteristically colonizes man-made structures near bodies of water, especially around artificial light in cities.[16] It was probably introduced into North America,[17] but its importation is undocumented. It was well established in South Carolina by 1847.[18]

I have observed the spider snaring insects at lamps on warm nights as early as March 12 and as late as October 22. Females stash their egg cases in recesses directly on lamp fixtures. The spider appears capable of overwintering and completing its life cycle without descending to the ground.

In Center City, buildings and bridges that attract bridge spiders also shelter its enemy: the black and yellow mud dauber (*Sceliphron caementarium*), a predator that specializes in preying exclusively on spiders. The wasp seizes, paralyzes, and carries them off to its nest to feed its young. In the first systematic study of the bridge spider in North America, Nicholas Marcellus Hentz obtained a diverse collection of bridge spiders from the nests of mud daubers.[19] Black and yellow mud daubers prefer to prey on spiders in flat webs, like those of bridge spiders, and they use chemotactic cues to find bridge spiders.[20]

Figure 9.13 Black and yellow mud dauber (*Sceliphron caementarium*) on a wild bean (*Strophostyles helvola*). Dedicating her prey exclusively to her offspring, she hunts in vegetation for spiders, but occasionally feeds on pollen or nectar. I have not seen mud daubers around municipal lamp fixtures.

Municipal lamps along the Schuylkill River distance bridge spiders from mud daubers in time and space: black and yellow mud daubers hunt for spiders in vegetation, and exclusively during the day.[21] In Center City, municipal lighting offers bridge spiders food, shelter, and protection from enemies.

SPOTLIGHT

DOGS

"Pee line," demarcating distribution of mosses and lichens on a tree trunk on 24th Street.

A field mark characteristic of Center City's trees is the "pee line," produced by dogs. In this photo the moss *Orthothrichum pumilum* and the bright yellow lichen *Candelaria concolor* grow above but not below this line. Pee lines are most conspicuous after rain in winter, when tree mosses and lichens brighten.

10

BLACK AND YELLOW MUD DAUBER

(*Sceliphron caementarium*)

The black and yellow mud dauber has inhabited Philadelphia since the beginning of the wasp's recorded history in North America more than two and a half centuries ago.

Figure 10.1 Black and yellow mud dauber collecting tidal mud by the Schuylkill River Trail at Walnut Street.

On April 25, 1745, members of the Royal Society of London heard a description of a "very curious" nest of a previously unknown North American wasp: the black and yellow mud dauber from Philadelphia. John Bartram had made the observations, which his patron, Peter Collinson, presented on his behalf to the Society. The paper also included Bartram's observations of the nest of a second mud dauber, the organ pipe mud dauber (*Trypoxylon politum*), also from Philadelphia.

Here is what the Royal Society heard, as published later in the *Philosophical Transactions* of the Royal Society of London:

> An Account of some very curious Wasps Nests made of Clay in Pensilvania
>
> By Mr. John Bartram: Communicated by Mr. Peter Collinson, F. R. S.
>
> Read April 25 1745.
>
> Mr. John Bartram a diligent Observer of natural Productions sent me, from Pensilvania, two Sorts of curious Wasps Nests made with Clay, which are commonly built against the Timber under the Roofs of Houses and Pales, to shelter them from the Weather. They feed as the Bees, on Flowers; but whether they sting like them I do not yet know.
>
> The plain Clay-Nest is fabricated by a small black Wasp, of the same Species of that in Fig.1. but less, that has a Speck or Stripe of Yellow in its Tail; and the Cells are made four or five together, joining Side by Side to each other. But the Clay-Nests that are so elegantly wrought are built by a purplish black Wasp such as is figured in Fig. 2.: After one Cell is formed, they stop it up, and join another to its End, and then add another to that; which makes these wrought Clay Fabrics longer than the plain ones.
>
> Their Method of Working is much alike, and it is very diverting to see them at it: Their Art and Contrivance is wonderful; and, as if it was given to cheer them at their Labours, they make a very particular musical Noise, the Sound of which may be heard at ten Yards Distance.
>
> Their Manner of Working is, to moisten Clay, and temper it up into a little Lump, of the Size of Swan-shot. This they carry to build with; they begin first at the upper End of the Cell, and work downwards, until it is long enough to contain the Nymph or Chrysalis: After they have spread out the little Lump in a proper Manner to form their little Fabric, they set up their musical Notes, and return to temper and work up more Clay for the next Course. Thus they continue alternately singing and working, until a Cell is finished; which is made delicately smooth withinside; then, at the further End of each cell, they lay an Egg; after this, by surprising Instinct, they go and catch Spiders, and cram the Cell full of them: But it is further wonderful to observe, that they only in some manner disable the Spiders, but not kill them; which is to answer two Purposes; first, that they should not crawl away before the Cell is finished; and next, that they may be preserved alive and fresh until the Egg hatches, which is soon.
>
> The Spiders, by wonderful Instinct, are provided for the Embryo to feed on: Having stor'd up sufficient for its Support, she very securely closes up the Cell, and then proceeds to build the next in the same Manner.
>
> The Maggot or Embryo, having eat up all its Provision, before October prepares for its Change, and spin itself up in a fine soft silken Case, in which it lies all the Winter in the Chrysalis-State, until the Spring, when it eats its Way out of its Clay Dwelling.
>
> April 3, 1745
>
> P. Collinson[1]

Figure 10.2 John Bartram's house in Philadelphia. In 1745 he described the black and yellow mud dauber nesting under the eaves of this house. It still nests here.

Figure 10.3 Arched entrance to Pennsylvania Hospital from 8th Street. A mud dauber is constructing her nest inside the archway.

Figure 10.4 Mud dauber in archway to Pennsylvania Hospital just after she landed on her nest. She carries a ball of mud, which she probably collected in the hospital's garden.

Figure 10.5 Close-up view of the ball of mud.

Figure 10.6 Daubing mud.

Figure 10.7 Tamping down.

Powers of reasoning attributed to the black and yellow mud dauber

In 1806 Benjamin Henry Latrobe— an architect and civil engineer who designed the Philadelphia Water Works (then located at the current site of City Hall[2]) and who was later appointed Architect of the United States Capitol[3]—added more observations and thoughts about the mud dauber. Latrobe weighed spiders in a series of cells in a mud dauber nest; he concluded that in each cell the wasp adjusted the number of spiders it stashed according to the size of the spiders in the stash.[4]

> This examination proves that the wasp distributes with much judgment the quantity of food necessary for its progeny; in most of the cellules, for instance, I ought to have found twenty-two or twenty-three spiders, and yet sometimes there are only five or six, but in this case they are very large ones.[5]

Latrobe accidentally damaged a few of the cells of a nest while the wasp was away gathering mud to make a new cell. He recorded what happened when the wasp returned and discovered the disturbance:

> In a short time the wasp arrived, loaded with a round lump of clay. It came merely for the purpose of making a new cellule; but seeing that its former works were deranged, it began to run rapidly over the cellules, apparently hesitating what to do. At last it deposited the clay upon the edge of one of the apertures, and began to spread it with its snout, pushing it before it, in the attitude of a sow digging in the ground. It emitted a shrill buzzing when at work. After having very properly replastered the work, it flew away. In four minutes it returned with a new load of clay which it deposited in the next aperture. It repeated its visits four times; and, after having finished the repairs and being convinced of the goodness of the workmanship by running over it several times, it flew off again and returned with a new load, with which it began to form a new cell.[6]
>
> If the faculty of modifying the conduct of an individual according to circumstances is one of the characteristics of reason, the fact I have now mentioned is surely a proof of reasoning in an insect. The wasp had remarked the unexpected derangement which had been made during its absence; the clay which it brought was intended for a new cellule; but observing the mischief done to the old ones, it repaired them before building any more.[7]

The black and yellow mud dauber as hunter

In 1890, Henry C. McCook described how mud daubers hunt spiders:

> On this errand she may be seen hawking over and near cobwebs of various sorts, venturing within the meshed and beaded snares that prove fatal to most incomers, and sometimes even to herself. She rarely fails in her errand. If the aranead occupant, expectant of prey, sallies forth to seize the intruder, it finds itself a captive, not a captor. For the wasp shakes the silken filaments from feet and wings, turns upon the spider, seizes and stings it, bears it to her cell, and thrusts it therein.
>
> She does not limit her hawking to cobwebs, but flutters over flowers, burrows among leaves, creeps with nervous, twitching tread along branches of trees, wherever spiders dwell or hunt, and with relentless cunning, zeal, and ferocity snatches those creatures away to add to the growing store within her egg nest. At last the cavity is filled, the circular opening sealed up, and the spiders left literally entombed alive within that clay sarcophagus.
>
> If one at this stage should break open the mud dauber's cell, he might dispute the statement that the imprisoned spiders are alive. To all appearances they are dead. In point of fact they are simply paralyzed. The effect of the poison injected by the wasp's sting within the tissues

> of her victim is such that all activity is at once and completely suspended, without destroying life. Thus, when the larval waspkin awakes to the pangs of hunger, it finds itself in the midst of a generous supply of the very food which Nature intended for it. The mother whom it is never to know, and who already perhaps has paid the last debt to Nature, had consumed her closing days in providing for the offspring which she was never to see. I have found these larvae, fat, white grubs, in the midst of their "preserved meats," feasting thereon, and have wondered at their enormous appetite and the greedy vigor with which it was satisfied.[8]

Today the black and yellow mud dauber still nests on John Bartram's House. It also nests on the Walnut Street Bridge, Pennsylvania Hospital, the Academy of Natural Sciences, Fairmount Water Works, and Boathouse Row.

Figure 10.8 The Academy of Natural Sciences of Drexel University. A black and yellow mud dauber nest is just barely visible as a light brown dot beside the Ionic scroll capping the left column.

Figure 10.9 Close-up view of figure 10.8. Mud dauber nest is the brown structure on the right.

Hunting around light

Urban lighting might support mud daubers just as it supports spiders. In 1847 when Nicholas Marcellus Hentz first reported finding the bridge spider, *Larinionides sclopetarius,* he reported finding it around the windows of houses; he considered the spider to be "domesticated."[9] He suspected domestication protected bridge spiders from mud daubers.

In 1970 William Eberhard at Harvard found just the opposite. He observed the black and yellow mud dauber preying on a different but closely related spider (*Larinioides cornutus*) during the day at windows that had been illuminated during the night. The mud dauber learned to look for spiders at windows, where these spiders hid in their silken retreats during the day and where the night before they had caught prey. Eberhard observed that this spider's daytime retreats around windows were conspicuous, whereas those in the field were impossible to find because they were hidden in curled-up leaves or under flakes of bark. Upon finding a silken retreat, the mud dauber would tear it apart and seize and sting the spider.[10] These observations suggest that mud daubers prey on spiders attracted to artificial lighting on buildings.

Mud daubers build nests near lamps that attract spiders. Eberhard noted that a black and yellow mud dauber foraging for spiders at a window had a nest only 5 meters away.[11] I found mud dauber nests on artificially illuminated ceilings and walls of buildings at Boathouse Row. At night these areas attract an abundance of spiders, and during the day the remains of their webs are conspicuous. In one instance, I noted a spider had constructed its retreat in an abandoned nest of a mud dauber.

Sparing spiders near mud dauber nests

Martin S. Obin, at the University of Florida, noted that mud dauber nests and spider webs commonly coexist in close proximity, but that mud daubers paradoxically ignore spiders near their nests—even though the spiders they ignore are the very species they capture and stash in their nests. Obin offered two hypotheses to explain the local spiders' immunity to attack by mud daubers. First, he observed that flies that parasitize or consume larvae of mud daubers gain access to mud dauber nests by lurking near spiders and trailing prey-laden wasps back to their nests. By ignoring spiders near their nests and attacking only those farther away, mud daubers increase the distance over which flies must track the wasps. Wasps would presumably be more likely to escape flies that had to trail them for longer distances, particularly since mud daubers take circuitous routes back to their nests. His second hypothesis is that spiders near mud dauber nests intercept parasitic flies approaching the nests. Spiders have been observed capturing flies approaching mud dauber nests.[12]

Figure 10.10 Hollenback House, fourth boathouse on Boathouse Row. Remnants of spider webs and their prey are distributed within the beam of the floodlight. In the rear is a clay nest of the black and yellow mud dauber, which provisions its nest exclusively with spiders.

Figure 10.11 Close-up view of a black and yellow mud dauber's nest surrounded by remains of spider webs, as shown in figure 10.10. Although mud daubers prey on spiders, they are reported to spare spiders close to their nests.

Longevity of populations of the black and yellow mud dauber

The black and yellow mud dauber has endured in Philadelphia's urban environment despite its dependence on water and mud for building material and on shelters protected from rain, which can ruin nests made of dried mud. The wasp is vulnerable to breaks in any of the links in its lengthy food chain. For example, an oil spill in the Schuylkill River could reduce populations of aquatic insects whose winged adults are food for spiders, which in turn are food for the larvae of mud daubers. Insecticides used to control mosquitoes and other urban pests could destroy black and yellow mud daubers by killing them or their larvae directly, or their spiders, or the prey of their spiders. Use of herbicides to destroy weeds destroys habitat for insects that support spiders consumed by larvae of black and yellow mud daubers. Herbicides also destroy wildflowers; black and yellow mud daubers feed on nectar and pollen of flowers rather than on spiders, which only their larvae eat.

The survival of black and yellow mud daubers despite such vulnerabilities in Philadelphia seems mysterious. In Center City, mud daubers have endured through the Industrial Revolution and dense urbanization. What might account for such resilience?

Reasons for the longevity of populations of black and yellow mud daubers in Philadelphia

Urban lighting is probably not responsible for the success of mud daubers in Center City. Although mud daubers build nests on walls and ceilings with lamps that attract spiders, they also build nests on walls far away from lamps. Mud daubers and spiders coexist probably because they both need structural support that is protected from sun and rain. An abundance of buildings with eaves and ledges suited to mud dauber nests, plus the availability of mud in gardens and riverbanks, are the most compelling reasons that mud daubers have endured centuries in Center City.

The harsh environment of downtown Philadelphia may work to the mud dauber's advantage. Not only do buildings provide mud daubers dry nesting sites rare in nature, but they also create habitats to which the mud dauber's numerous enemies may be poorly adapted. These enemies include flesh flies (Sarcophagidae), bee flies (Bombyliidae), velvet ants (Mutillidae), cuckoo wasps (Chrysididae), ichneumon wasps (Ichneumonidae), and chalcid wasps (Melittobia).[13]

Reason for scarcity of organ pipe mud daubers in Philadelphia

In Center City the black and yellow mud dauber has fared better than the organ pipe mud dauber, the other species of mud dauber that nested on John Bartram's house in Philadelphia in the eighteenth century. A study of urban wasp nests in Piscataway, central New Jersey, reported that people destroyed almost two thirds of nests that organ pipe mud daubers had built on apartment buildings. These wasps had built their nests on walls near doors of porches. However, people spared nests of black and yellow mud daubers, which had built their nests out of reach in eaves.[14] Organ pipe mud daubers guard their nests and buzz loudly—behavior considered musical by Bartram, but probably intimidating by today's homeowner. Although neither kind of mud dauber is aggressive, the organ pipe mud dauber's buzzing and guarding probably incites people to destroy its nests.

Prehistoric links to human houses

Mud daubers in North America have nested on buildings since the beginning of their recorded history. Recently, James J. Krakker at the National Museum of Natural History of the Smithsonian Institution reported finding black and yellow mud dauber nests associated with human habitation in Missouri in the middle Holocene. The age of the nests was radiocarbon-dated to 5,500 to 6,200 years. Krakker concluded that mud daubers built these nests on houses.[15] The success of mud daubers in Philadelphia over the past two and a half centuries may be an extension of an even longer association of the wasp with buildings.

Figure 10.12 Old nest of organ pipe mud dauber (*Trypoxylon politum*) near Fairmount Water Works. It is on a retaining wall inaccessible to people.

Destructive impact of building maintenance on mud dauber nests

I recently looked for mud dauber nests on buildings around the Water Works and Boathouse Row, where the nests had been particularly common. The whole area had been spruced up, with more landscaping, new walkways, and new and restored gazebos high on the cliff overlooking the river. The facades of most of the buildings looked scrubbed and freshly painted. The restaurant in the Water Works was busy. Black and yellow mud dauber nests hidden in ornamental nooks beneath the eaves survived, but nests in more exposed locations had disappeared. All traces of the organ pipe mud dauber on buildings were gone. I found the remains of an old nest of an organ pipe mud dauber on an inaccessible stone retaining wall remote from buildings.

Buildings, along with mud and spiders, have contributed to the surprising survival of populations of mud daubers in Philadelphia during the past two and a half centuries. But in Center City, only one of the two species of mud daubers that nested on John Bartram's house is thriving; the other had the misfortune of selecting nest sites vulnerable to people who do not admire them as did Bartram. The contrasting fates of these two species of mud daubers illustrate how persecution of animals in downtown Philadelphia may be narrowly focused, quirky, and, from the perspective of the persecutors, effective.

11

YELLOWJACKETS

(*Vespula* species)

In the late 1990s a plague of yellow-jackets in Center City made outdoor dining hazardous. They have since become scarce here.

Figure 11.1 Eastern yellowjacket (*Vespula maculifrons*) on porcelainberry (*Ampelopsis brevipedunculata*) just off Martin Luther King Drive in Fairmount Park, September 2012.

In 1887, Ezra Townsend Cresson of the Academy of Natural Sciences of Philadelphia completed a 350-page monograph on North American Hymenoptera, which includes bees, ants, and wasps. Cresson wrote that unlike paper wasps, which construct "nests on trees, or in corners of buildings, or under the roofs of outbuildings, yellowjackets build their nests underground, as most country boys know by painful experience."[1]

The German yellowjacket

Ninety years later, yellowjackets started nesting in attics and walls of buildings. To investigate this phenomenon, entomologists at Cornell University in Ithaca used a newspaper notice to solicit information from local residents about any wasp nests. Readers identified twenty wasp nests—seventeen in houses, two in vegetation, and one underground. The predominant species nesting in houses was the German yellowjacket (*Vespula germanica*), an introduced species that until recently had been rare. Only one of the seventeen nests in houses was colonized by what had previously been Ithaca's most common species of yellowjacket, the eastern yellowjacket (*Vespula maculifrons*), which is native to North America.[2]

The same year that the German yellowjacket was recognized as the dominant species in houses in Ithaca, it was discovered to be the principal species in houses in northern Delaware.[3] Five years later it was found to be more common in urban than rural areas in northern New Jersey, although still less common overall than the eastern yellowjacket, according to surveys using traps baited with fish-flavored cat food.[4] In the 1980s, the German yellowjacket spread to the Midwest,[5] and homeowners' complaints to pest control operators about yellowjackets increased.[6]

Pestiferous yellowjackets in Center City

By the 1990s, swarms of yellowjackets attracted to food and sweet beverages made eating outside in Center City unpleasant and dangerous. Yellowjackets would typically appear at the dinner table within ten minutes of food and drink being served. I recall interrupting dining to move inside to escape them. Yellowjacket traps, then popular items in local hardware stores, killed yellowjackets but offered no relief.

Figure 11.2 Parc Restaurant, Bistro, and Cafe, at Rittenhouse Square, April 2012. The disappearance of yellowjackets has improved the quality of outdoor dining.

In addition to fruit, honey, molasses, soft drinks, and fruit juices, their culinary preferences include fine meats and fish: roast beef, roast lamb, boiled ham, cooked beefsteak, bacon rind, cooked calf's liver, wienerwursts, salmon, rabbit, and chicken.[7] In order to assess baits, one study systematically tested yellowjackets' preferences for processed meats: it found canned chicken and canned fish superior to freeze-dried chicken.[8] Yellowjackets feeding on meat attract more yellowjackets.[9] Yellowjackets use odors to locate food high in sugar content. When they return to their nests, they carry these odors with them; other yellowjackets leaving the nest learn these olfactory cues and use them to guide them as they forage.[10]

Despite what would appear to have been a threat to public health, the plague of yellowjackets in Philadelphia unfolded without any systems in place for monitoring it. Philadelphia has no quantitative record of yellowjacket populations, just as it has no objective documentation of the outbreaks of insects that defoliated the city's shade trees in the mid-nineteenth century. The technical difficulties of tracking populations of pollinators such as bees[11] exemplifies problems that would complicate attempts to measure populations of yellowjackets, whose abundance may vary from block to block and day to day, depending on many variables, including use of insecticides.

Medical documentation of yellowjackets in Philadelphia

To uncover evidence of this plague, I searched the medical literature for all reports of victims of insect stings in Philadelphia at any time in the past. I found four publications—all dated from 1996 to 1999. Thomas Jefferson University Medical Center recorded 449 visits to the emergency room for stings from 1991 to 1996, peaking during September and October, a period the author dubbed "yellow jacket delirium."[12] At the Hospital of the University of Pennsylvania, gastroenterologists' endoscopic examination of a woman with painful esophageal obstruction revealed she had ingested a yellowjacket.[13] Over a period of five years, eight children were treated at Children's Hospital for anaphylaxis (a life-threatening allergic reaction) due to insect stings.[14] At the Hospital of the Veterinary School of the University of Pennsylvania, two dogs were treated for massive envenomation due to yellowjacket stings; one survived.[15]

Causes of yellowjacket outbreaks

When the urban yellowjacket problem surfaced in the 1970s, Harry G. Davis of the U.S. Department of Agriculture concluded that urban sprawl had displaced yellowjackets from their traditional nesting sites underground, while at the same time providing new sources of food for species like German and eastern yellowjackets, members of a group of *Vespula* (yellowjacket) species that are scavengers. He reviewed the global dissemination of pestiferous species of yellowjackets, especially the German yellowjacket in New Zealand, Tasmania, South Africa, Chile, Europe, and North America. Facilitating the spread of the German yellowjacket may have been the evolution of new genotypes that produced large colonies nesting in buildings.[16]

Danger from yellowjackets

Urban yellowjackets are more dangerous than mud daubers. Mud daubers spend most of their time hunting spiders, visiting flowers, collecting mud, and constructing nests. The male organ pipe mud dauber may guard its nest, but it does not sting. Mud daubers are not attracted to human food. The wasp nests themselves house only eggs and larvae—no adult wasps until the moment they hatch. The wasps that build these nests live alone, not in groups, and they are not aggressive.

Yellowjackets, in contrast, live in colonies. When a hive is disturbed, its members sting in swarms, like those of Africanized ("killer") bees.[17] Members produce alarm pheromones that attract other members and induce them to attack.[18] Roger Simon and Allen Benton at Pennsylvania State University studied a nest of the eastern yellowjacket in a barn. They determined that it contained over 5,000 adult yellowjackets and over 10,000 cells. The wasps remained active into December.[19] Yellowjackets become more aggressive and are more likely to sting in the fall, when their natural sources of food disappear.[20] Nationwide, deaths due to attacks by animals are reported to the U.S. Department of Health and Human Services. From 1979 through 1990, the average number of deaths each year due to attacks was forty-four by bees, wasps, and hornets; sixteen by dogs; six by snakes, and four by spiders.[21] Yellowjackets are classified as wasps.

Yellowjacket stingers are barbed and may get stuck in victims, particularly the stingers of eastern yellowjackets. When a victim brushes off a yellowjacket, the stinger apparatus may separate from the wasp and stay lodged, where its venom contains pheromones that guide other yellowjackets to the victim even while he or she is attempting to flee.[22] The loss of the stinging apparatus in the body of the victim results in the insect's death, which has been cited as an example of ultimate self-sacrifice, or altruism—from the perspective of the hive. The selective forces that favor such altruism, however, are selfish from the perspective of the genes of the stinging insect, whose defense of the hive serves to propagate its own genes through those of its close relatives.[23]

Decline in numbers of yellowjackets

Although yellowjackets in urban settings are pestiferous, elsewhere they provide important ecological services, particularly as predators of other insects, including agricultural pests. Declines in populations of yellowjackets could harm agriculture. Recent reports have suggested that such a decline may be taking place. Analysis of stings reported per year to the Illinois Poison Center found a reduction of 50 percent after 2005. Almost all of the stings took place around victims' homes in urban areas. The authors suggest that declines in reported stings are symptomatic of declines in populations of Hymenoptera generally—including not only yellowjackets, but also pollinators such as honeybees.[24]

In Center City, yellowjackets have disappeared. Over the last three years (2011 to 2013), I found none, even after attempting to lure them with bait such as soda, juice, fruit, and meat in September and October. Bees, on the other hand, remain conspicuous in community gardens and on flowering shrubs and trees. Bees easiest to observe include bumblebees, honey bees, carpenter bees, and sweat bees.

Wasps other than yellowjackets are also common, including paper wasps, which, like yellowjackets, live socially. The European paper wasp (*Polistes dominula,* also referred to as *Polistes dominulus*) is yellow and black and easily mistaken for a yellowjacket. Common in Philadelphia, it is an introduced species whose range in North America has been expanding rapidly.[25] Populations of some local bees and wasps may have declined, but the disappearance of yellowjackets is so distinctive that it likely represents a separate phenomenon.

In the summer of 2012 I found a yellowjackets' nest in the ground at Bartram's Garden. The species was the native eastern yellowjacket (*Vespula maculifrons*), the most common pestiferous yellowjacket in the eastern half of the country prior to the introduction of the German yellowjacket.[26]

Figure 11.3 Bumblebee (*Bombus impatiens*) on chicory (*Cichorium intybus*). Unlike yellowjackets, bumblebees are still common in Center City. They do not bother people eating outside.

Figure 11.4 Male eastern carpenter bee (*Xylocopa virginica*) waiting in midair for female to emerge from the wood of this bench behind the Philadelphia Museum of Art. It resembles a bumblebee, except its abdomen is smooth rather than hairy. It is uninterested in the bottle. Like bumblebees, it continues to be common in Center City.

Figure 11.5 European paper wasp (*Polistes dominula*) on porcelainberry. Its orange antennae distinguish it from yellowjackets, whose antennae are black. It has recently become common in Center City. Unlike yellowjackets, it does not pester people dining outdoors.

Figure 11.6 Nest of European paper wasp. It is suspended under an eve at the Fairmount Water Works.

Figure 11.7 Bald-faced hornet (*Dolichovespula maculata*) at 23rd and Panama Streets in Center City. In Center City it does not harass people dining outdoors.

Figure 11.8 Bald-faced hornet's nest on Kater Street.

Causes of the decline

What might account for the disappearance of yellowjackets in Center City? Just as no quantitative measure documented the population explosion of yellowjackets here in the 1990s, no systematic measure has documented the population crash. Declines in populations of bees galvanize popular interest because bees are agriculturally important pollinators, but yellowjackets attract interest only when they are pestiferous. Records of the collapse of Center City's populations of yellowjackets reside exclusively in people's memories, which can recall yellowjackets at dinner tables better than they can specify the years and extent of yellowjacket abundance. The decline in abundance has yet to be delineated geographically, although I have noted such a decline in Narberth, a suburb of Philadelphia. I remember infestations of yellowjackets around garbage cans a decade ago in Fairmount Park. Searching for them in Fairmount Park during the past three years in late summer and fall, I found no yellowjackets around trash cans. However, in 2012 I identified German yellowjackets infesting an attic in a home in rural Churchtown, Lancaster County.

Populations of yellowjackets may have gone through cycles of boom and bust like those of bagworms and ailanthus moths. Populations of German yellowjackets, in particular, may have proliferated soon after their establishment, only to crash as parasites or pathogens caught up with them.

Many conditions could theoretically have contributed to the decline in numbers of yellowjackets. Tighter construction in buildings discourages yellowjackets from nesting in exterior walls and attics. Neonicotinoid insecticides, whose use by homeowners is unregulated and unmonitored, are toxic to insects even in low doses.[27] The increase in crowds of people in Center City aggravates soil compaction, reducing ground suitable for subterranean nests of yellowjackets. Improved collection of garbage in Center City deprives yellowjackets of food.

European paper wasp

One hypothetical cause for the decline of yellowjackets in Philadelphia is the establishment of the European paper wasp, which was first reported in Pennsylvania in 1990.[28] It is now the most common wasp in Center City. Its black and yellow stripes make it easy to confuse with yellowjackets, but its antennae are orange, in contrast to yellowjackets' antennae, which are black. The European paper wasp does not congregate around soda cans and human food as do yellowjackets, but it may compete with yellowjackets for prey. It has been blamed for declines in populations of native paper wasps, such as *Polistes fuscatus,*[29] still present in Philadelphia.

Wasp years

Yellowjackets in England and in the Pacific Northwest have shown mysterious cyclical population spikes, called "wasp years." The spikes recur at irregular intervals, sometimes decades apart, and simultaneously encompass more than one species of yellowjacket.[30] Many hypotheses have attempted to explain these jumps in abundance, but none is convincing.[31] If past plagues of yellowjackets in Center City reflect such population cycles, they may recur.

SPOTLIGHT

COMMON MORNING GLORY

Common morning glory (*Ipomoea purpurea*) growing in curbside leaf litter that has decomposed into soil around a gutter on Naudain Street.

Accumulation and decomposition of leaves from street trees is sufficient during summer to produce soil overlying asphalt pavement around intakes of sewer drains. The common morning glory shown in the photo may be an escapee from a nearby garden.

12

COMMON EASTERN FIREFLY

(*Photinus pyralis*)

The common eastern firefly thrives in Center City despite the light pollution.

Figure 12.1 Male and female common eastern fireflies on the rear wall of our house. They were on top of each other before I disturbed them to take this photograph.

In 1774 Baron Charles De Geer described and named an American firefly sent to him by Israel Acrelius, a Swedish clergyman in Christina (now Wilmington), Delaware. At the time Delaware was under the jurisdiction of the governor of Pennsylvania, and De Geer named the beetle *Lampyris pensylvanica.*[1]

In 1851 John L. Leconte was the country's authority on fireflies. Referring to De Geer's firefly as *Photuris pennsylvanica,* he wrote in the *Proceedings of the Academy of Natural Sciences of Philadelphia* that the species is abundant in every part of the United States.[2] This firefly has since been designated the official state insect of Pennsylvania.[3] It is absent in Center City, where another firefly thrives, the common eastern firefly, *Photinus pyralis.*

Deciphering firefly flashes

Frank Alexander McDermott succeeded Leconte as the regional expert on fireflies. In 1911 he published his findings on the flashes of *Photinus pyralis.*[4] He hypothesized that the flashes of light emitted by males signal females. To test this, he lit safety matches in the evening near females. During the flare of ignition, he swung the matches in imitation of the characteristic arc made by the flash of a male of this species in flight. He blew out the matches as soon as the flares ended. He described his findings:

> In each instance the flash of light from the match was followed, within two to five seconds, by the flashes of females of pyralis in the surrounding grass and weeds. Most of them flashed at the end of about four seconds. They did not flash in the intervals between the lighting of matches, except in response to the flash of a passing male.[5]

He then repeated his experiment, except this time he used an electric lamp to simulate the answering flash of females, which flash while at rest, typically on a blade of grass.

> If the male is in a position to see the light of the bulb, he will almost invariably drop, and repeating the process will bring him up to the bulb; usually he will crawl around and over it excitedly for a few minutes, and then fly away. Sometimes males would crawl up grass stems above the bulb, and apparently looking over the edge of the blade, hold perfectly still for a moment, and then flash; the instant the bulb was flashed in answer they would commence to wave their antennae rapidly, and crawl quickly down the blade and toward the bulb. Early in the flying period of an evening, as many as a dozen males have been thus attracted in a few moments.[6]

McDermott reproduced these findings for two other species of *Photinus.* He demonstrated that each of the three species has its own flash code, which he simulated with his electric light. He tried to do the same for *Photuris pennsylvanica,* but failed:

> Although a quite close watch has been kept on *Photuris pennsylvanica* Deg. for a considerable number of nights, nothing definite can be said as to the possible relation of its light emission to its reproductive life. A large number of these insects fly about in the trees and bushes, emitting their light in the various ways that have been described for it, and yet apparently paying no attention to each other.[7]

Discovery of femme fatale fireflies

Six years after McDermott's paper, Francis X. Williams, working in New England, unlocked the mystery surrounding the flashes of *Photuris* fireflies. He observed *Photuris* females eating males of another species of firefly. The victims were members of the genus *Photinus*, which includes Center City's common eastern firefly.[8]

> One evening six females were disturbed at such meals. The fact that the victims were always males, though the females were nearly as abundant in this locality, and that the feeders were invariably females, strongly suggests that the weak *Photinus* males were drawn to their untimely ends by the lure of the greenish-yellow light of the female *Photuris*. When bottled up with *Photinus* they would readily devour the latter, despite its active exudations.[9]

Half a century later, James E. Lloyd at Cornell University extended these findings in a classic paper, "Aggressive Mimicry in *Photuris*: Firefly Femme Fatales."[10] Lloyd's findings were serendipitous. In the course of his research on fireflies, Lloyd needed to collect *Photinus* females, so he searched for them by signaling with a flashlight that mimicked flashes of *Photinus* males. He was essentially applying McDermott's methods as a tool to locate female *Photinus* fireflies. He discovered *Photuris* females flashing in answer to his flash simulating male *Photinus pyralis*, the common eastern firefly. He then observed that *Photuris* females not only answered the flashes of *Photinus* males, but they lured them in, seized them, and ate them.[11] Later he showed that *Photuris* females have repertoires of flashes; they match the particular flashes of different species of *Photinus*, depending on which is available as potential prey.[12] Lloyd reported that *Photuris* males (which do not eat) mimic the flashes of the prey of *Photuris* femme fatales; he speculated that *Photuris* males use this mimicry to seduce *Photuris* femme fatales.[13] He and Steven Wing showed that *Photuris* females hunt *Photinus* males not only by luring them, but also by directly attacking them in midair, guided by their prey's flashes.[14] The measures and countermeasures that fireflies use to signal each other have been called an evolutionary "arms race."[15]

Discovery of firefly poisons

The ecological toxicologist Thomas Eisner and his colleagues showed that *Photinus* fireflies synthesize defensive poisons that *Photuris* fireflies do not produce; *Photuris* females acquire chemical protection by eating *Photinus* males and sequestering their poisons.[16] *Photuris* females themselves avoid the toxicity of the poisons they eat while endowing their eggs with high concentrations of these poisons.[17] Selective pressure due to *Photuris* predation may be responsible for an evolutionary switch from nocturnal to diurnal behavior in some members of the firefly family, Lampyridae. The light-producing organs in these diurnal "fireflies" are only vestigial.[18]

Revision of firefly taxonomy based on flashes

In retrospect, McDermott's initial bafflement over the function of the flashes of *Photuris pennsylvanica* can be appreciated in the context of femme fatales. Female *Photuris* fireflies respond to the flashes of male *Photuris* fireflies only before mating. After mating, they become femme fatales, ignoring the flashes of male *Photuris* fireflies.[19] McDermott continued to delve into *Photuris* mating signals for fifty years and discovered cryptic species, based on differences in mating flashes and other traits not detectable in museum specimens. He and an associate, H. S. Barber, concluded that *Photuris pennsylvanica* is a complex of cryptic species distinguished by their flashes.[20] McDermott surmised that the *Photuris* species that Baron Charles De Geer named *pensylvanica* was probably another species, *Photuris versicolor,*[21] which today in Delaware is more common.[22] In describing the flashes, Acrelius, the pastor who sent De Geer the firefly, had told him only that "they glow and appear to viewers as thousands of sparks, but they shine even more when they fly"[23] (my translation of the French). Acrelius's specimens are lost.[24]

Photinus pyralis in Center City

During the latter half of June and early July in Center City, I have watched fireflies flashing in Rittenhouse Square, Fitler Square, and Schuylkill Park. They are abundant in our courtyard garden and the College of Physicians garden, which was first planted in 1914 and later converted to a medicinal herb garden.[25] From mid-June to early July, they start flashing around dusk, and in twenty minutes the flashing ends, except for rare stragglers. The J-shaped arc of the flash of males about a meter or less above the ground is characteristic of *Photinus pyralis.* In Center City I have yet to see *Photuris* fireflies flashing, although in a suburban garden just outside of Center City I have seen them flashing high in trees late at night.

Figure 12.2 Benjamin Rush Medicinal Plant Garden of the College of Physicians of Philadelphia, where fireflies are common. Cultivation of this site began almost a century ago. Photographed July 1, 2012, just before renovations.

The abundance of *Photinus pyralis* in Center City is surprising, even if one concedes that their flashing may make the fireflies appear more numerous than they actually are. They thrive here despite streetlights and light pollution. At night they fly to door lamps; in the day they rest exposed on doors and walls.

Diet of *Photinus pyralis* fireflies

What accounts for their abundance in Center City is unknown. Their larvae are predaceous and live underground, but their subterranean habits have proved difficult to study. Parasites of larval *Photinus* fireflies include mites and maggots from two families of Diptera (Tachnidae and Phoridae).[26] Conceivably Center City protects *Photinus* from parasites that do not tolerate urban conditions.

In 1868 early American entomologists deduced that *Photinus pyralis* larvae eat earthworms:

> It lives in the ground where it feeds on other soft bodied insects. At times these "fire-fly" larvae must subsist almost entirely on young earth-worms, for we have found them abundantly in soil, on which no vegetation had grown for at least one year, and where in consequence there was scarcely another animal to be found, besides these two— the "fire-fly" larva feeding upon the earth-worm, and the latter subsisting on the earth itself.[27]

McDermott reared *Photinus* larvae on earthworms.[28] The population density of earthworms in soil has been found to be higher in urban than rural and suburban forests,[29] and to increase as urban parks age.[30] In our backyard, two introduced species are common, the rosy-tipped earthworm (*Aporrectodea rosea*) and the common nightcrawler (*Lumbricus terrestris*). The rosy-tipped earthworm has developed tolerance to contamination of soil with lead[31] and zinc.[32]

Figure 12.3 Earthworm under a log in our backyard. Populations of earthworms take time to build up in soil. Center City's nineteenth-century row houses are well endowed with old gardens and earthworms, prey of larvae of the common eastern firefly.

Photinus fireflies flash in large numbers over grass in Schuylkill River Park's older sections, which were completed several decades ago, but not in newer sections along the bike path, completed only six years ago. The lawn that hosts an abundance of fireflies consists of a mixture of weeds and grasses that have not been subjected to pesticides and herbicides. I do not find fireflies flashing over perfect carpets of weed-free grass, probably because of the chemicals required to achieve such perfection. Center City's rich legacy of nineteenth-century courtyards with gardens may have contributed to its abundance of fireflies. Flashes of fireflies, like songs of tree crickets, may be favorable indicators of environmental health.

Figure 12.4 Turf containing white clover (*Trifolium repens*), plantain (*Plantago*), and diverse grasses in Schuylkill Park, where fireflies are common. Fireflies are rare over "perfect" lawns—grass monocultures—dependent on pesticides and herbicides.

Light pollution as protection against femme fatale (*Photuris*) fireflies

Center City may have an abundance of *Photinus* fireflies because, among other reasons, it affords them safe haven from their primary enemy, *Photuris* femme fatales, which are absent here. Such an urban safe haven may benefit *Photinus* fireflies in much the same way as urban refuges benefited ailanthus silkmoths before their enemies moved in. *Photuris* larvae, unlike subterranean *Photinus* larvae, can be found on the ground surface.[33] Perhaps in Center City, pedestrian trampling takes a heavier toll on *Photuris* than on *Photinus*.

I presented this thought to Sara Lewis, one of the country's leading experts on fireflies. I noted that *Photinus* fireflies in Center City flash at dusk near artificial lights. She offered alternative hypotheses:

It's quite unusual to find such high firefly density in an urban environment. I'm intrigued by your idea that some environmental condition might differentially affect *Photuris* predators, and thus indirectly increase *Photinus* abundance.

It seems like there are a couple of possibilities here: soil conditions (as you suggest) or artificial lighting. As for soil conditions, I'm not convinced that *Photinus* & *Photuris* larvae are all that different in their habitat requirements (although we know remarkably little about this because they're so hard to raise in captivity). Both groups live & pupate underground, although *Photuris* larvae do forage more actively on the surface at night. Based on my own anecdotal observations, *Photuris* larvae are more generalist scavengers—for example, they'll eat cat food. In contrast, *Photinus* larvae appear to specialize on eating earthworms. Any urban gardens nearby?

Another possibility is they may have different reactions to artificial lighting —I assume these are streetlights? Because for many *Photuris* species, courtship takes place when it's fully dark, artificial lighting might be more disruptive for these than for crepuscular *Photinus* species.[34]

The possibility that light pollution might contribute to the abundance of fireflies is counterintuitive. James E. Lloyd has pointed out multifarious ways that nocturnal artificial lighting harms fireflies, particularly by confounding and whiting out their flash signals and disrupting their nocturnal navigation systems.[35] Light pollution may constitute another example of "creative destruction," in which habitat degradation benefits one species at the expense of another. Urban lighting acted this way in the case of the bridge spider, *Larinioides sclopetarius,* which feeds on insects attracted to municipal lights.

A group of firefly investigators who trained under Professor Lewis at Tufts University has collaborated with the Museum of Science in Boston to engage volunteers in the study of fireflies. It provides protocols for volunteers to collect data on firefly abundance and behavior.[36] Such an undertaking may clarify whether artificial lighting favors *Photinus* at the expense of *Photuris.*

Fireflies as an artifact of urbanization

In Center City, artificial lighting may protect common eastern fireflies while old gardens support their larvae. In his poem "Philadelphia," Rudyard Kipling viewed fireflies as a natural heritage:

If you're off to Philadelphia this morning,
And wish to prove the truth of what I say,
I pledge my word you'll find the pleasant land behind
Unaltered since Red Jacket rode that way.
Still the pine-woods scent the noon; still the catbird sings his tune;
Still autumn sets the maple-forest blazing.
Still the grape-vine through the dusk flings her soul-compelling musk;
Still the fire-flies in the corn make night amazing!
They are there, there, there with Earth immortal
(Citizens, I give you friendly warning).
The things that truly last when men and times have passed,
They are all in Pennsylvania this morning![37]

Although fireflies may evoke the primordial past, neither Kipling's cornfields nor Center City's gardens nor artificial lights belong to that past. And in Pennsylvania, less than a quarter of the species of earthworms, which *Photinus* fireflies feed on, are native.[38] One would have to conclude that the abundance of fireflies in Center City is at least in part a legacy of man-made disturbance.

13

LAND PLANARIAN

(*Bipalium pennsylvanicum*)

A new species of flatworm was discovered three decades ago in a suburb of Philadelphia. Introduced into North America from Asia, it specializes in preying on earthworms.

Figure 13.1 Land planarium (*Bipalium pennsylvanicum*) found consuming an earthworm in a garden near Spring City, Chester County, Pennsylvania. Identification was based on papers by Robert E. Ogren and Joseph K Sheldon, who first discovered this species.[1]

In 1978 Joseph K. Sheldon at Eastern College in St. Davids, a suburb of Philadelphia, discovered a wormlike animal with a head shaped like a hammerhead. It has three stripes down its back and grows to a length of 80 millimeters (3 inches). Sheldon found it beneath logs around a vegetable garden and under stones on the college campus. He also found it in his home garden 2.4 kilometers from the campus. Four years later a population was discovered on the campus of Ursinus College in Collegeville, Pennsylvania, after a student walking across the grassy campus noticed one on his shoe.[2]

Robert E. Ogren of Wilkes University in Wilkes-Barre, Pennsylvania, determined that the animal was a terrestrial planarian, or flatworm (Phylum Platyhelminthes), that had not previously been described. Ogren deduced that it had been introduced into this country from Asia, where its genus, *Bipalium,* is endemic. He named the new species *Bipalium pennsylvanicum.* It is the second member of this genus living outdoors in southeastern Pennsylvania.[3] The first is *Bipalium adventitium,* widely distributed in North America[4] and common in New York,[5] especially in New York City.[6]

How flatworms attack earthworms

Both planarians, sometimes called turbellarians in reference to their taxonomic class (Turbellaria), are specialized predators of earthworms. *B. adventitium* tracks earthworms by following their mucous trails, and pursues them into their tunnels, where it blocks their escape.[7]

It digests its prey alive before consuming it:

> Upon contact, the turbellarian immediately crawled onto the earthworm's body. The earthworm did not react until the broad translucent pharynx of the *Bipalium* was extended over the segments. At that instant the annelid moved tortuously and excessive exudations of mucus resulted. Occasionally turbellarians were forced off the body by this action, particularly with larger earthworms. Segments over which the pharynx was extended were always found to be liquefied and soon after swollen. Once permanent attachment by the pharynx was attained liquefaction continued until small earthworms (100–200 mg) were almost completely consumed. Large holes a centimeter wide were formed as portions of larger earthworms *(Lumbricus terrestris)* were digested. As extracellular digestion occurred, streams of material could be seen passing through the pharynx into the predator's digestive tract causing the anterior portion of the *Bipalium* to swell. While feeding, the turbellarian hung flaccidly as the earthworm became motionless. No earthworm recovered.
>
> Feeding by *Bipalium* lasted an average of 45 minutes on earthworms weighing less than 600 mg. Multiple sites were attacked on 4–5 g earthworms and feeding was completed in an average of six hours…The mean weight gain by Bipalium during feeding was 89 mg +/- 12 mg, or 82% of body weight. After feeding, the Bipalium crawled away from the victim, curled into a knot-like position, and remained somewhat motionless for several days.[8]

In North America *B. adventitium* has been reported to feed on fourteen species of earthworms,[9] including individuals fifty-five times its own weight.[10] Two of these species (*Lumbricus terrestris* and *Aporrectodea rosea*) are common in our garden in Center City. *B. pennsylvanicum* also feeds on earthworms, and not on arthropods or gastropods (slugs and snails).[11]

Exotic flatworms depleting populations of earthworms in Europe

Introduction of an exotic planarian onto the Faroe Islands and Northern Ireland has reduced populations of earthworms.[12] *Bipalium* in North America has not been shown to deplete populations of earthworms, but its full impact may take time to develop. Populations of exotic planarians predatory on earthworms are spreading in North America and Europe. Typically the planarian is first detected in or near gardens, and then spreads to surrounding areas.[13] Ogren concluded that transfer of soil used in horticulture distributed the worms in Pennsylvania.[14]

I have yet to find planarians in gardens in Center City, but the animals are easy to miss. Peter K. Ducey, who has studied the natural history of *Bipalium,* suggests looking for them feeding on earthworms on paved walkways in city parks and university campuses at dawn after rain.[15]

North American terrestrial flatworms first appeared in Center City

Exotic terrestrial planarians had an auspicious beginning in Pennsylvania. In 1851 Joseph Leidy discovered the first one in North America. He found it under a flowerpot in his garden at his home at 1302 Filbert Street,[16] currently the location of Philadelphia's Criminal Justice Center, a block from City Hall. Leidy described the species and named it *Rynchodemus sylvaticus.* He also found it in woods along the Schuylkill River and Wissahickon Creek, now part of Fairmount Park.[17] This species, also present in Europe, was likely introduced into North America.[18] It has since been reported in five states,[19] but the worm, which feeds on soil insects rather than on earthworms,[20] has yet to become abundant.[21]

Figure 13.2 Philadelphia's Criminal Justice Center, site of Joseph Leidy's nineteenth-century home and the discovery of North America's first land planarian, *Rhynchodemus sylvaticus,* under a flowerpot in his garden.

Infestations of planarians theoretically could take a toll on *Photinus* fireflies, whose larvae, like *Bipalium,* are specialized predators of earthworms. In the laboratory, *Bipalium* did not prey on beetle larvae.[22] Conceivably, firefly larvae prey on *Bipalium,* which is soft-bodied like earthworms.

Barriers to observing planarians and their ecological impact

Defining the relationship between *Photinus, Bipalium,* and earthworms is problematic. Measurement of populations of animals thinly distributed in soil is imprecise. Techniques for monitoring their interactions underground have yet to be developed. Expertise that encompasses all three animals would require a team of dedicated specialists.

American robins *(Turdus migratorious)* commonly hunt earthworms in turf in Schuylkill River Park. Fireflies flash profusely in June and July. Earthworms are abundant in gardens here. If *Bipalium* flatworms have infested Center City, they have yet to take an obvious toll.

Figure 13.3 American robin (*Turdus migratorius*) grabbing an earthworm in Schuylkill River Park.

Postscript

Since I completed writing this chapter, *Bipalium pennsylvanicum* was discovered in Philadelphia. In September 2014, Heather Rinehart sent me a photograph of this worm, which she found at Bartram's Garden, approximately three kilometers from Center City.

14

AMERICAN ROBIN

(*Turdus migratorius*)

In Pennsylvania, populations of robins have increased over the past decade, while those of other common urban birds have declined.

Figure 14.1 American robin. (Hand-colored engraving, "Turdus pilaris migratorius. The Fieldfare. Aristolochia. The snake-root," plate 29 in Mark Catesby's *The Natural History of Carolina, Florida and the Bahama Islands,* vol. I, published in 1731.)

In 1731, over a century before the publication of Audubon's *Birds of America,* the English naturalist Mark Catesby produced the first volume of *The Natural History of Carolina, Florida and the Bahama Islands.*[1] It includes an engraving of an American robin on its back on a tree stump that resembles a chopping block. The bird's feet are askew up in the air, and its neck hangs down over the edge of the stump. In the accompanying text, Catesby offers no clue as to why he chose to depict the bird as if it were ready to be carved up.

Delicacy

Alexander Wilson offers a clue in the first volume of his *American Ornithology,* published in Philadelphia in 1808:[2]

> So fond are they of gum-berries, that, wherever there is one of these trees covered with fruit, and flocks of Robins in the neighborhood, the sportsman need only take his stand near it, load, take aim, and fire; one flock succeeding another, with little interruption, almost the whole day: by this method, prodigious slaughter has been made among them with little fatigue.
>
> ...When fat, they are in considerable esteem for the table, and probably not inferior to the Turdi of the ancients, which they bestowed so much pains on in feeding and fattening. The birds are frequently and easily raised, bear the confinement of the cage, feed on bread, fruits, &c., sing well, readily learn to imitate parts of tunes, and are very pleasant and cheerful domestics.[3]

Pokeberries

The slaughter of robins might have continued were it not for pokeberries. Wilson, whose ornithological career began in Philadelphia,[4] describes how pokeberries protected robins:

> Sometimes they will disappear for a week or two, and return again in greater numbers than before, at which time the cities pour out their sportsmen by scores, and the markets are plentifully supplied with them at a cheap rate. In January 1807, two young men, in one excursion after them, shot thirty dozen. In the midst of such devastation, which continued many weeks, and, by accounts, extended from Massachusetts to Maryland, some humane person took advantage of a circumstance common to these birds in winter, to stop the general slaughter. The fruit called poke-berries (Phytolacca decandra, Linn) is a favorite repast with the Robin, after they are mellowed by the frost. The juice of the berries is of a beautiful crimson, and they are eaten in such quantities by these birds, that their whole stomachs are strongly tinged with the same red color. A paragraph appeared in the public papers, intimating that, from the great quantities of these berries which the Robins had fed on, they had become unwholesome, and even dangerous food; and that several persons had suffered by eating of them. The strange appearance of the bowels of the birds seemed to corroborate this account. The demand for, and use of them, ceased almost instantly and motives of self-preservation produced at once what all the pleadings of humanity could not effect.[5]

Figure 14.2 Pokeberry (*Phytolacca americana*), a common wild herbaceous perennial plant in Center City.

Warnings about eating robins appeared in a dissertation submitted by Benjamin Shultz for a degree of doctor of medicine at the University of Pennsylvania in 1795. Noting that robins eat pokeberries, he describes the berries:

> These, when perfectly ripe, are extremely smooth, and of a dark reddish colour. They of course are very tempting to eat, but this gratification can seldom be enjoyed to any degree, without great inconvenience. Man is not the only animal to whom these berries are unfriendly; many birds are observed to be purged by them. The flesh of those who eat them acquires a high red colour, a disagreeable flavour, and is destitute of adipose substance. [6]

Outside of Philadelphia the mass shooting of robins continued well into the nineteenth century. In the text accompanying *Birds of America*, published in 1834, Audubon describes the slaughter with pleasure:

> In all the Southern States...their presence is productive of a sort of jubilee among the gunners, and the havoc made among them with bows and arrows, blowpipes, guns, and traps of different sorts, is wonderful.[7]

Attraction to human habitation

Despite the shooting of robins in the eighteenth and nineteenth centuries, the birds were conspicuous around people's homes. In 1749 Peter Kalm, the Swedish naturalist who visited Philadelphia, wrote: "It sings very melodiously, is not very shy, but hops on the ground, quite close to the houses."[8] Two centuries later Witmer Stone, Philadelphia botanist and ornithologist, noted that the robin had recently become the most abundant bird at Cape May. He attributed the population increase to "the steady increase in dwellings with gardens and shrubbery and well-kept lawns."[9] The attraction of robins to human residential development appears to be characteristic of the species rather than to a particular place. In Alberta, increases in robins correlated with settlement of the prairie and creation of gardens, which also introduced earthworms.[10]

Over the past decade populations of robins have increased in Pennsylvania, according to the Breeding Bird Survey.[11] The robin is the most widely distributed bird in the state.[12] Among populations of urban birds in Pennsylvania, only robins have increased, while house finches (*Haemorhous mexicanus*), house sparrows (*Passer domesticus*), and starlings (*Sturnus vulgaris*) have decreased.[13] I have seen robins foraging in Center City throughout the year.

Urban habitat

Why have robins fared so well in Center City? On the surface, cities would appear to offer them few advantages: cities have only recently become part of robins' evolutionary history. Street pigeons (*Columba livia*) and house sparrows, by contrast, have had millennia to evolve in Europe and the Middle East, where fossils of house sparrows date back 65,000 years,[14] and evidence of domesticated pigeons goes back 4,800 years.[15] Center City does not offer robins, which are indigenous to the region, oceanic barriers that protect them from native enemies; bird feeders downtown provide seed to house sparrows, but robins ignore such offerings; buildings in the city offer potential nesting sites, but robins here typically nest in trees.

Leonard A. Eiserer of Franklin and Marshall College in Lancaster, Pennsylvania, observed that lawn mowing attracts robins. He found that the time robins spent on lawns decreased as the height of grass increased. Given a choice, robins foraged on lawns whose grass was short rather than long. The act of lawn mowing itself, independent of height of grass, drew robins. Eiserer surmised that robins fed on insects that lawn mowing exposed.[16] Mowing grass theoretically may nurture earthworms: mowed fragments of blades of grass are just the right size for earthworms to pull into their burrows. Mowing, grass clippings, and mulch increase populations of earthworms or their castings.[17] I suspect shorter grass improves robins' mobility on the ground and their capacity to spot and seize prey.

In Center City, sprinklers attract robins in parks and in our garden. The water brings earthworms to the surface, as it does after rain.[18] In our backyard in the summer, robins are quick to forage after I have gardened and exposed freshly turned earth. Invertebrates other than earthworms are diverse and abundant in urban and suburban gardens.[19] Robins have been found to eat invertebrates in ninety-one families.[20]

Short grass and bare ground may protect robins from exposure to ectoparasites such as lice, louse flies, mites, and ticks. Robins have been found to harbor fifteen species of these arthropods.[21] Over a third of robins sampled using mist nets in Lyme, Connecticut, carried *Ixodes scapularis,* the tick that transmits the spirochete causing human Lyme disease. Just under a third of robins in this survey were coinfected with both nymphs and larvae of these ticks.[22] Robins harbor not only the Lyme disease tick, but also the Lyme disease pathogen, *Borrelia burgdorferi*.[23] In Maryland, the number of *Ixodes* ticks per robin declined with increasing urbanization.[24]

Precolonial contact with agriculture

The attraction of robins to yards and gardens dates back to our earliest records of this bird's behavior; it persisted even in the face of systematic hunting. Red-tailed hawks, in contrast, were reclusive in the face of hunting; only after protections were enforced did they habituate to people. The affinity of robins for human habitation likely preceded European settlement.

Native Americans presented robins with opportunities to forage in gardens long before the arrival of Europeans.[25] Crops found in prehistoric archaeological sites in Pennsylvania include squash (*Curcurbita pepo*), maize (*Zea mays mays*), tobacco (*Nicotiana* sp.), sunflower (*Helianthus annuus*), bean (*Phaseolus vulgaris*), and goosefoot (*Chenopodium* sp.). The age of the oldest remnants of crops dated by radiocarbon in this region is over 2,000 years.[26] Native Americans once gardened across the continent,[27] including in Arizona and New Mexico, where archaeological excavation of abandoned pueblos turned up bones of robins.[28] The American robin evolved in Central America and later colonized North America from coast to coast.[29] Robins likely foraged in pre-Columbian Native American gardens just as they did in gardens of European settlers, and as they do in gardens in Center City today.

Berries

In Center City's old courtyards, wild and ornamental trees and shrubs offer robins abundant berries, especially in winter. Vines that produce berries include Virginia creeper (*Parthenocissus quinquefolia*), Japanese honeysuckle (*Lonicera japonica*), English ivy (*Hedera helix*), and oriental bittersweet (*Celastrus orbiculatus*).[30] Shrubs and trees presenting food to robins include hawthorn (*Crataegus*), holly (*Ilex*), crabapple (*Malus*), mountain ash (*Sorbus*), mulberry (*Morus*), hackberry (*Celtis*), dogwood (*Cornus*), juniper (*Juniperus*), sumac (*Rhus*), and cherry (*Prunus*).[31]

In the spring, suburban robins have been found to begin breeding earlier than rural robins.[32] One possible explanation is the abundance of ornamental berries available during the winter around suburban homes and parks.

Figure 14.3 Virginia creeper (*Parthenocissus quinquefolia*) on the wall of the municipal swimming pool building on Taney Street in Center City. Robins eat its berries.

In his *Compendium Florae Philadelphicae,* published in 1818, William P. C. Barton described the city's flora, including vines with berries lasting into winter. One example is native bittersweet, *Celastrus scandens:*

> A climbing plant frequently reaching the tops of trees, twenty or thirty feet high. Flowers yellowish white, small. Berries a bright orange-red. Said to possess medicinal virtues. In hedges and among small trees and shrubs on rocky ground. Frequent near Mendenhall's tavern on the east bank of the Schuylkill, not far from the falls along the fences; and in the stony and hilly copices back of Powelton, abundant.[33]

Among the most common native woody vines to colonize successional habitats in Philadelphia outside of Center City today is poison ivy (*Toxicodendron radicans*), whose berries are eaten by robins.[34] Barton called it "Poison Vine" and described it as follows:

> No plant is more generally known than this. It is extremely poisonous. Berries white… Particularly common along fences…are possessed of medicinal virtues.[35]

In Center City, robins' supply of berries is more secure than the supply of small seeds that sustain house finches and house sparrows. Berries are abundant on vines, shrubs, and trees, whereas small seeds are available mostly in bird feeders and herbaceous weeds. Property owners apply herbicides to wild herbaceous plants, which in Center City typically grow in cracks at the base of buildings and in pavement. Destruction of these plants may in part explain why populations of house sparrows and house finches have declined compared to those of robins.[36]

Viruses, raptors, cats, and other enemies

Center City may buffer robins from predators and competitors less well adapted to downtown. Despite the conspicuous presence of red-tailed and Cooper's hawks, most species of raptors known to prey on robins[37] are not common downtown. Arborial snakes, which are nest predators, are absent. Starlings, which have been observed stealing worms from robins,[38] are common but have declined in numbers in Pennsylvania.[39]

In Middle Atlantic and Northeastern states, West Nile virus has caused declines in populations of both robins and its nest predators, including American crows and blue jays.[40] In 2010 the Philadelphia Department of Public Health reported mosquitos infected with West Nile virus in every section of the city surveyed, including Center City. It also reported thirteen human cases of illness caused by West Nile virus infection.[41] West Nile virus paradoxically could have boosted populations of robins if their immunity to infection by the virus exceeded that of their nest predators.[42]

Free-ranging domestic cats in the United States have been estimated to kill over a billion birds annually, but mortality of robins due to cats in Center City is unknown.[43] Also unknown is the mortality due to squirrels, which are nest predators of robins.[44]

Cowbirds

Robins are well defended against one common enemy here: the brown-headed cowbird *(Molothrus ater)*. Alexander Wilson described the behavior of cowbirds in 1810:

> The most remarkable trait in the character of this species is the unaccountable practice it has of dropping its eggs into the nests of other birds, instead of building and hatching for itself; and thus entirely abandoning its progeny to the care and mercy of strangers.[45]

Figure 14.4 Brown-headed cowbird *(Molothrus ater)* on a lawn off Martin Luther King Drive, Philadelphia.

Figure 14.5 Intact, speckled egg on brick sidewalk of 2400 block of Waverly Street in Center City. The setting and appearance of this egg suggest it is an ejected egg of a cowbird.

In experiments conducted in Michigan and Connecticut, robins rejected artificial and real cowbird eggs, usually by ejecting the eggs from their nests.[46] Robin eggs are clear blue, in contrast to cowbird eggs, which are white with spots; cowbird eggs are smaller than robin eggs. When artificial cowbird eggs were experimentally deposited in robin's nests, robins accepted cowbird eggs painted clear blue, but rejected cowbird eggs painted blue with spots, or clear white.[47] Song sparrows, in contrast, do not reject cowbird eggs, and their populations have increased when local cowbirds were trapped and removed.[48]

How did robins evolve defenses against brood parasitism by urban cowbirds? The brown-headed cowbird, like the American robin, evolved in Central America and dispersed over most of North America.[49] It inhabited North America for at least a million years and likely peaked in abundance 15,000–20,000 years ago, when North America supported its greatest diversity of large mammals, including bison, oxen, horses, llamas, camels, mammoths, and mastodons.[50]

In 1799 Benjamin Smith Barton, professor of materia medica, natural history, and botany at the University of Pennsylvania, described how the brown-headed cowbird foraged in Pennsylvania:

> It follows cows and horses, pulling asunder their excrements, in order to get at the seeds. It alights on their backs, eating flies and other insects from them. In some parts of Pennsylvania, it is best known by the name of Cow-Bird.[51]

By the time horses and cows disappeared from North American cities, cowbirds had become urbanized. Brown-headed cowbirds have recently been found to be more abundant in urban than rural areas.[52] Robins and cowbirds forage together on lawns in Center City.

Wherever the two species coexisted, brood parasitism in cowbirds would have exerted selective pressure on robins.[53] The clear blue color of robin eggs may have evolved in response to cowbirds, or it may have evolved initially in response to other selective pressures, such as predatory attacks on robin eggs[54] and then secondarily as a defense against brood parasitism.[55] Whatever the sequence, robins had to cope with cowbirds long before the two met in Center City.[56]

Latitude, altitude, and temperature

The earliest systematic records of robins in Philadelphia date to 1802, when William Bartram began a twenty-year log of observations on weather and natural history at his home, about 3 kilometers from today's Center City. Entries in his log indicate that he saw robins in the month of January for eight years, including eleven days in January 1821, when he noted that the temperature fell to −8°F, the ice on the Schuylkill River was 12–14 inches thick, and the ground was covered with 3 inches of snow. He observed that traffic on the ice across the Delaware River was constant, and included carts and sleds drawn by teams of six horses transporting hay and wood.[57]

The mortality of robins under such icy conditions is unknown, but the caloric intake required for maintaining body temperature in winter can exceed robins' capacity to consume berries.[58]

To explain the overwintering of robins in Philadelphia, Alexander Wilson offered one hypothesis, which he linked to the name *Turdus migratorius:*

> The name of this bird bespeaks him a bird of passage, as are all the different species of Thrushes we have; but the one we are now describing being more unsettled, and continually roving about from one region to another, during fall and winter, seems particularly entitled to the appellation. Scarce a winter passes but innumerable thousands of them are seen in the lower parts of the whole Atlantic states, from New Hampshire to Carolina particularly in the neighbourhood of our towns; and from the circumstance of their leaving, during that season, the country to the northwest of the great range of the Alleghany, from Maryland northward, it would appear that they not only migrate from north to south, but from west to east, to avoid the deep snows that generally prevail on these high regions for at least four months in the year.[59]

The *Canadian Atlas of Bird Banding* has documented robins migrating southeast from Ontario across the Allegheny plateau and southeastern Pennsylvania to southern New Jersey and Maryland.[60] Philadelphia offers overwintering robins refuge from cold associated with higher latitude and altitude.

Heat island

Measured by satellite infrared imaging, Center City is typically warmer than surrounding suburbs by 1.7 to 3.3°C (3–6°F); on clear, calm winter nights, it is warmer than nearby rural areas by 5.6 to 11°C (10–20°F).[61] Satellite spectroradiometric imaging has monitored the onset of "greenup" (vegetation leafing out) in the spring along the Washington–Philadelphia–New York corridor. Greenup was 8.7 days earlier within urban cores compared to 8 to 10 kilometers outside. The rise in surface temperature along a gradient from rural areas to urban cores paralleled the advance of greenup. The growing season in urban core areas was 15 days longer than in outlying rural areas.[62] Thermal mapping of Center City shows variation in elevated surface temperatures according to the landscape; for example, Rittenhouse Square stands out as relatively cool, corresponding to its tree canopy.[63]

Center City's heat island is a result of pavement and buildings that trap solar energy during the day and radiate it at night, along with heat generators such as cars, air conditioners, and power plants.[64] However, William Bartram's records demonstrate that overwintering of robins in Philadelphia preceded the city's heat island.[65]

In Columbus, Ohio, an urban heat island effect was demonstrated in the case of the northern cardinal (*Cardinalis cardinalis*), which nested and formed clutches 7 to 10 days earlier in urban compared to rural habitats. The difference was found to be due to higher urban temperatures and not to greater availability of food.

Climate warming

Climate warming has advanced seasonal behavior of robins. By the spring of 2000, robins were arriving 14 days earlier than they had in 1981 at their breeding grounds in the Rocky Mountains in Colorado.[66] Advanced arrival times of robins in spring in Maine have correlated with higher temperatures in Maine and New England.[67] Comparable studies have not been done in Philadelphia, where robins newly arrived from the south are indistinguishable from those that have overwintered here from the north.

Light pollution

American robins typically start singing at dawn, but in Center City in early spring, I have heard them start by 2 a.m. Mark W. Miller of the U.S. Geological Survey hypothesized that light pollution advances the time when robins begin singing. In Schuylkill County, Pennsylvania, he found that robins exposed to light pollution of high intensity began singing as early as 1:10 a.m., the earliest ever reported. On average, they began singing 107 minutes earlier than did robins exposed to no light pollution, and 68 minutes earlier than did robins exposed to moderate light pollution.[68]

Similar findings have been reported for dawn singing in four species of European songbirds. The consequences of shifting the onset of singing to earlier times in the morning included disturbances in the timing of reproductive behavior.[69]

Noise pollution

Noise pollution has shifted the timing of bird songs. The European robin (*Erithacus rubecula*) typically sings during the day; but in noisy neighborhoods in Sheffield, England, it sings at night. An analysis of noise levels in over a hundred locations in the city concluded that daytime noise induced the birds to sing at night. The birds sang at night even when levels of artificial light were low.[70]

Traffic noise has shifted the spectral characteristics of the American robin's song.[71] Spectral shifts in response to noise vary not only with emission of noise but also with nearby impervious surfaces, which cause sounds to reverberate.[72] In theory, noise and light pollution could act synergistically on robins' song.

Relationships with people

The robin strips trees and shrubs of ornamental berries and harbors West Nile virus; it carries Lyme disease ticks and the Lyme disease pathogen, *Borrelia burgdorferi;* yet it is exempt from the kind of persecution endured by pigeons and starlings. Alexander Wilson considered the affection that people reserve for robins and concluded that it transcends the beauty of its song. He suggested that people associate its song with spring, and the name "robin" with the European robin, a species much admired:

> This song has some resemblance to, and indeed is no bad imitation of the notes of the Thrush or Thrasher *(Turdus rufus);* but if deficient in point of execution, he possesses more simplicity; and makes up in zeal what he wants in talent; so that the notes of the Robin, in spring, are universally known, and as universally beloved. They are as it were the prelude to the grand general concert that is about to burst upon us from woods, fields and thickets, whitened with blossoms, and breathing fragrance. By the usual association of ideas, we therefore listen with more pleasure to this cheerful bird than to many others possessed of far superior powers, and much greater variety. Even his nest is held more sacred among schoolboys than that of some others; and while they will exult in plundering a Jay's or a Catbird's, a general sentiment of respect prevails on the discovery of a Robin's. Whether he owes not some little of this veneration to the well known and long established character of his namesake in Britain, by a like association of ideas, I will not pretend to determine. He possesses a good deal of his suavity of manners; and almost always seeks shelter for his young in summer, and subsistence for himself in the extremes of winter, near the habitations of man.[73]

In Louisville, Kentucky, a study measured the distance robins allowed people to approach before they took flight. Robins tolerated people at closest distances when people approached them on paths and did not look at them. These responses appeared to be learned: in all trials, adult robins allowed people to approach closer than did young robins.[74] A study in Seattle measured flight responses of birds in neighborhoods where people reported that they repelled birds: American crows (*Corvus brachyrhynchos*) and European starlings (*Sturnus vulgaris*)—but not American robins—exhibited exaggerated flight responses.[75]

Alexander Wilson noted two centuries ago that people are less likely to persecute robins than other birds. Popular regard for robins has endured, endowing robins with rich rewards: berries, worms, lawns, water, nesting sites, and protection from ectoparasites and predators. In Pennsylvania, it may account, at least in part, for increasing numbers of robins.

15

CHINESE MANTID

(Praying mantis; *Tenodera sinensis*)

Over a century ago a Philadelphia nursery accidentally introduced the Chinese mantid into North America, and it spread through most of the eastern United States. In Center City, the smaller Carolina mantid *(Stagmomantis carolina)* is replacing it.

Figure 15.1 Chinese mantids (*Tenodera sinensis*) from Mt. Airy, Philadelphia, where the first Chinese mantid in North America was discovered. From Philip Laurent's collection, deposited in the Academy of Natural Sciences of Drexel University. (Courtesy of the Academy of Natural Sciences of Drexel University)

On October 16, 1897, Joseph Hindermyer encountered a Chinese mantid *(Tenodera sinensis)* on a tomato vine in his garden in Mt. Airy, Philadelphia.[1] Never before seen on this continent, it was the largest carnivorous insect in North America. His neighbor, Philip Laurent, an entomologist, reported what happened next:

> Mr. H., not being familiar with the insect's harmless nature, was afraid to touch it, but at last managed to secure it in a paste-board box, in which condition it was brought to me... Although a careful search was made in the vicinity in which the specimen was found, no others were discovered. Learning later that the native habitat of the insect was China and Japan, I made inquiry among those having nurseries and conservatories in the neighborhood where the specimen was captured, regarding the importation of plants from the above named countries. At the nursery of Thomas Meehan & Sons—the largest nursery in the vicinity of where the insect was captured—I was informed that they were constantly receiving plants from all parts of the world, so that it is more than likely that the insect was introduced through this channel.[2]

In March 1898, Ella Jacobs, visiting Meehan's Nursery in Germantown, found six specimens of what she thought were galls. Unable to identify them, she took them to her office and kept them around to see what might develop:

> About the end of May, as I went to my office, the janitor greeted me with the pleasant news that my room was full of "bugs." Rather startled, I proceeded to investigate, and discovered several hundred insects on the wall, over pictures and desk. I examined closely and decided it was the fault of my unnamed specimen. I noticed that it was broken open in ridges; I placed it in a box and in an hour I saw several of the insects emerge. The curious part is, that these insects appeared to be the Praying Mantis. A visit to Dr. Skinner, at the Academy, confirmed this fact.
>
> It seems rather a coincidence to have found these in this locality so soon after the report of Mr. Laurent's find of a somewhat similar character.
>
> I greatly regret that we took all of the cases we saw, six of them, as I know now that their contents would have been a valuable acquisition to the nurseries as these carnivorous insects would have eaten other insects injurious to the plants.[3]

Importation into North America

In the archives of the McLean Library of the Pennsylvania Horticultural Society is a Meehan nursery catalog dated 1858—39 years before Hindermyer found his Chinese mantid. It offers for sale Chinese magnolias and Japanese paulownias.[4] Imported nursery stock could have introduced the Chinese mantid into Meehan's nursery decades before it surfaced on Hindermyer's tomato plants. Discoveries of accidentally introduced species of insects typically occur after a lag during which their populations grow to the point that they come to people's attention. The lag may have been shorter than usual in the case of the Chinese mantid because its 10-centimeter body and 2.5-centimeter egg case are so conspicuous.

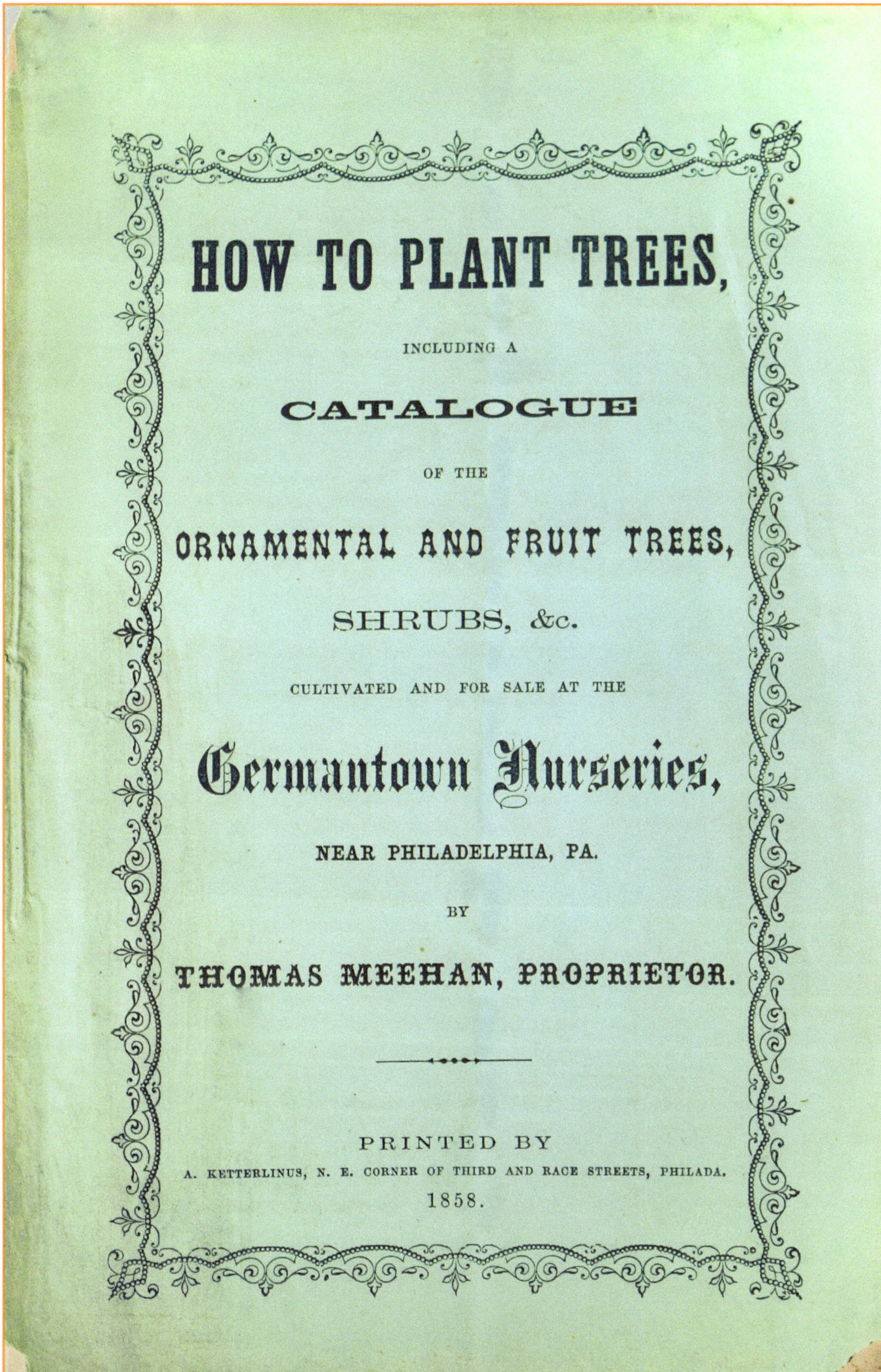

Figure 15.2 Cover of the catalog of the Germantown Nurseries of Thomas Meehan, 1858. Offerings include species from China and Japan, likely sources for the accidental importation of the Chinese mantid, *Tenodera sinensis,* into North America. (Courtesy of the McLean Library of the Pennsylvania Horticultural Society)

Figure 15.3 Chinese mantid egg case (ootheca) found in Bartram's Garden, Philadelphia, in 2012.

Beneficial or harmful?

In 1926, Walter R. Thierolf attempted to determine whether this predator was economically beneficial. His investigation was the basis of his thesis at the University of Pennsylvania, and is reminiscent of studies that attempted to determine whether the house sparrow was helpful or harmful. He collected fifty egg cases in Glenside, Pennsylvania, and released an estimated 10,000 individuals into the neighborhood around his home. He identified their prey by observing mature mantids eating and by dissecting their alimentary tracts.

Thierolf's verdict on the economic value of the species was cautious, in part because he found that the mantids avidly ate honeybees. He diplomatically concluded, "The fact that the insect victims of mantids are so extremely varied would appear to make them worthy agents in nature's plan to retain a normal balance in insect life."[5] In the course of releasing 10,000 mantids into new territory, Thierolf likely expedited the mantid's dispersal, beneficial or not.

Means of dispersal

By 1950, according to an annual report of the Smithsonian Institution, the Chinese mantid had been found along the Atlantic Coast from Connecticut to Virginia and in scattered locations elsewhere, including California.[6] Today, it is found in almost every state east of the Mississippi and in several western states, plus Ontario and Quebec, as documented by a website that posts photographs of insects that people submit for identification.[7]

This dispersal is greater than expected based on the female mantid's low mobility. The female does not fly and is stopped by common barriers such as streams or highways.[8] Vendors of garden supplies market egg cases of the Chinese mantid for control of insect pests and, along with nurseries, have undoubtedly accelerated the mantid's dissemination.[9] (In advertisements, garden supply businesses offer the egg cases for sale under the name *Tenodera aridifolia sinensis,* but the nomenclature recently has reverted back to *Tenodera sinensis,* its name at the time of its discovery in Philadelphia.[10])

Collecting egg cases

In Center City the Chinese mantid is a casualty of its celebrity status as a "beneficial" insect. I have not found one for several years downtown. Gardeners collect their egg cases and place them in their gardens, where squirrels, birds and mice prey on them, or they bring them indoors, where they hatch prematurely in the spring. The last egg case I found in Center City was on a sapling transplanted from nursery stock; the egg case disappeared a few days after I discovered it. Egg cases in Fairmount Park just outside of Center City are still common in brambles in old fields.

A "native" praying mantid introduced into Center City

The Chinese mantid is one of four species of mantids in this region. Two were introduced from Asia, one from Europe, and one, the Carolina mantid (*Stagmomantis carolina*), from just south of Pennsylvania.[11] The Carolina mantid ranges from northern Brazil and Ecuador into Venezuela, Colombia, Central America, Mexico, and the southern United States.[12] Along the east coast of the United States, its northern limit was Chestertown, Maryland, according to an analysis of specimens in the Academy of Natural Sciences of Philadelphia in 1937.[13] I have found both sexes of the Carolina mantid attracted to door lamps on our row house on Pine Street in Center City. Like bridge spiders, Carolina mantids capture prey attracted to electric light.

The Carolina mantid is smaller than the Chinese mantid, but still among the biggest insects one is likely to find in Center City. Body lengths of adults reach 5–6 centimeters (2–2.4 inches), compared to 8–10 centimeters (3–4 inches) for the Chinese mantid. The female Carolina mantid has short stubby wings and is usually green. The male has fully developed wings, and its body is usually brown, but color in either sex can be predominantly brown or green.

Egg cases of Carolina mantids are better camouflaged than those of Chinese mantids. Unlike the bulky, conspicuous globular egg cases of Chinese mantids, egg cases of Carolina mantids are smaller, elongated, and oriented so they blend in with the twigs to which they are attached. Crowds of people in Schuylkill Park ignore them.

Figure 15.4 Egg case (ootheca) of Carolina mantid (*Stagmomantis carolina*). Its linear profile is smaller and less conspicuous than the bigger and more globular egg case of the Chinese mantid. (Photographed at Bartram's garden)

Figure 15.5 Female Carolina mantid (*Stagmomantis carolina*) attracted to the porch light of our house at night in Center City. The black spot on its wing distinguishes it from other species.

Figure 15.6 Male Carolina mantid on our house.

Figure 15.7 Carolina mantid eating a moth fly (Family: Psychodidae) beside our front door.

Mystery of the dispersal of the Carolina mantid into Philadelphia

How the Carolina mantid colonized Center City is unknown. Mantids have low "vagility," meaning the distances they crawl or fly in the course of their lives are short. In Northern Delaware the lifetime distance traveled by adult Chinese mantids that were marked and recaptured was 70 meters maximum, and usually less than 20 meters.[14] The Carolina mantid, which is about half the size of the Chinese mantid, has no obvious means of dispersal other than its legs; the female's wings are too short for powered flight. Despite limited mobility, this species has managed to populate a vast geographic expanse from Brazil to Maryland. Fossil mantids dating back 90 million years have been found in New Jersey amber,[15] so mantids have had a long time to disperse. Still, in the absence of human assistance, how might this species have traversed barriers like rivers?

The answer to this question is not known. Mantids have been found on top of the Empire State Building in Manhattan;[16] conceivably females crawled up trees or cliffs and used their stubby wings to glide. Six species of birds have been reported to feed on mantid egg cases;[17] in principle birds might excrete and disperse viable mantid eggs embedded in the tough matrix of their cases, but this hypothesis has yet to be investigated. Floods might have carried egg cases or adults on debris across rivers; but this too is conjecture.

Freight trains and boats as possible agents of dispersal

The first record of a Carolina mantid in Pennsylvania was in 1862, after ootheca (egg cases) imported from Maryland produced populations that reproduced for two or three generations in Lancaster City.[18] In 1899 Philip Laurent reported identifying a Carolina mantid collected on a wharf in Philadelphia.[19] Nursery stock could disperse the Carolina mantid like the Chinese mantid; egg cases of both are offered for sale online.

Lawrence Hurd of the University of Delaware reported finding ootheca (egg cases) of the European mantid (*Mantis religiosa*) attached to the undercarriage of railroad freight cars, which he concluded transported this mantid to fields along train tracks in Northern Delaware.[20] In Center City freight trains frequently park along tracks bordered by wild vegetation. I found an egg case of a Carolina mantid attached to a pine tree in Schuylkill Park less than 50 meters from these tracks.

Figure 15.8 Vegetation along the CSX railroad tracks in Center City. The view faces south near the Schuylkill River and Interstate 676. In Delaware, egg cases of mantids have been found attached to the undercarriages of freight trains, which may disperse them.

Center City's heat island

In the northeastern United States, the distribution of many species of southern insects, particularly butterflies, has been moving northward due to temperatures rising from climate change.[21] The northward advance of the distribution of the Carolina mantid is part of this trend, even though people have introduced it. Center City's urban heat island may have encouraged establishment of Carolina mantids.

Pressure for smallness downtown

Carolina and Chinese mantids coexist in Bartram's garden, Philadelphia, just outside of Center City. Here open fields provide abundant prey, and the remote location protects them and their egg cases from destruction by people. In contrast, in Center City, the larger (Chinese) mantid and its egg case have become rare. Persecution of these big attractive targets may explain their rarity downtown, but another hypothesis is worth considering. In *Why Big Fierce Animals Are Rare,* Paul Colinvaux argues that scarcity of prey limits the abundance of big predators such as lions and tigers.[22] Perhaps scarcity of insect prey downtown contributes to the scarcity of Center City's biggest carnivorous insect.

16

PILLBUGS

(Isopods; *Armadillidium*)

Pillbugs thrive downtown despite vulnerability to predators, parasites, pathogens, and desiccation. They have gained safety in numbers.

Figure 16.1 Pillbug, *Armadillidium nasatum*, rolled into an imperfect ball, with a gap on the right. Rolling into a ball (conglobation) protects pillbugs from desiccation and predators.[1]

In the first volume of the *Journal of the Academy of Natural Sciences of Philadelphia,* published in 1818, Thomas Say presented "An Account of the Crustacea of the United States." Crustacea are arthropods such as lobsters, crabs, shrimp, barnacles, and fourteen-legged creatures called isopods. Terrestrial isopods include familiar garden animals known by many colloquial names, such as woodlice, sowbugs, roly-polies, and pillbugs. Say noted that one species, currently named *Armadillidium vulgare,* "is very common in moist places, under stones, in decaying wood, &c."[2] This species inhabits our garden in Center City.

Figure 16.2 Our Center City row house garden, habitat for a diverse community of exotic animals, including six species of isopods, such as pillbugs.

Introduction of pillbugs

Unlike the Chinese mantid, *A. vulgare* in North America left no obvious clues to its place of origin. A genetic study of 10,000 of these pillbugs in 157 populations in Europe and North America concluded that this species was introduced from northern Europe.[3] Root balls in imported horticultural and agricultural stock could have carried it in, or dirt used in ship ballast dumped near American ports could have transported it here.

All species of terrestrial isopods in the northeastern United States have been introduced except for the few endemics that inhabit caves or seashores, which protected them from Pleistocene glaciation and permafrost.[4] Pleistocene permafrost on the mid-Atlantic Coastal Plain, which includes Center City, extended at least as far south as southern Delaware and southern Maryland.[5]

A century after Thomas Say's report on crustaceans, Henry Fowler of the Academy of Natural Sciences of Philadelphia noted that *A. vulgare* "is of world-wide distribution, living mostly in moist places, as under stones or logs, in crevices or rocks, about greenhouses, cellars, under boards, etc." He named four neighborhoods in Philadelphia where he collected it.[6] Now, almost two centuries after Thomas Say's paper on crustaceans, those he described are still here, and new ones have arrived and established themselves.

Figure 16.3 Pillbug, *Armadillidium vulgare*, introduced from Europe and common in this region by 1818. Specimen from our garden.

Figure 16.4 Pillbug, *Armadillidium nasatum*, a more recent introduction and the most abundant isopod in our garden. Specimen from our garden.

In 1984, Philip F. Ganter and Wilma Kane Hanton at the University of North Carolina reported that more than 80 percent of pillbugs (*A. vulgare*) in local populations were female. Referring to electron micrographs revealing bacteria in pillbug ovaries, they hypothesized that an intracellular bacterial parasite skewed sex ratios by transforming male pillbugs into females.[7] In 1991 Thierry Rigaud and colleagues in France identified the bacterium as *Wolbachia*,[8] named after the Harvard microbiologist S. Burt Wolbach, who, with Marshall Hertig, first described the bacterium in the house mosquito in 1924.[9]

Wolbachia, transmitted maternally through pillbug eggs, transforms genetic male pillbugs into functional females that produce viable eggs and offspring. By converting genetic male pillbugs into reproductive females, the microbe commandeers the reproductive machinery of the pillbug and subordinates it to the bacteria's own benefit. *Wolbachia* behaves like a selfish gene, reprogramming its host's reproduction to maximize its own.[10]

In one survey *Wolbachia* infected almost half of species of terrestrial isopods,[11] but the actual proportion is likely greater.[12] It infects about two thirds of all species of insects,[13] plus an indeterminate number of nematodes, spiders, scorpions, and mites.[14] Reproductive effects of *Wolbachia* on infected hosts differ depending on host species. In some cases, it kills developing males; in others, it creates females that reproduce asexually,[15] permanently eliminating males from propagation unless the insects are treated with antibiotics.[16]

Wolbachia's reproductive gain is not necessarily the pillbug's loss. Spread of a similar maternally transmitted infection to populations of the sweet potato whitefly increased the fly's fitness.[17] Elimination of *Wolbachia* infection from the bedbug (*Cimex lectularius*) retarded the bedbug's growth and induced sterility.[18]

Wolbachia infection may have helped pillbugs colonize our garden. Normal males are poor agents for dispersal: a solitary male that has dispersed into a new habitat cannot alone found a colony; but males that *Wolbachia* has transformed into females can mate, disperse, and then establish populations in new habitats. Radioactively labeled pillbugs tracked in the field dispersed as far as 25 meters.[19] *Wolbachia* infection, at least theoretically, helps pillbugs cope with severe fragmentation of habitat, such as that in downtown Philadelphia.

The capacity of a normal male pillbug to inseminate many females compensates for depletion of males in populations after the males have been transformed into females,[20] but this transformation has its costs: normal males prefer real (i.e., genetic) females,[21] and they exhaust their supply of sperm after mating multiple times.[22] The pillbug may be either *Wolbachia's* beneficiary or its victim, depending on the value of the traits that *Wolbachia* transmits. This value may differ under different environmental conditions, as in the case of *Wolbachia* infection in mosquitoes.[23] Populations of pillbugs can eliminate *Wolbachia* infection,[24] possibly by evolving resistance.[25]

Parasites: spiny-headed worms

Another infection commandeers pillbugs, but unlike *Wolbachia*, its life cycle is not confined to pillbugs. The infectious agent is a parasite, a spiny-headed worm (*Plagiorhynchus cylindraceus*) in the phylum Acanthocephala, which includes over a thousand named species and is unrelated to earthworms, flatworms, or roundworms. *P. cylindraceus* has been found in birds on five continents. The only study of its prevalence in North American birds found it in 62 percent of robins, 56 percent of grackles, 42 percent of starlings, and lower frequencies in blue jays, house sparrows, and brown-headed cowbirds.[26]

The parasite was first discovered in North America in 1918 by Harley J. Van Cleave,[27] who later looked for it but did not find it in Joseph Leidy's collection of Acanthocephala preserved in the University of Pennsylvania and the Academy of Natural Sciences of Philadelphia.[28] Van Cleave reported the parasite in birds in Pennsylvania, New Jersey, Maryland, New York, and Washington, DC, and inferred that it is an exotic species introduced only recently.[29]

In 1929 D.T. Sinitsin, a Russian parasitologist who had fled the Soviet Union,[30] reported finding a 4-millimeter worm in a pillbug (*Armadillidium vulgare*) that he had collected near the Chesapeake and Ohio Canal in Washington, DC. He identified the worm as a young *P. cylindraceus*.[31] In 1964 Gerald D. Schmidt and O. Wilford Olsen at the University of Colorado harvested eggs of gravid *P. cylindraceus* worms located in the small intestine of robins. They presented the eggs of the worm first to beetles, which ate and digested them. They then presented the eggs to pillbugs, which ate the eggs but did not digest them; the eggs hatched and developed into infectious cysts inside the pillbugs. When birds consumed pillbugs containing parasitic cysts, the proboscis of the worms pierced the cysts' walls and attached to the birds' guts, where the worms developed to maturity and produced eggs. The eggs exited in the birds' feces, where they became available to pillbugs, completing the cycle.[32] The parasitic infection has been found to lower birds' metabolic rate and weight, but effects on mortality and fitness have not been studied.[33]

Reported prevalence of these worms in populations of pillbugs is paradoxically low —for example, only 1 in 1,500 individuals examined in one study.[34] How could the worm be so abundant in birds yet vanishingly rare in pillbugs? In 1982 Brent B. Nickol and Glen E. Dappen at the University of Nebraska showed that mature pillbugs were relatively resistant to infection by the parasite, which infected predominantly young individuals.[35] The next year, Janice Moore at the University of New Mexico showed that the worm changed the behavior of the pillbugs it infected such that its victims exposed themselves to predators such as robins.[36]

Like *Wolbachia*, the worm commandeers the behavior of pillbugs for its own benefit. Unlike *Wolbachia*, it offers pillbugs nothing in return. The worm's diversion of pillbugs to predators explains the paradoxical rarity of worm-infected pillbugs compared to worm-infected birds. Worm-infected pillbugs promptly become food for birds such as robins.

Pillbug aggregation: vulnerability to enemies

Pillbugs in our garden aggregate, piling up on each other beneath stones or logs. Why would they clump together and risk exposing themselves to predators and pathogens? Most aggregations of pillbugs in our garden consist of *A. nasatum*, a pillbug susceptible to infection by both *Wolbachia*[37] and spiny-headed worms[38] and probably also iridovirus.[39] This pillbug initially inhabited greenhouses in North America, but by the 1950s it had established populations outside greenhouses.[40] A genetic polymorphism produces different color forms.[41]

Aggregations of *A. nasatum* would appear to be easy targets for the European centipede (*Cryptops hortensis*) and the pillbug hunter *Dysdera crocata*, an introduced spider that specializes in preying on terrestrial isopods.[42] This spider has long fangs that it uses like pincers to seize isopods, which it secures with one pincer clamping down on the armored top and the other puncturing the soft underside,[43] which it injects with venom that can kill within seven seconds.[44] It hunts at night without use of a web to snare prey.[45]

Figure 16.5 Pillbug hunter (*Dysdera crocata*) under a log in our garden. Remains of its victims are scattered about.

Figure 16.6 Pillbug hunter. I collected this individual in our garden.

Figure 16.7 Pillbug hunter's formidably long fangs work like pincers. The top fang holds the victim and the bottom fang skewers it on the soft underside and injects the venom.

Pillbug aggregation: avoidance of dessication

In 1931 Warder Clyde Allee published *Animal Aggregations: A Study in General Sociology.* Citing his own research,[46] he suggests why isopods bunch:

> Land isopods (Allee, 1926) tend to collect in aggregations in the hot, dry summer and in the cold, and often physiologically dry, winter. These aggregations are frequently such as might result when shelter is limited, provided there is a tolerance for the presence of other similar animals; but at times these animals collect in much closer units than can be entirely explained on this basis. That is to say, the isopods do not occupy all the available and apparently equally desirable space, but clump together in one part of this.
>
> When a drop of water was introduced on a dry background, the isopods tended to occupy all of that favorable location regardless of whether or not they were in contact. The bunching in close physical contact came later, and might take place as a thigmotropic reaction, perhaps modified by chemical stimuli, or might have been conditioned by the drying of the small moistened region.[47]

Fifty years later Naokuni Takeda at Toho University in Japan showed that the pillbug (*A. vulgare*) produces a pheromone that promotes aggregation, which in turn reduces desiccation and increases growth. In other species of isopod, aggregation pheromone was shown to prolong survival.[48] Mark Hassall and his colleagues at the University of East Anglia in England concluded that aggregation can protect isopods from climate warming,[49] which might apply to our garden, located as it is in an urban heat island. Cédric Devigne and his colleagues in France have shown experimentally that changes in temperature and humidity alone do not fully explain why isopods aggregate; they concluded that unidentified social benefits favor aggregation.[50]

Pillbug aggregation: Avoidance of pillbug hunters

In 1971 William D. Hamilton published his iconoclastic "Geometry of the Selfish Herd," presenting the evolution of aggregation behavior as a selfish response to predators. He refers to birds, fish, frogs, ungulates, and insects—but not isopods:

> This paper presents an antithesis to the view that gregarious behavior is evolved through benefits to the population or species…Gregarious behaviour is considered as a form of cover-seeking in which each animal tries to reduce its chance of being caught by a predator. It is easy to see how pruning of marginal individuals can maintain centripetal instincts in already gregarious species...Besides this, simply defined models are used to show that even in non-gregarious species selection is likely to favour individuals who stay close to others.[51]

Pillbug aggregations—like those of ungulates, birds, and fish—buffer members in the inside from attack by predators on the outside. Pillbug hunters hide under logs and stones, as do pillbugs. I have not found the hunters within aggregations of pillbugs. Pillbug hunters could, in theory, exploit pillbug aggregations by grouping inside these aggregations, much as bridge spiders exploit concentrations of prey at electric lights along the Schuylkill River; but, unlike bridge spiders, pillbug hunters are solitary, and they dwell outside aggregations of pillbugs. The pillbugs most vulnerable to pillbug hunters would be expected to be those on the periphery of aggregations or outside them—as Hamilton's theory predicts. Protection against predators may be the primary benefit of pillbug aggregation, or it may be secondary, after protection against desiccation.

Figure 16.8 Bunching of pillbugs (*Armadillidium nasatum*) on the underside of a paving stone in our garden. Aggregation protects individuals from desiccation and predators. A genetic polymorphism contributes to differences in color among individuals.

Hamilton contended that even nongregarious species gain safety in numbers. In our garden his conclusion sheds light on pillbugs that happen to live outside of aggregations. Typical of territorial spiders[52] pillbug hunters in our garden space themselves apart and keep their population densities low. Searching in pillbug hunters' favorite shelters such as under logs, I usually find no pillbug hunters. When I do find one, it is solitary. In our garden, low densities of pillbug hunters and high densities of pillbugs keep ratios of pillbug hunters to pillbugs low; these low ratios also keep the odds of an attack on any individual pillbug, even those outside aggregations, low.

Abundance of any one kind of isopod in our garden would be expected to contribute to the protection of others. Pillbug hunters specialize in all kinds of isopods, not just pillbugs. Our garden has six species of isopods, all introduced. Two are pillbugs (*A. vulgare* and *A. nasatum*), which defend themselves by rolling into a ball, and the rest are runners that defend themselves by fleeing.*

Slugs

Like isopods, slugs aggregate, or "huddle," a behavior that also protects them from desiccation.[53] All three species of slugs in our garden have been reported to huddle,[54] but I have observed huddling in only the most common, the threeband gardenslug (*Lehmannia valentiana*).[55] For this slug, high population densities facilitate huddling and protection from desiccation—another instance of safety in numbers.†

Figure 16.9 Threeband gardenslugs (*Lehmannia valentiana*), huddling under a paving stone. Huddling protects slugs from desiccation. (A recent synonym for its scientific name is *Ambigolimax valentianus*.)

Impact of exotic animals on our garden's ecology

Daniel Simberloff coined the phrase invasional meltdown for destruction of native ecosystems by introduced species. The term meltdown, an allusion to nuclear power plants, refers to positively reinforcing interactions among the immigrants.[56]

Safety in numbers of slugs and isopods in our garden exemplifies positively reinforcing interactions— but not an invasional meltdown. These animals have enriched an urban habitat stripped long ago of most native plants and animals. They decompose organic debris, amend the soil, and obviate my need to dispose of leaf litter. They are members of a community of diverse animals including native species, such as the common eastern firefly, *Photinus pyralis*, and the American robin.

* In our garden the four isopods that are runners are: common pygmy woodlouse (*Trichoniscus pusillus*), common striped woodlouse (*Philoscia muscorum*), *Porcellionides pruinosus*, and *Hyloniscus riparius*.

† The two other slugs are the giant garden slug (*Limax maximus*) and the grey field slug (*Deroceras reticulatum*).

17

COMMON MILKWEED

(*Asclepias syriaca*)

Common milkweed thrives just outside Center City but not inside, despite wind-blown seeds that disperse downtown.

Figure 17.1 Common milkweed (*Asclepias syriaca*) along railroad tracks by the Schuylkill River at Locust Street in 2008. This plant was part of a sprawling colony destroyed in 2011 by railroad maintenance crews.

In 1803 Benjamin Smith Barton described entrapment of houseflies in flowers of common milkweed, which he concluded is a kind of "muscipula," or flytrap. Barton, professor of materia medica, natural history, and botany at the University of Pennsylvania, presented his findings to the American Philosophical Society in Philadelphia:

> In the summer of 1801, I discovered a vegetable muscipula in the vicinity of Philadelphia. Having collected some branches, in flower, of the Asclepias syriaca, or Syrian Swallow-wort, well known in the United States by the names of Wild-cotton, cotton-plant, &c; with the view of making some experiments with the milky juice of this plant, I was not a little surprised to find in the course of a few hours, a number of the common houseflies strongly attached to the flowers; being secured, some by their proboscis, and others by their legs: the greater number, however, by their legs. I, at first, imagined, that the flies were merely retained by the viscous juice of the flowers of this Asclepias: but I soon found, that this was not the case. They were detained by the small valves of the flower, and I observed, that the irritability of the valves seemed to reside exclusively in one particular spot, not larger than the point of a common sized pin. Neither in this spot nor in any other part of the valve, could I observe the least vestige of a glutinous or viscous quality. I think it sufficiently evident, that the valve is endued with the irritable principle.
>
> In the genus Asclepias, the valves which I have noticed, are ten in number, being situated in pairs, so as to form five little foviae, the structure and uses of which are not sufficiently known to botanists.
>
> A considerable number of flies, not less perhaps than sixty or seventy, which alighted upon the flowers of my Asclepias, were detained in the manner I have mentioned...Many of the flies, particularly the larger ones, were enabled, after some time, to disengage themselves from their prison, without the loss of any of their limbs or organs, or any perceptible injury whatever. Many others effected their escape, not however, without the loss of one or more of their legs, or their proboscis. Not a few, after making long and repeated efforts to regain their liberty, perished in their vegetable prisons.[1]

Milkweed pollination

Barton's "valves" are *pollinia,* or agglutinated masses of pollen. Paired pollinia linked together are *pollinaria*. Barton was the first to observe pollinia adhere to the legs and mouthparts of insects visiting flowers of common milkweed, but he misconstrued their functional significance as entrapment akin to that of Venus flytraps (*Dionaea muscipula*) and sundews (*Drosera*).

From the perspective of bees, Barton correctly perceived milkweed's behavior as less mutualistic than that of, say, clover, which rewards bees with nectar and pollen. Milkweed offers bees only nectar. Bees visiting milkweed do collect pollen—in the form of pollinia stuck to their feet and other body parts—but they are unable to use such pollen as food. Douglass H. Morse at Brown University found that pollinia of common milkweed slowed down the foraging of bumblebees by 25 percent; pollinia entangled their mouthparts and their appendages, and caused loss of body parts, including claws and segments of legs,[2] much as Barton described for houseflies.

By the end of the nineteenth century European botanists demonstrated that pollinia contain pollen. They showed that pollinia in common milkweed flowers adhere to insects' legs and feet and other body parts as they take nectar, and that milkweed flowers later snag pollinia off the insects when the insects incidentally insert the pollinia into flower chambers containing the stigma, the flower's receptive female structure.[3]

Figure 17.2 Leg of honeybee (*Apis mellifera*) entangled by orange pollinia containing milkweed pollen.

Figure 17.3 Honeybee struggling to extricate herself from milkweed flowers that have snagged pollinia stuck to her feet.

Scarcity of milkweed downtown

In 1818, William P. C. Barton, a relative of Benjamin Smith Barton, published the first systematic inventory of plants in the vicinity of Philadelphia. About common milkweed, his *Compendium Florae Philadelphicae* states:

> The commonest species of this genus in this neighbourhood. From three to four feet high. On the banks of the Schuylkill, Delaware, and all our creeks, very frequent. Perennial. June.[4]

Common milkweed currently blooms in scattered colonies along the west bank of the Schuylkill River beside Martin Luther King Drive just outside Center City. Here I watched a honeybee struggling to extricate herself from a common milkweed flower that had snared her in a tangle of pollinia. After five minutes she finally liberated herself. She flew about 10 centimeters away only to turn around and return to the same cluster of flowers that had just trapped her. She drank nectar for a minute until, still free, she flew away.

Milkweed thrives in Philadelphia despite over two centuries of urbanization. The plant's success, however, has been uneven. In Fairmount Park just outside Center City, it makes sprawling, multistemmed perennial colonies with abundant flower heads and seed-bearing pods. By contrast, in Center City it typically produces just a few stalks that last only one or two seasons and yield no pods. For the past several decades in Center City, I have found it to be absent or rare. In 2011 along tracks by the Schuylkill River, railroad maintenance crews destroyed the last big colony here. Its scarcity in Center City is surprising, since it tolerates a broad range of conditions, including drought and soil ranging from alkaline to acidic.[5]

Figure 17.4 Rare example of common milkweed in Center City. It is on the edge of a parking lot near 22nd and Sansom Streets. It never produced seedpods, and was gone the following year. Concrete prevented the roots from spreading and establishing a colony. Photographed August 31, 2008.

Figure 17.5 Colony of common milkweed off Martin Luther King Drive along the Schuylkill River about a kilometer from Center City. Space sufficient to accommodate such a large colony is scarce in Center City.

Baker's Law

What might account for the absence of a plant that is generally regarded as a weed? In 1955 Herbert George Baker observed that plants that disperse long distances typically reproduce vegetatively, or by self-fertilization.[6] This observation was later dubbed Baker's Law.[7] Baker's Law offers a possible explanation for the rarity of common milkweed in Center City. Seeds of common milkweed disperse on strands of silk blown by the wind. In Center City I have watched common milkweed seeds floating in the air far from milkweed patches. Common milkweed violates Baker's Law in the sense that it is a long-distance colonizer that reproduces predominantly by outcrossing.[8] Perhaps the reason common milkweed is rare in Center City is its violation of Baker's Law.

Common milkweed does make vegetative clones from its roots, but these clones do not disperse over long distances unless the roots are broken into pieces that can travel as independent propagules, as in agricultural fields that have been tilled.[9] Self-pollination in common milkweed does produce seedpods, but the rate is low—only 4 percent in experiments in which milkweed was pollinated by hand.[10] In theory, common milkweed's poor compliance with Baker's Law might account for its rarity in Center City: a milkweed that colonized Center City would be far away from potential mates required for outcrossing.

Figure 17.6 Seed of common milkweed about to disperse from a colony of milkweed along railroad tracks in Center City. Wind can carry these seeds long distances. The red insect is a young nymph of the large milkweed bug (*Oncopeltus fasciatus*).

Milkweed's trouble obtaining mates downtown would constitute an Allee effect, named after the same Allee who investigated aggregation of pillbugs. Allee effects occur when low population densities impede sexual reproduction. As distances separating members of a population increase, their access to mates decreases, potentially causing reproductive failure and collapse of the population.[11] In theory, the proportion of Center City covered with concrete and asphalt is so high that it depresses population densities of milkweed and impedes sexual reproduction.

Tatyana Livshultz, pollination biologist and botanist at the Academy of Natural Sciences of Drexel University in Philadelphia, and her colleagues at the Royal Botanical Gardens, Kew, found evidence that Allee effects due to drought and thinning of milkweed populations in Africa influenced the evolution of milkweed flowers, particularly of pollinia.[12] Pollinia improve chances that pollen grains carried by pollinators reach their destination—the stigma of a milkweed plant of the same species.[13] In addition, pollinia package pollen in quantities optimized for reproduction.[14]

The evolutionary experience of milkweeds coping with Allee effects in Africa may have prepared common milkweed for Allee effects in Center City. Before railroad crews destroyed it, the colony of milkweeds growing along the railroad tracks produced abundant seedpods. Common milkweed has demonstrated that in Center City it can overcome reproductive barriers due to low population densities. Pollinia and self-fertilization both may have contributed to this success.

Light pollution

Pollinia protect pollen from becoming fodder for bees, and they facilitate pollination; but they may be vulnerable to light pollution. In 1957, Stuart W. Frost at Penn State University found milkweed pollinia attached to 290 banded tussock moths (*Halysidota tessellaris*) that had flown into light traps, mostly during the first few weeks of July. He reported that no milkweed grew near the light traps.[15] The following year he replicated these findings. He noted that the pollinia could have come from any of three milkweed species, including common milkweed.[16] I have found the banded tussock moth and its larvae in Center City. Specimens probably collected in Philadelphia are included in Titian Ramsey Peale's nineteenth-century moth collection housed at the Academy of Natural Sciences in Philadelphia.[17] The larvae feed on many kinds of shade trees, shrubs, and vines.[18]

Urban light pollution might undermine milkweed reproduction by diverting moths carrying milkweed pollinia. Such diversion would disrupt pollination and deplete stores of pollinia. Common milkweed has been shown to produce four times more nectar and double the amount of sugar during the night compared to the day.[19] Per visit to a milkweed flower, nighttime pollinators are twice as likely to produce a milkweed pod than are daytime pollinators.[20]

Despite these findings, light pollution does not explain the rarity of milkweed in Center City. In a study comparing pollination during the day and night, flowers of common milkweed exposed only to daytime pollinators produced eight times more pods than did flowers exposed only to nighttime pollinators.[21] The greater abundance of daytime pollinators of milkweed more than offsets their lower efficiency of pollination.[22]

Ozone

In theory, air pollution could disrupt milkweed pollination. Ozone destroys volatile floral hydrocarbons that attract pollinators.[23] Common milkweed in the laboratory begins to develop purple stipling when exposed to concentrations of ozone below ozone concentrations measured in Philadelphia.[24] Detrimental effects of ozone on plants, however, have been found to be greater outside core urban areas than inside.[25] The absence of purple stipling on the common milkweed that I have observed in Center City suggests that ozone is not a cause for the failure of this species to establish itself in Center City.

Pollinator scarcity

Scarcity of pollinators downtown could prevent common milkweed from making pods, especially since the railroad intensified suppression of weeds. Common milkweed requires insect pollinators even for self-pollination; neither wind nor gravity is capable of depositing milkweed's pollinia into its flowers' stigmatic chambers.

Gerald A. Mulligan and Judy N. Findlay at the Canada Department of Agriculture placed bags over flower heads of common milkweed to exclude insect pollinators; these flower heads produced no seed. They obtained similar results for a handful of other widespread weedy species, none of which is common in downtown Philadelphia.

In contrast, they did obtain seeds from bagged flowers of many of Center City's most abundant weeds. Examples are: dandelion (*Taraxacum officinale*), pigweed (*Amaranthus retroflexus*), lambsquarters (*Chenopodium album*), horseweed (*Erigeron canadensis*), fleabane (*Erigeron annuus*), purslane (*Portulaca oleracea*), smartweeds (*Polygonum persicaria, P. aviculare, P. lapathifolium,* and *P. pensylvanicum*), groundsel (*Senecio vulgaris*), foxtails (*Setaria viridis* and *S. glauca*), chickweed (*Stellaria media*), bull thistle (*Cirsium vulgare*), St. John's wort (*Hypericum perforatum*), evening primrose (*Oenothera biennis*), broadleaf plantain (*Plantago major*), and common mullein (*Verbascum thapsus*).[26]

Among the weeds Mulligan and Findlay tested, dandelion is one of the most conspicuous in Center City, where it grows in cracks in pavement and in almost any kind of soil. Dandelion makes seed asexually without insect pollinators, even though it produces showy flowers containing nectar and pollen.[27] Dandelion and common milkweed are both perennials that disperse seeds on fine strands of silk blown by wind. At first glance, dandelion's success compared to milkweed's failure in Center City might be attributed to dandelion's capacity to produce seed without pollinators.

Figure 17.7 Dandelion, like milkweed, uses silk strands to disperse seeds long distances.

Figure 17.8 Dandelion blooming in sidewalk crack on Locust Street in Center City. It produces seed asexually, eliminating dependence on insect pollinators.

On closer examination, pollinator scarcity in Center City does not offer a compelling explanation for the scarcity of common milkweed, as plants dependent on insects for pollination do produce seed in Center City. A conspicuous example is northern catalpa (*Catalpa speciosa*);[28] ailanthus is probably another, despite claims of pollination by wind.[29] Male and female ailanthus flowers reside on separate trees.[30] One survey found that a diverse group of insects, especially bees and flies, pollinate ailanthus;[31] another identified a soldier beetle (*Chauliognathus marginatus,* Cantheridae) as a principal pollinator.[32] This soldier beetle is common in Center City. I have observed it taking nectar at milkweed blossoms in Fairmount Park just outside Center City.

Figure 17.9 Ailanthus altissima in bloom. In one study, a soldier beetle was found to be a principal pollinator of ailanthus. Figure 17.10 shows this beetle on flowers of common milkweed.

Figure 17.10 Soldier beetle (*Chauliognathus marginatus*) on common milkweed off Martin Luther King Drive in Fairmount Park. Pollinia from milkweed are attached to tarsi of two legs. This species also pollinates ailanthus trees.

In Center City white clover (*Trifolium repens*) is another wild plant that produces seed only after insect pollination.[33] Its flowers here attract bumblebees and honeybees.

Scarcity of space

By increasing the number of its flower-bearing stalks, common milkweed can increase the number of flowers that it presents to pollinators; but lack of growing space in Center City limits such expansion. A milkweed seedling normally spreads vegetatively before it flowers. During its initial year of growth, it does not flower, but first sends off horizontal roots that produce new shoots from root buds. By the time it flowers the second year, it has already established vegetative clones with many stems, each producing three to seven heads of flowers. During four years, one common milkweed seedling produced fifty-six stalks vegetatively and ninety-six seedlings in an area of 9 square meters.[34] In Center City, space big enough to accommodate such reproductive sprawl is rare.

Ailanthus and catalpa, as trees, can offer pollinators nectar and pollen on a large scale.

White clover, spreading vegetatively in lawns, produces carpets of flowers, presenting pollinators with bountiful offerings of food. In contrast, lack of space constrains the mass of flowers that common milkweed can muster for pollinators in Center City compared to areas just outside, as in Fairmount Park.

Figure 17.11 Honeybee with pollen basket filled with pollen from white clover (*Trifolium repens*). Pollen baskets of honeybees visiting milkweed remain empty.

Figure 17.12 A carpet of white clover on a lawn offers pollinators a bountiful source of nectar and pollen. Flowers grow just below the height of lawn mower blades.

Persecution

Concrete and asphalt are the primary physical barriers to formation of big colonies of common milkweed in Center City, but other barriers may be more important in impeding its establishment downtown. Seedlings of common milkweed take two growing seasons to mature, typically in full sun.[35] Pod-bearing stems grow to a height of a meter or more. To produce seed, the conspicuous stalks of common milkweed must, for two consecutive years, escape the scrutiny of hostile property owners, landscape maintenance crews, and others who would regard them as unsightly intruders. In Center City, I can think of few places where common milkweed could escape persecution.

Other wild herbaceous plants in Center City are better adapted for evading detection. Seedlings of tall annual weeds like pigweed (*Amaranthus hybridus*) mature in half the time that it takes milkweed, and biennials (which take two years to mature) typically keep a low profile their first year, when they form rosettes, as in the case of Queen Anne's lace (*Daucus carota*). White clover is a perennial like milkweed, but in Center City it keeps its flower heads below the height of lawn mower blades. Lambsquarters (*Chenopodium album*) produces seed on intrusive-appearing plants a meter tall, like common milkweed, but in Center City lambsquarters also makes seed on plants whose height is a tenth of a meter. Grasses, plantains (*Plantago*), and other denizens of pavement in Center City hide below pedestrians' sight lines, and they tolerate trampling. Lawn pennywort (*Hydrocotyle sibthorpioides*), a perennial like milkweed that produces flowers in umbels and spreads vegetatively, completes its life cycle completely within cracks between the bricks of our front sidewalk.

Herbicide

In 1999 Robert G. Hartzler at Iowa State University found that common milkweed was present in 51 percent of fields of corn and soybean in Iowa. Ten years later, the number had dropped to 8 percent. Hartzler blamed the declines on glyphosate, an herbicide whose use increased after introduction of corn and soybeans genetically engineered to resist this herbicide.[36] John M. Pleasants and Karen S. Oberhauser, also at Iowa State University, found that declines in populations of common milkweed coincided with an 81 percent decline in Midwestern production of monarch butterflies, whose larvae feed on milkweed. They concluded that widespread agricultural use of glyphosate reduced populations of monarch butterflies and made them more vulnerable to other threats.[37] In 2011 railroad crews used herbicide to kill Center City's only seed-producing colony of milkweed.

In 1784 John and William Bartram sent common milkweed to a European patron,[38] presumably for a garden. Currently in Center City, gardeners cultivate tropical milkweed (*Asclepias curissavica*), swamp milkweed (*Asclepias incarnata*), and butterfly weed (*Asclepias tuberosa*); common milkweed would probably thrive in Center City if left alone with space to grow.

18

PURPLE-STEMMED CLIFFBRAKE

(*Pellaea atropurpurea*)

Purple-stemmed cliffbrake grows in masonry of the Eastern State Penitentiary just outside Center City. It is absent downtown.

Figure 18.1 Purple-stemmed cliffbrake growing in the north wall of Eastern State Penitentiary. Brown structures containing spores line the edges of the undersurface of the leaves.

In 1822 the Commonwealth of Pennsylvania began construction of Eastern State Penitentiary from locally quarried Wissahickon gneiss and schist. The construction of the outer walls took seven years to complete. These walls are 10 meters high, over 3 meters thick at ground level, and over 200 meters long on each of four sides, covering 4 hectares (10 acres). The prison closed in 1970 and in 1994 reopened as a historic site.[1]

Figure 18.2 Eastern State Penitentiary, north wall, viewed looking east. At this distance, ferns growing in the wall are practically invisible.

Figure 18.3 Purple-stemmed cliffbrake on the north wall, viewed looking up through a telephoto lens.

The prison walls are habitat to purple-stemmed cliffbrake (*Pellaea atropurpurea*), a native fern that festoons the top of the wall and grows out of cracks between stone blocks. Accompanying it but less abundant is another fern: ebony spleenwort (*Asplenium platyneuron*). Unlike most ferns, purple-stemmed cliffbrake likely evolved in dry rocky habitats resembling a desert. The center of dispersal for its taxonomic group is the southwestern United States and adjacent Mexico. It is found on limestone ledges and cliffs from Guatemala to Vermont.[2]

Figure 18.4 Ebony spleenwort (*Asplenium platyneuron*) growing in the penitentiary's north wall.

Distribution outside Center City

Like common milkweed, purple-stemmed cliffbrake grows wild just outside Center City but not inside. Unlike milkweed, it requires little space and thrives on vertical surfaces. It tolerates heat, cold, and drought and flourishes in otherwise barren habitats inhospitable to most other plants. In addition to populating Eastern State Penitentiary, it grows in masonry retaining walls around the Fairmount Water Works. Just south of Center City, I found it thriving on the wall of a rundown brick industrial building, since demolished. Buildings downtown offer this fern seemingly infinite opportunities to colonize masonry walls, but paradoxically, it has failed to do so.

The plant has lived in the vicinity of Philadelphia for as long as records exist. In 1793 Henry Muhlenberg recorded it in Lancaster County,[3] and in 1837 William Darlington reported it in Chester County.[4] It was omitted in the first flora of Philadelphia, by William P. C. Barton in 1818,[5] but included in one published in 1905.[6] Its long tenure in the region would appear to have given the plant ample time to colonize buildings downtown.

Tolerance of dry habitat and desiccation

In 1911, William Nicholas Steil at the University of Wisconsin discovered that purple-stemmed cliffbrake produces spores asexually,[7] a trait present in only 5 to 10 percent of ferns.[8] The life cycle of ferns usually begins with spores germinating and producing minute plants, called prothallia (gametophytes), which produce male and female organs. Sperm swim to female organs containing eggs, which, when fertilized, develop into ferns (sporophytes) that make spores. Steil observed that prothallia of purple-stemmed cliffbrake did not produce sex organs; ferns developed directly from sexually undifferentiated prothallia. By bypassing sexual development and fertilization, purple-stemmed cliffbrake eliminates the need for water as a medium for sperm to swim to eggs. The plant can complete its life cycle on dry rock, like the wall of Eastern State Penitentiary.[9]

In 1931 Fermen Layton Pickett at Washington State University showed that purple-stemmed cliffbrake tolerates desiccation during active growth phases of its life cycle.[10] This phenomenon is rare among vascular plants, which encompass all higher plants such as ferns, flowering plants, and gymnosperms (including conifers). Only 0.15 percent of all vascular plants tolerate desiccation outside of dormant stages such as seeds and spores.[11] Pickett showed that prothallia of purple-stemmed cliffbrake that had been air-dried in the laboratory for five years grew when rehydrated. He found that prothallia remained viable after repeated exposure to periods of air drying lasting three to four weeks. He discovered similar tolerance of desiccation for ebony spleenwort.[12] Pickett's findings suggest how these two species of fern survive heat and drought on penitentiary masonry high above ground.

Figure 18.5 Effects of drought and heat on two ferns. Purple-stemmed cliffbrake (the taller fern) looks healthy compared to ebony spleenwort, whose leaves have turned brown at the tips. Photographed July 29, 2012.

Favorable ingredients in mortar

In 1920, Edgar Wherry, who became one of Philadelphia's foremost experts on ferns, published the results of studies on the acidity of soil around the roots of ferns that grow in rocks. He had invented a method for measuring acidity of soil in the field, and he applied his new technology to ferns in different habitats. In Pennsylvania he found purple-stemmed cliffbrake growing on limestone, sandstone, schist, and shale. It thrived in soil of neutral pH, but became stunted in acidic soil low in calcium. Wherry concluded that the primary determinant of the suitability of habitat for this fern is type of soil rather than type of rock.[13] Applying his findings to Eastern State Penitentiary, one might hypothesize that calcium carbonate in the wall's mortar endows the wall's scanty soil with calcium and buffer needed to maintain a neutral pH, especially helpful in acid rain.

Mysterious absence in Center City

The question remains why purple-stemmed cliffbrake does not colonize masonry downtown. Matthew Wild and Daniel Gagnon at the University of Quebec in Montreal recently investigated the rarity of this fern in Canada. They evaluated habitats where the fern grows and compared them to habitats nearby where the fern does not grow. They could find no significant difference between occupied and unoccupied habitats. They suggested that the fern's rarity is due to constraints on dispersal rather than habitat.[14]

Buildings, rivers, and pavement in Center City theoretically are barriers to dispersal, but they would not be expected to block dustlike spores blown by wind. Spores of ferns have been recovered in the jet stream, and they are resistant to the cold and ultraviolet radiation expected at high altitudes. They are a primary reason ferns have colonized remote oceanic islands more often than have flowering plants, and they were the means by which ferns recolonized the island of Krakatau after volcanic destruction of the island's vegetation.[15] The maximum diameter of spores of purple-stemmed cliffbrake is 60 microns,[16] about the thickness of human hair,[17] and typical for spores of ferns.[18] The dryness of the fern's habitat on the wall of Eastern State Penitentiary would be expected to facilitate release of its spores as aerosols. The way urban heat islands pull in surrounding air[19] should draw spores downtown.

Center City might harbor the fern's enemies, such as insects. I have found scale insects on the leaves of purple-stemmed cliffbrake growing on a retaining wall at the Fairmount Water Works. Insects, however, infrequently eat ferns compared to flowering plants,[20] and the scale insects on the purple-stemmed cliffbrake caused no sign of injury. Edgar Wherry was able to cultivate this fern in a system of nested flower pots,[21] and the plant has been recommended for rock gardens.[22] One might expect that a plant so easily cultivated could establish colonies downtown.

Figure 18.6 Yellow scale insects on leaf of purple-stemmed cliffbrake growing on a stone retaining wall at the Fairmount Water Works. Despite the infestation, the fern looks healthy and has produced spores, located in the brown granular material along the lower margin of the leaf.

Weathering required for colonization

All of the habitats of this fern just north and south of Center City are distinctive in the same way: they are old and weathered. Reports outside of Philadelphia have called attention to purple-stemmed cliffbrake colonizing historic structures, such as an old wall in Washington, DC,[23] a prison in Carlisle, Pennsylvania,[24] and a brick kiln in Fairfax, Virginia.[25] In 1899 Ellsworth Jerome Hill, observing this species in a quarry, noted the importance of weathering:

> No *Pellaea* was seen on any of these artificially made exposures, though various mosses and other forms of vegetation were well established. The fern, wherever found, grew upon rocks weathered to a dark gray, and with an exposure doubtless of many centuries' duration, or dating back to the time when a glacier carved out the rock bed of the river, its face only changing by the slow process of disintegration.
>
> It is not easy to account for this preference of the fern for the old weathered surface. There is noticeable, however, a marked difference in the color of the recently exposed stone and that long subjected to weathering. Some chemical change is produced by atmospheric agencies, for the freshly exposed surfaces are soon stained with yellow or drab due to the presence of iron-oxide. This color is not seen on surfaces long exposed. The absence of the *Pellaea* may not be due to the presence of certain metallic ingredients in excess, but they suggest a possible or partial cause of it.[26]

Weathering may include colonization by lichens, algae, fungi, or bacteria that purple-stemmed cliffbrake may require for extraction of minerals and nutrients from rocky substrate. Microbial biofilms coating rock are themselves complex ecosystems,[27] and how the fern might engage them has not been studied. The rough texture of weathered surfaces may trap spores and contribute to purple-stemmed cliffbrake spore banks, analogous to seed banks.[28]

In Britain until the late nineteenth century, mortar was made of lime, sand, ash, loam, straw, and dung. It weathered quickly. After 1870, the constituents of mortar shifted to cement and sand, and sometimes furnace slag, which weathered more slowly; these walls must age 40 to 80 years before they host flowering plants. In London, saxifrage occurs only on walls that are at least 150 years old.[29] American mortar shifted to Portland cement around 1880.[30]

Habitat loss due to property maintenance

The establishment of this fern in masonry in Philadelphia may require weathering or old mortar or both. In Center City property owners view weathering of masonry as a sign of structural deterioration, which induces them to institute sandblasting, painting, or pointing. This fern is absent from Center City probably because people want neither weathering nor plants on the facades of their buildings. Purple-stemmed cliffbrake takes three to four years to mature and produce spores;[31] this long period affords property owners ample opportunity to intervene before the plant has completed its life cycle.

Figure 18.7 Purple-stemmed cliffbrake on a crumbling brick wall of an old industrial building just south of Center City. Fern on left is unidentified. Buildings with masonry in such poor repair are rare in Center City. This building has since been torn down.

The suggestion that lack of weathered masonry excludes the fern from Center City deserves qualification. Some weathering of masonry in Center City is tolerated, or even encouraged, as evidenced by silvergreen bryum moss (*Bryum argenteum*), which grows on impermeable surfaces including rock, brick, mortar, asphalt, and concrete. It also populates soil and tree trunks. Unlike purple-stemmed cliffbrake, however, it keeps a low profile and presents an inconspicuous target; its ability to fill the interstices of crevices makes it hard to eradicate; and its growth between bricks on patios and walkways is viewed as decorative rather than destructive.

Figure 18.8 Silvergreen bryum moss (*Bryum argenteum*) with spore capsules. In Center City it contributes to weathering that conditions brick or stone for colonization by other plants.

Purple-stemmed cliffbrake bears ecological similarities to common milkweed and the organ pipe mud dauber. All three are common outside but not inside Center City. All are vulnerable to destruction through property maintenance. All stand out as attractive targets for persecution. In Center City, persecution has been effective in eradicating each of them.

SPOTLIGHT

EASTERN GARTER SNAKE

Blow fly (Family Calliphoridae) laying eggs inside mouth of dead eastern garter snake (*Thamnophis sirtalis*) at Bartram's Garden, Philadelphia.

Maggots hatching from this fly's eggs may feed on the snake; however, just after I took this photograph ants began scavenging in the snake's mouth and, perhaps, preying on these eggs. I have found eastern garter snakes near the Fairmount Water Works in Center City. The first published reference to "garter" snakes in the vicinity of Philadelphia was in 1743, more than a decade before the systematic description and naming of *T. sirtalis,* by Carl Linnaeus.[1]

19

MUGWORT

(*Artemisia vulgaris*)

In the latter half of the nineteenth century, mugwort spread rapidly from sites of introduction at ports, especially in the Philadelphia-Camden area. These sites were dumps where ships discarded ballast.

Figure 19.1 Mugwort (*Artemisia vulgaris*) surrounded by tufts of green foxtail (*Setaria viridis*) in the median strip at Broad and Bainbridge Streets, facing City Hall.

By the time the first compendium on the flora of Philadelphia was published in 1818, most of the species of weeds common today in Center City were well established. One notable exception is mugwort (*Artemisia vulgaris*), an introduced species the compendium omits.[1] The oldest specimen of this species in the United States was collected by Thomas Nuttall on a botanical trip from Philadelphia to Delaware in 1809, and the next oldest was collected in Camden, New Jersey, in 1837;[2] but the plant had naturalized in New England[3] and Canada long before. Merritt Lyndon Fernald concluded that Jesuit missionaries introduced mugwort into southeastern Canada in the sixteenth century.[4]

Medicinal herb

The species was cultivated as a medicinal herb and used as a panacea, as described in William Salmon's *Botanologia,* published in London in 1710:

> It prevails powerfully against the Poyson and Malignity of Poppies and Opium: rectifies the Stomach, stops Vomiting, and causes a good digestion. It allays Vapors, opens all sorts of Obstructions of the Bowels, and cures the Rickets in Children: It likewise cleanses the Reins and Bladder of Tartarous Mucilage. Dose from twenty to sixty Drops or more, according to the quantity of the Vehicle it is taken in: it may be given in Canary, or other Generous sort of Wine, two, three, or four times a day.[5]

Botanicum Officinale, published in London in 1722, reported additional uses:

> The Leaves of Mugwort, are chiefly used, and principally against, Distempers incident to the Female Sex, being of great Service in promoting the menstrual evacuations, both given inwardly and used outwardly in Baths and Semicupia; they strengthen the Head and Nerves, and are very good against hysteric Fits or Vapours.[6]

Evolution in North America

Jacob Barney at Cornell University systematically recorded the date and location of dried specimens of mugwort in historic collections in herbaria. He used the data to track the species' dispersal in North America. He found that the range of the plant had been stable in the United States until around 1860, when it started to expand rapidly outside of its established centers of distribution in New England and Canada.[7]

He and his colleagues cultivated mugwort from populations native to Europe and compared its growth with that of mugwort from populations naturalized in North America. Compared to European mugwort, American mugwort was shorter and germinated earlier; it produced more vegetative clones (ramets), more biomass, and higher ratios of roots to shoots. When cultivated with goldenrod (*Solidago canadensis*), a native species with which it competes, it suppressed goldenrod more effectively than did European mugwort. Barney and colleagues concluded that after its introduction into North America, mugwort evolved adaptations that promoted its rapid spread.[8]

Ships' ballast

How did mugwort adapt? The center of distribution of the genus *Artemisia* is located in the cold arid steppes of central Asia.[9] In their *Flora of North America* published in 1841, John Torrey and Asa Gray noted "a dozen varieties of this polymorphous and widely diffused species," which then included four named varieties in North America.[10] Barney found that rapid expansion of mugwort in North America began geographically with mugwort first colonizing sites created by the dumping of rocks and earth in ships' ballast, as recorded on labels of historic specimens of mugwort in herbariums. He concluded that importation of mugwort in ships' ballast was the primary source for this species' explosive geographic dissemination starting in the latter half of the nineteenth century. He mapped ballast sites where the plant was collected on the East, West, and Gulf Coasts, and found that ballast in Philadelphia and Camden counties together accounted for eleven possible introductions—more introductions than all the others combined.[11]

The concentration of herbarium specimens of mugwort from dumps of ballast in Camden and Philadelphia could be an artifact of the dedication of Philadelphia's botanists,[12] but such a bias would not change the observation that colonization of ballast heralded local geographic spread of mugwort. Repeated introduction of mugwort through the ports of Philadelphia and Camden brought together foreign strains that had previously been geographically isolated. In ballast dumps near the city's docks, the same breezes that brought sailing ships into port would have helped cross-pollinate these geographically disparate strains of mugwort, which is wind pollinated.[13] Hybridization likely contributed to mugwort's dramatic spread in the latter half of the nineteenth century.

Figure 19.2 Port along the Delaware River, south Philadelphia, 1870. Just south of here ships dumped ballast of rocks and soil, introducing exotic plants from around the world. They also dumped ballast in Camden, directly across the river. (Photo from Free Library of Philadelphia. Courtesy of Free Library and PhillyHistory.org, a project of the Philadelphia Department of Records.)

Figure 19.3 Mugwort (*Artemisia vulgaris*) collected June 13, 1897, from ballast ground, Kaighns Point, Camden, New Jersey, directly across the river from south Philadelphia. (From the herbarium of the Academy of Natural Sciences of Philadelphia, now the Academy of Natural Sciences of Drexel University. Courtesy of the Academy)

Dissemination

Historical collections of mugwort in herbariums show that, after mugwort appeared in ballast dumps in ports, it appeared along highways and railroads, indicating that rails and roads dispersed mugwort after its initial introduction. Air turbulence from passing traffic and trains presumably carried the seeds, which are only 1–2 millimeters in diameter. At the same time, horticultural trade spread mugwort as ornamental and herbal plantings; nurseries also dispersed it accidentally.[14] In Philadelphia, ships, trains, motor vehicles, and nurseries acted together in mugwort's importation and dissemination.

Figure 19.4 Population of mugwort growing as a dense, continuous monoculture along railroad tracks. The view is looking north from Walnut Street Bridge in Center City.

The first publication documenting the presence of this species in our region was William Darlington's *Flora Cestrica,* a 640-page treatise published in 1837. About mugwort he wrote:

> The A. vulgaris, or common Mugwort,—with pinnafid leaves, green above, and whitish, tomentose beneath—is occasionally to be found about old gardens; but can hardly, in strictness, be considered either as naturalized, or cultivated for any useful purpose. It is certainly not a native, here.[15]

In 1945, Hugh E. Stone produced a monumental sequel to Darlington's flora. It was published in two volumes, totaling 1450 pages. Stone found no evidence of mugwort in Chester County since Darlington's report.[16] The failure of mugwort to establish itself in the first half of the nineteenth century in Chester County makes sense: the plant's evolutionary transformation in North America took place in the second half of the nineteenth century.

Mugwort has been ranked among the ten worst weeds of nurseries in the eastern United States.[17] Ann Fowler Rhoads and William M. Klein, presenting results from the Pennsylvania Flora Database of herbarium specimens, reported that the plant inhabits all counties in southeastern Pennsylvania, including Chester County.[18] Its North American distribution, once concentrated in Canada,[19] now encompasses almost all of the eastern United States, including Florida.In Washington, DC, its pollen has become a common aeroallergen.[20] It inhabits all continents except Africa and Antarctica.[21]

Figure 19.5 Mugwort leaves, broad and toothed.

Figure 19.6 Mugwort leaves, deeply cut.

Figure 19.7 Mugwort leaves, inconspicuous on upper stems.

Figure 19.8 Mugwort flowers in September.

Future of mugwort in Philadelphia

Today in Center City, mugwort grows in cracks in pavement and masonry, along curbs, and in gardens and any kind of soil. It thrives especially along railroad tracks, despite frequent application of herbicide, which suppresses it temporarily. Mugwort is resistant to many herbicides.[22]

Although widely distributed downtown, mugwort is easy to overlook. In August, its height ranges from 2 meters to 0.1 meter, depending on location. By the time the plant has grown to eye level, its diagnostic leaves are hidden in its base or withered. The leaves vary in size and shape on different plants and on the same plant, and they may resemble those of other species, including chrysanthemum, common ragweed (*Ambrosia artemisiifolia*), and white heath aster (*Symphyotrichum ericoides*). The white, wooly texture that Darlington pointed out on the undersurface of its leaves helps distinguish mugwort from species with leaves of similar shape.

Philadelphia catalyzed the evolution of mugwort into a successful urban colonizer. Its ballast dumps brought together genotypes of mugwort from around the world. Its industrial transformation subjected mugwort to novel selective pressures. Its highways, rails, and ports endowed mugwort with diverse routes of dissemination. In Center City today, the morphological variability of mugwort continues to present targets for natural selection.

SPOTLIGHT

HOUSE FINCH

House finch (*Haemorhous mexicanus*) eating a sunflower seed at our feeder in May.

Native to southwestern United States, house finches proliferated in the East after a pet store released caged individuals in New York City in 1939. The birds appeared for the first time in Philadelphia in the late 1950's, when they behaved as winter migrants. Populations here later expanded to include year-round residents.[1]

After their release in New York City, house finches evolved changes in structure, physiology, and coloration. Presumably, evolution combined with bird feeding contributed to the house finch's success in adapting to new conditions.[2]

Despite their abundance, house finch populations in Pennsylvania in the last two decades declined by over half.[3] Lethal bacterial conjunctivitis spread to house finches congregating at bird feeders,[4] but other forces may have taken a toll, as discussed in the case of house sparrows in Chapter 3.

20

FRESHWATER SPONGE

(*Spongilla* [*Eunapius*] *fragilis*)

Spongilla fragilis **is a freshwater sponge first described from a specimen found in the Schuylkill River in Philadelphia in the nineteenth century. Freshwater sponges still inhabit the river here.**

Figure 20.1 Fairmount Dam, regarded in the late nineteenth century as one of the richest places for collecting freshwater sponges, including *Spongilla fragilis*. It diverted water to the Fairmount Water Works (middle of photo), which pumped water up to a reservoir whose site is now occupied by the Philadelphia Museum of Art, built after pollution of the river made the reservoir obsolete. In the foreground is the Fairmount fish ladder, which allows migrating fish to surmount the dam.

In 1851, at a meeting of the Academy of Natural Sciences in Philadelphia, Joseph Leidy announced his discovery of a new species of freshwater sponge. He found it growing on the underside of stones below the low water mark in the tidal Delaware and Schuylkill Rivers in Philadelphia. He reported its diameter as one to two inches. Noting that most of the sponge disintegrates after it dies, he named it *Spongilla fragilis.* It has no common name.[1] The current name for the genus is *Eunapius* instead of *Spongilla,* but in this account I have retained *Spongilla.*

In 1870 he reported finding this sponge living in association with other invertebrates in the Schuylkill River below the Fairmount Dam. The other animals included ciliated polyps, mollusks, rotifers, protozoans, bryozoans, and polychaete worms.[2] He discovered one of these, a polyp he named *Urnatella gracilis,* the same year he discovered *S. fragilis.*

Sponges on Fairmount Dam, 1887

In 1887 Edward Potts, Leidy's colleague at the Academy, published the last account I could find of this sponge in the Schuylkill River in Philadelphia. In a monograph on freshwater sponges, he reported finding it on the timbers of the Fairmount Dam, which he ranked as one of the richest places in the world for freshwater sponges. At the time the dam powered turbines that drove pumps in the Fairmount Water Works, which pumped water to a reservoir on top of Fairmount Hill, now occupied by the Philadelphia Museum of Art. On the west side of the dam, across the river from the water works, was a navigation canal with locks, currently the site of a fish ladder. Potts gained access to the dam during summer months, when water spilling over the dam subsided. On one visit to the dam he found six species of freshwater sponge, including *S. fragilis.*[3]

Potts described a dense population of *S. fragilis* on the walls of the canal:

> Upon one occasion when the water was withdrawn from the canal basin at the head of the locks at Fairmount Dam, Philadelphia, the exposed, perpendicular walls of dressed stone were seen to be lined with them, probably hundreds in number; some of minute size, but many covering two or three square feet of surface. They were rarely much more than an inch thick near the middle and shaded off all around to filmy edges. They had no apparent preference for the comparatively rough surface of the stones, for some of the finest specimens were found upon the timbers of the gates, from which they were easily removed.[4]

Lower Schuylkill riverbed void of life, 1876

At the time Potts reported rich populations of *S. fragilis,* pollution in the lower Schuylkill River was already severe. A decade earlier, Josua Lindahl, secretary of the Swedish Commission, offered Joseph Leidy an opportunity to collect specimens from the Schuylkill riverbed, which a small steamer under Lindahl's command was prepared to dredge experimentally. Expecting a trove of small animals from the bottom of the river, Leidy accepted Lindahl's invitation, only to be disappointed:

> No living thing whatever was drawn up, as the mud and sand were black and saturated with bituminous oil. This latter fact was unexpected, and would appear to illustrate the mode of formation of more ancient bituminous shales. The refuse of the city gas-works, and probably of some coal-oil refineries, run into the river. The oils appear to have an affinity for the particles of clay carried down the river, and, precipitating, become bituminous sediments at the bottom.[5]

Pollution of the city's water supply

A year before the dredging, a commission of engineers appointed by the mayor submitted its report on pollution of the city's water supply in the Schuylkill River:

> For many years, and to within a recent period, the Schuylkill water has been remarkably pure and wholesome; but it has been impaired by impurities, accompanying the growth of population and the extension of industries. The contamination of this stream is not alarming, yet it is believed that unless a remedy be applied it will ultimately be rendered unfit for domestic uses. The principal causes of deterioration are, the sulphuric acid from the coal mines, and the refuse and the sewage from population and from the numerous manufactories which drain into the Fairmount pool.[6]

The commission reported that, since 1842, sulfur in the Schuylkill River at Fairmount, the source of the city's drinking water, had increased eightfold. It found that the causes of the pollution extended far beyond the city itself:

> The region drained by the river is estimated at something over 1,800 square miles, of which 1,200 square miles is below the mining region, a large portion of which is a highly cultivated, populous, and thriving region. Several cities and towns, numbering about 20, occupy closely its banks, and many of them have become centers of manufacturing interests, and are estimated to contain at present an industrious population of over 100,000, Reading alone containing nearly 40,000 people; and, in addition to the enormous coal mining operations of the upper Schuylkill, iron banks have been opened, and iron furnaces built upon its margin; cotton factories, carpet and dye works, woolen and hat manufactories, paper mills, tanneries, chemical and gas works, breweries, and indeed the advantages of the location are so obvious, that almost every branch of manufacture has found a convenient location on its banks, and some of them on a scale not exceeded by any in this country; nor should the recognized influence of railroads on the banks of rivers be lost sight of, in estimating the future probable distribution of the population and industries of this valley.[7]

Figure 20.2 Wharf on the east bank of the Schuylkill River at Walnut Street, 1888. Twelve years earlier, Joseph Leidy had found no living plants or animals in sediment dredged from the riverbed. (Photo courtesy of PhillyHistory.org, a project of the Department of Records of the City of Philadelphia)

The commission recommended new sewers and pumps and modification of reservoirs and freshwater intake pipes.

None of these recommendations addressed pollution from coal and oil, or discharge of waste upstream in the river's vast watershed.[8] At the time Leidy discovered *S. fragilis,* coal shipped to Philadelphia by rail and boat from mines along the Schuylkill watershed totaled 15 million tons per year and was increasing.[9] These mines dumped silt and culm, consisting of fine particles of coal, directly into streams and onto stream banks.[10] Locally, the Philadelphia Gas Works manufactured illuminating gas from coal in a factory on the riverbank about 100 meters downstream from the water works.[11] Ruth Patrick, limnologist at the Academy of Natural Sciences, recalled seeing the river colored black from pollution from coal in the first half of the twentieth century.[12] Coal dust dumped into the river must have challenged the river's populations of sponges, which are filter feeders that consume bacteria and algae they sieve from the water. Fine sediment suspended in water clogs their pores.[13]

Figure 20.3 "East bank of the Schuylkill, below Spring Garden Street Bridge. The banks have long been a disgrace to Philadelphia." Photo and caption from *The Redemption of the Lower Schuylkill* by John Frederick Lewis, published by the City Parks Association, Philadelphia, 1924.

Schuylkill River Project, 1951

In 1951 the Commonwealth of Pennsylvania and the United States Army Corps of Engineers issued their final report on the Schuylkill River Desilting Project,[14] a monumental effort to dredge and remove 38 million tons of culm along 208 kilometers of river.[15] After the desilting project ended, Patrick noted that the river no longer turned black, but lack of funding prevented completion of the project:

> Although the removal of the sediments improved the quality of water in the Fairmount Park area of the Schuylkill River, conditions were not ideal for aquatic life. In our 1958 studies, we found many organisms on the banks and substrates that were above the riverbed. The rocks that protruded from the surface of the bed often supported a fair amount of aquatic life, whereas very little was found in the bed of the river itself.
>
> One tangible evidence of improvement in aquatic life was the finding of a bryozoan described by Joseph Leidy in 1851 from the Schuylkill. Its name is *Urnatella gracilis*. Leidy wrote in 1870 that it was abundant, but by 1883 it had vanished because of the city's sewage and industrial pollution. I am sure he would have been pleased to know that it was again established in 1958.[16]

Figure 20.4 Schuylkill River desilting project discharging dredged materials into an impounding basin upstream at Stouds Ferry, Berks County. (From *The Schuylkill River Desilting Project, Final Report of the Schuylkill River Project Engineers, 1 July 1951*[17])

The City of Philadelphia has since constructed three sewage treatment plants and introduced new sewage treatment technology.[18] It installed steel bulkheading along the Schuylkill shoreline in Center City and transformed the riverbank into a landscaped park.[19] Erection of a fish ladder enabled fish to migrate over the Fairmount Dam.[20] Dredging to reduce Schuylkill culm resumed.[21] Coal mining decreased,[22] as did the city's population.[23] Federal legislation, especially the Clean Water Act of 1972 and its amendments, improved monitoring and management of waste.[24]

Reduction in pollution

Based on monitoring from 2001 to 2005, water quality of the Schuylkill River in Philadelphia was rated "good" with respect to suspended sediment, and "improved" with respect to phosphorus, but still "poor" for phosphorus and nitrogen.[25] Sodium and chloride concentrations have increased, especially in the winter, due to road salt and suburban sprawl,[26] while sulfur near the mines in the upper Schuylkill decreased.[27]

Surveys of fish sampled by electrofishing in the spring in the tidal Schuylkill from 2002 to 2006 found thirty-three species; four additional species were found by video monitoring inside the fish ladder.[28] A survey of macroinvertebrates in the lower Schuylkill in 1975–1976 identified twenty-two genera, including mollusks, insects, crustaceans, and the polychaete worm *Manayunkia speciosa*.[29] Joseph Leidy discovered this worm in the Schuylkill River in 1858 and named the genus *Manayunkia* in reference to an Indian name for the Schuylkill River. This worm was part of the faunal community Leidy associated with *S. fragilis* below the Fairmount Dam in 1870.[30]

Figure 20.5 September 2012, same location as in figure 20.3.

Whether populations of *S. fragilis* have survived in the lower Schuylkill River has not been reported. The distribution of the species is cosmopolitan. By 1884, *S. fragilis* had been found from Florida to Nova Scotia, and from the Great Lakes west to the Columbia River.[31] The species has since been reported from all continents, in climates ranging from tropical to subarctic, and in diverse freshwater habitats, including caves.[32]

Pollution could have transiently eliminated this sponge, as in the case of *Urnatella gracilis.* Sponges can be transported as minute dormant propagules, called gemmules. The gemmules of *S. fragilis* are minute (up to 1 mm in diameter),[33] and tolerate salt, desiccation, anoxia, freezing and thawing, and long periods of inactivity.[34] Ships theoretically could have reintroduced *S. fragilis* in the form of gemmules on wooden hulls[35] and in ballast tanks.[36] The mystery of its cosmopolitan distribution has generated speculation about its dissemination, such as by wind, insects, birds, and mammals, including people.[37]

A search for *Spongilla fragilis*

I recently hunted for *S. fragilis* just below the Fairmount Water Works at low tide in the early fall, the season when the size of the sponge is largest. Leidy found the species in this location a century and a half ago. I waded into the river and inspected stones and logs, looking for sponges and encrusted gemmules. Exposed to light, sponges can look green due to symbiotic algae, which can confound identification. I looked for sponges on stones' undersurfaces, which were shielded from light. I did not identify any sponges, but green encrustations were common and may have been sponges coated with algae.

Up the river about a kilometer I found a grapefruit-sized gelatinous ball floating just below the surface in an inlet. I suspected it might be a sponge, but Richard J. Horwitz at the Academy of Natural Sciences identified it as the bryozoan *Cristatella* (*Pectinatella*) *magnifica,* which Leidy discovered in Philadelphia and named the same year he discovered and named *S. fragilis.*[38] Like sponges, bryozoans are filter feeders, but they have microscopic tentacles.[39]

Discovery of spongillaflies

Although I failed to identify *Spongilla fragilis* in the river, I discovered spongillaflies (*Climacea areolaris*) attracted to a black light in our backyard a few blocks from the river. The larvae of these spongillaflies are aquatic and feed exclusively on freshwater sponges (family Spongillidae), including *Spongilla fragilis*.[40] Adults fly and feed on nectar and, under laboratory conditions, live for two to three weeks.[41] They resemble brown lacewings and belong to the same order (Neuroptera) of insects. They are weak fliers, so they likely emerged from the Schuylkill River close by.

Figure 20.6 Spongillafly (*Climacia areolaris*), attracted to black light in our backyard a couple of blocks from the Schuylkill River. Its larvae feed exclusively on freshwater sponges (members of the family Spongillidae).

The Schuylkill River is part of the Delaware River basin, home to 835 documented species of aquatic invertebrates, including 10 species of freshwater sponge.[42] The sponges inhabiting the lower Schuylkill today have yet to be systematically surveyed. Viewed in the context of a river black from coal dust less than a century ago, the presence of even one species of sponge would appear to be evidence of the river's resilience.

A long history of corrective action contributed to the restoration of the health of the river. The report of the city's engineers who documented the Schuylkill River's pollution[43] preceded Rachel Carson's *Silent Spring*[44] by almost a century. The city protected its supply of potable water primarily to serve its citizens, not aquatic wildlife, but its actions served both.

Tolerance and intolerance of pollution

Credit for the presence of freshwater sponges in the Schuylkill River may belong less to the remediation of the river than to the toughness of some freshwater sponges, especially *Spongilla fragilis.* Edward Potts reported it thriving on the Fairmount Dam[45] one decade after Joseph Leidy had discovered that industrial pollution had destroyed all life in the lower Schuylkill riverbed[46] and city engineers had determined that refuse, sewage, and sulfuric acid had polluted the pool behind the dam.[47] On the Fairmount Water Works in 1884, freshwater sponges were so numerous that Potts considered them to be *causes* of pollution.[48]

A review of studies on the tolerance of *Spongilla fragilis* to pollution found that this sponge is practically insensitive to hydrogen ion concentration and siltation. Even though it is a filter feeder, healthy colonies have been found growing on substrates submerged in mud; it has been collected in water with coliform counts of 24,500 colonies/ml; and it tolerates pollution in the form of nitrates, phosphates, sulfates, and many other contaminants.[49]

Spongilla fragilis tolerates pollution better than does its enemy the spongillafly, *Climacea areolaris.*[50] It likely tolerates pollution better than do many of its other enemies. Animals known to feed or live on freshwater sponges include fish, crayfish, mites, nematodes, protozoans, rotifers, bivalves, oligochaetes, and insects (dipterans and trichopterans as well as neuropterans).[51] The protection that pollution offers *Spongilla fragilis* may account for this sponge's paradoxical abundance in polluted water.

SPOTLIGHT

LAMBSQUARTERS

Young lambsquarters (*Chenopodium album*) on sidewalk of South 23rd Street, Center City.

In 1818 an account of plants growing wild in Philadelphia described lambsquarters (*Chenopodium album*): "This weed in its young state is eaten at our tables. It attains the height of five or six feet. In wastes, in dunghills, near rubbish, and in gardens, everywhere very common."[1]

Ten years later a report concluded that importation of agricultural seed contaminated with seed from *C. album* had introduced this plant into the United States from Europe.[2]

Lambsquarters' association with people is ancient. In northern Syria archeological excavation found seeds of *C. album* with artifacts of human habitation dating back 10,000 years.[3] In Jutland, Denmark, seeds of *C. album* were recovered from the gut of Grauballe Man, whose corpse was found submerged in a state of partial preservation in a peat bog approximately 2,400 years after his death.[4] In Alberta, Canada, prehistoric Native Americans harvested seeds of this species.[5]

21

BROWN BULLHEAD

(*Ameiurus nebulosus*)

The brown bullhead was first described from Philadelphia in the early nineteenth century, when it was common and savored. Recent fish surveys in Center City have not detected it.

Figure 21.1 Brown bullhead (*Ameiurus nebulosus*) caught in Driscoll Pond, Haddonfield, New Jersey, by Leo Sheng. (Photo by Leo Sheng)

In 1819, Charles Alexandre Lesueur, a member of the Academy of Natural Sciences of Philadelphia, described and named a species of catfish new to science. He reported that the species, now designated as *Ameiurus nebulosus,* was very common in Philadelphia ("tres nombreuse à Philadelphie") and that people fished for it and had high regard for its white flesh.[1] According to Thaddeus Norris, an expert on fish culture at the time, this species inhabited ponds, ditches, and creeks, including tidal water.[2] He contrasted it with the bigger catfish (*Ameiurus catus*) also native to Philadelphia's waters:

> If these smaller species were not so common they would be more generally esteemed. These are far better fish for the pan; their flesh is firm and sweet, and resembles that of the trout or the breast of a young chicken, more than the flesh of any other fish. "Catfish and coffee" at the Falls of Schuylkill was formerly and to some extent is still an "institution."[3]

Historic abundance in the Schuylkill River

In 1914 Henry Fowler, ichthyologist at the Academy, noted that fish in the Schuylkill River were more common than one might have expected, given the severity of the river's pollution:

> For many years the tidal reaches of the Schuylkill River to the Fairmount Dam in Philadelphia have been greatly polluted, suggesting the impression that they support little or no fish life. I have recently received a number of fishes from this region, through Mr. W. E. Meehan, the Director of the Philadelphia Aquarium, besides notes on others not sent.[4]

He listed thirteen species that he ranked as common, including *Ameiurus nebulosus.*[5]

The "Schuylkill cat," as *A. nebulosus* was called in the early twentieth century,[6] or the "brown bullhead," as it is known today, is distinctive for its tolerance of pollution. A guide to game fish published in 1905 offers a description alleged to be by Henry David Thoreau:

> They stay near the bottom, moving slowly about with their barbels widely spread, watching for anything eatable. They will take any kind of bait, from an angleworm to a piece of tomato can, without hesitation or coquetry, and they seldom fail to swallow the hook.[7]

Diet of nonbiting midges (chironomids)

An evaluation of the contents of stomachs of brown bullheads from a lake in New York found that they selectively ate chironomid larvae—wormlike aquatic stages of nonbiting midges, which are flies. These larvae live in sediment and belong to the largest and most ecologically diverse family of aquatic insects.[8] In Lake Erie, where brown bullheads are common, numbers of chironomids increased fourfold from 1930 to 1961, a period when pollution increased;[9] chironomid abundance subsequently decreased when pollution abated.[10] Compared to brown bullheads in Lake Erie's less polluted tributaries, those in the most polluted waters grew larger and produced more eggs per female.[11]

The abundance of chironomids in polluted water helps explain their possible contribution to the brown bullhead's tolerance of pollution. In the course of an evaluation of an outbreak of chironomid midges, the population density of chironomid larvae in mud samples taken from a lake bottom polluted from runoff from the Twin Cities was 7,000 individuals per square yard.[12] Chironomid midges around bodies of polluted water in urban areas have been treated as pests.[13] In the Delaware River basin, which includes the Schuylkill River, 18 percent of the genera of all aquatic invertebrates belong to the family Chironomidae.[14]

Figure 21.2

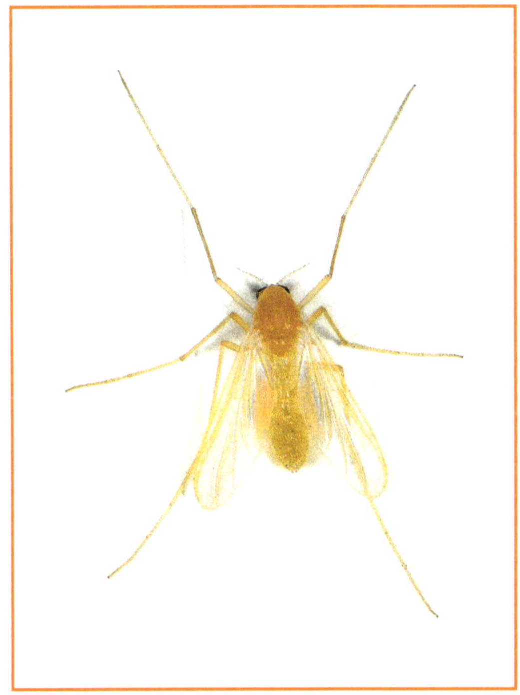

Figure 21.3

Figure 21.4

Figure 21.5

Figure 21.6

Figure 21.7

Figures 21.2–21.7 Chironomids, or nonbiting midges. Their aquatic larvae are favorite prey of brown bullheads. These midges were attracted to black light in our backyard, several blocks from the Schuylkill River. They belong to the most species-rich family of aquatic animals.

Diet of worms (oligochaetes)

Other aquatic fauna may have contributed to the bullhead's pollution tolerance that Henry Fowler observed in the tidal Schuylkill. Like chironomids, oligochaetes have proliferated in rivers with increasing pollution.[15] Oligochaetes are segmented worms in the same taxonomic class as earthworms.[16] When oligochaetes in a polluted river outnumbered chironomid larvae, brown bullheads ate more oligochaetes than chironomids, even though the fish favored chironomids.[17] In a survey of macroinvertebrates on the bottom of the tidal Schuylkill River in 1975 and 1976, oligochaetes were the most abundant animals, numbering over 6,000 per square meter. By weight and by numbers, they constituted over 98 percent of the macroinvertebrate fauna on the bottom of the river.[18]

If oligochaetes and chironomids were the reason that brown bullheads tolerated pollution in the Schuylkill River, why did Joseph Leidy not find them in the sediment dredged from the bottom of the Schuylkill River in 1876?[19] Freshwater invertebrates, especially oligochaetes, were one of his specialties, particularly the genus *Limnodrilus*,[20] which constituted 99 percent of the oligochaetes identified in the Schuylkill River survey.[21] Perhaps the bituminous sediment that Leidy found was distributed unevenly in the riverbed, which supported aquatic life in sections with less contamination.

Diet of sewage

Alternatively, pollution may have supplied brown bullheads with food other than oligochaetes and chironomids. The brown bullhead's diet in polluted sections of the Monongahela River in West Virginia suggests what this mysterious other food might have been. In a study of the stomach contents of brown bullheads in the Monongahela, the food brown bullheads ate in greatest volume was not prey but sewage, and the second greatest was detritus. Although they did consume oligochaetes and chironomids, by volume the fraction of the brown bullhead's diet consisting of sewage and detritus was 70 percent.[22] In Philadelphia in 1876, brown bullheads may have been able to compensate for scarcity of prey by eating sewage and detritus.

Tolerance of pollution

Compared to other fish, brown bullheads are better able to tolerate extreme conditions associated with pollution. These include water that is acidic (pH 3.3),[23] hypoxic (oxygen 0.5–1 mg/liter),[24] and warm (temperature 40°C [105°F]).[25] In an impoundment in the upper Schuylkill, they have lived with sediments contaminated with lead, cadmium, chromium, copper, and zinc, and they showed no gross pathology.[26] Brown bullheads living in a tidal creek contaminated with heavy metals in North Carolina showed no histologic, biochemical, or hematologic abnormalities.[27]

Benefits of pollution for brown bullheads

Like the freshwater sponge *Spongilla fragilis,* the brown bullhead may tolerate pollution better than its enemies. Brown bullheads defend their eggs and young from predators such as minnows and sunfishes, which they chase away.[28] The U.S. Environmental Protection Agency has classified as tolerant of pollution only a quarter of species of minnows (cyprinids) and less than 10 percent of species of sunfish (centrarchids).[29] The native redbreast sunfish (*Lepomis auritus*) had become uncommon in the polluted tidal Schuylkill River according to Fowler's report in 1914.[30]

In 1999 Anthony C. Steyermark at Drexel University and his colleagues found that tapeworms (cestodes; *Proteocephalus* sp.) parasitized all brown bullheads from a pond in a residential area in New Jersey, whereas in the urban industrialized Schuylkill River, cestodes parasitized no brown bullheads. The cestodes attacked the fishes' hearts, livers, kidneys, and gonads, which carried high parasite burdens; one fish harbored 314 cestodes. Fish from the pond were stunted compared to those from the Schuylkill River. Steyermark et al. suggested that contamination in the Schuylkill River protected brown bullheads from these cestodes, whose larvae require crustaceans as intermediate hosts.[31]

Pollution may protect brown bullheads from consumption by fisherman. The Pennsylvania Department of Environmental Protection issues annual guidelines on the safety of eating fish caught locally. It tests fish for two contaminants: mercury and PCBs (polychlorinated biphenyls). Because of PCBs, it recommends limiting consumption of fish caught in the tidal Schuylkill River to one meal a month for all fish it tested except carp and eels, which it recommends never be eaten.[32] Although the guidelines are intended to protect people, they may also protect fish.

Figure 21.8 Fishermen near the 30th Street train station. The boat in the background belongs to the Philadelphia Water Department, which monitors the quality of both the water and fish.

The presence of PCBs in fish caught in the Schuylkill River does not necessarily deter fishermen from eating them. One study found that, compared to non-Hispanic white fishermen, non-Hispanic black fishermen were more likely to fish in watersheds with high PCB contamination and more likely to consume catfish. It suggested that consumption of contaminated catfish caught by fishermen is the reason levels of PCBs are higher in non-Hispanic blacks than in non-Hispanic whites.[33]

Contamination of brown bullheads may suppress reproduction in fish-eating birds. Contamination of fish with chlorinated hydrocarbons was first shown to suppress reproduction in birds in the case of bald eagles and the pesticide DDT,[34] but other chlorinated hydrocarbons (such as PCBs) have caused similar effects and have involved other fish-eating birds, including herons,[35] cormorants,[36] and ospreys,[37] all of which I have observed preying on fish in the Schuylkill River in Philadelphia. In 1984, chlordane, DDT, dieldrin, and PCBs in the Schuylkill River in Philadelphia were found at four trophic levels, exemplified by green algae, snails, minnows, and largemouth bass.[38] Oligochaetes and chironomids in riverbeds ingest PCBs from the sediment and pass them on to fish, which in turn pass them on to their predators. This transmission up the food chain concentrates PCBs, which are lipid soluble and accumulate in animal fat.[39]

Figure 21.9 Great blue heron (*Ardea herodias*) fishing at Boathouse Row. Like brown bullheads, it prefers shallow, quiet water.

Figure 21.10 Double-crested cormorants (*Phalacrocorax auritus*) on a log grounded at the Fairmount Dam. This diving bird fishes in the lower Schuylkill in Center City. The posture with spread wings is typical.

Figure 21.11 Great egret (*Ardea alba*) at Columbia railroad bridge over the Schuylkill River, Philadelphia.

Costs of pollution to brown bullheads

For brown bullheads, the benefits of tolerance of pollution may have costs, such as exposure to carcinogens. In 1941 Balduin Lucké and Hans G. Schlumberger at the University of Pennsylvania School of Medicine and Wistar Institute described tumors on the lips of 166 brown bullheads from the Schuylkill and Delaware Rivers around Philadelphia:

> This neoplasm usually occurs as solitary or multiple, large, red, fleshy masses upon the lips or dental plates, and by reason of its size, may prevent closure of the mouth...The larger growths frequently invade adjacent normal tissues and force their way into vessels where they are found as emboli. The clinical course of the tumor is one of relatively slow but progressive growth.[40]

The tumors as they described them bear features typical of malignancy, and later authors classified them as such (squamous cell carcinoma).[41] In 2004 liver cancers were found in 26 percent of brown bullheads in Darby Creek in Philadelphia's John Heinz National Wildlife Refuge at Tinicum, a habitat so highly contaminated that the U.S. Environmental Protection Agency designated it a hazardous waste Superfund site.[42] In a study sponsored by the Delaware River Basin Commission, brown bullheads collected at various sites in the Delaware River were found to have lip tumors and liver lesions.[43]

The evidence of a causal relationship between chemical pollutants and tumors was initially compelling.[44] Prevalence of tumors in brown bullheads was found to be high in contaminated industrial sites compared to uncontaminated sites in widely scattered locations in the United States, especially in the East and Midwest.[45] In the laboratory, brown bullheads dosed with extracts of sediment containing industrial pollutants (polycyclic aromatic hydrocarbons, or PAHs) developed liver and skin tumors indistinguishable from tumors they developed in the wild.[46] Prevalence of liver tumors in brown bullheads in the Black River in Michigan dropped after a coking plant shut down and the river was dredged, reducing PAH contamination.[47]

Despite strong evidence incriminating pollution, brown bullheads in uncontaminated reservoirs and ponds in New York State were found to have a prevalence of tumors reaching 100 percent for skin and 30 percent for liver or bile ducts.[48] In the South River on the Chesapeake Bay, prevalence of tumors in brown bullheads was high despite the absence of high concentrations of known carcinogens.[49] Descendants of brown bullheads from the Delaware River estuary that fisheries personnel introduced into ponds developed tumors, suggesting transmission of an infectious or genetic carcinogen from river to pond.[50] The cause of tumors in brown bullheads remains enigmatic.

Disappearance of brown bullheads

Populations of brown bullheads have been in decline in the Schuylkill River in Philadelphia. Fish surveys here from 2002 to 2006 identified no brown bullheads among 44,000 fish identified. These surveys identified locally introduced game fish, including 3,499 channel catfish (*Ictalurus punctatus*), a competitor of brown bullheads, and 469 flathead catfish (*Pylodictis olivaris*), a predator of brown bullheads.[51] The surveys sampled fish populations by electrofishing in the Schuylkill River below the Fairmount Dam, and by video observation in the fish ladder at the Fairmount Dam. In contrast, brown bullheads were abundant in samples obtained by trawling and electrofishing in the tidal Schuylkill River from 1971 to 1976.[52] In 1979 they were observed in large numbers in the "turn-pool" at the base of the fish ladder.[53] A fisherman told me he recently caught brown bullheads with rod and reel in Center City at night in the summer.[54]

Figure 21.12 Channel catfish (*Ictalurus punctatus*), the most common catfish reeled in from the Schuylkill River in Center City. Its introduction as a game fish likely contributed to declines in brown bullheads, which are native to the Schuylkill River.

The scarcity of brown bullheads has been attributed in part to decreasing pollution.[55] The manufacture of PCBs, for example, was banned in 1979 by the U.S. Environmental Protection Agency under the federal Toxic Substances Control Act.[56] As pollution decreased, populations of brown bullheads lost a comparative advantage over predators and competitors less tolerant of pollution. A less toxic Schuylkill may have destroyed a safe haven for brown bullheads in Center City. Brown bullheads, however, thrived in the Schuylkill before the advent of industrial pollution. Introduction of game fish such as the flathead catfish (*Pylodictis olivaris*)[57] and channel catfish (*Ictalurus punctatus*)[58] may have taken a toll that compounded that caused by reduced pollution. Joseph Perillo, aquatic biologist at the Philadelphia Water Department, has confirmed the recent introduction and establishment of another fish-eating predator, the northern snakehead (*Channa argus*),[59] whose prey includes brown bullheads.[60]

In addition to introduced game fish and cleaner water, Schuylkill bulkheads made out of concrete, wood, and steel may have adversely affected populations of brown bullheads. Bulkheading deprives brown bullheads of shallow water—which they prefer—along riverbanks, and no streams or ponds connect to the banks of the

Schuylkill in Center City. Aggravating the adverse effects of bulkheading, brownish flocculent material (visible in shallow water at low tide) obscures the riverbed and potential prey such as oligochaetes, chironomids, and crustaceans. Brown bullheads feed on green plants when they are available, but aquatic vegetation other than algae was absent in a study of the riverbed of the tidal Schuylkill in the late twentieth century.[61]

Power plants' cooling water intake pipe

The Exelon Generation Company operates a cooling system that pumps water from the Schuylkill River to a cluster of three power plants located on the east bank of the Schuylkill at Christian Street, a few blocks south of Center City.[62] The three plants have a combined generating capacity of 421 megawatts,[63] which is about 20 percent of the capacity of the Limerick Nuclear Power Plant on the Schuylkill River upstream in Montgomery County.[64] The cooling systems of the three plants draw cooling water from the river through an intake pipe 34 meters in length and 3 meters in diameter. From 2001 to 2005, pumps drew water through the intake pipe at an average rate of 170,000 liters per minute. After water traverses the length of the pipe, it passes through a traveling screen that diverts fish and debris into a trash trench; from here the fish are transported for disposal offsite. Eggs, larvae, fish, and debris tiny enough to pass through the screen mesh (3/8 inch, or about 1 centimeter) are pumped with the cooling water around the power plants and eventually discharged back into the river.[65]

Figure 21.13 Cluster of three power plants on the Schuylkill River at Christian Street, a few blocks south of Center City. View is from the Schuylkill Expressway. A cooling water intake structure along the shoreline draws water from the river. Fish sucked through a pipe strike a screen that diverts them into a trash trench for disposal offsite.

In an Exelon-sponsored study in 2005 and 2006, samples of fish impinged against the screen at the end of the water intake pipe included a single brown bullhead,[66] compared to 109 brown bullheads in 1975 and 192 in 1976.[67] Perhaps mortality of brown bullheads sucked into the water intake pipe over many decades contributed to their disappearance. Brown bullheads, which spawn in water 2 meters deep or less,[68] may have selectively positioned themselves near the opening to the cooling water intake structure, which is located on the shoreline. Also potentially harmful to brown bullheads is the heat of the effluent that the power plants' cooling system discharges into the river.[69]

The magnitude of the risk that the cooling water intake pipe poses to brown bullheads may be considered in the context of the magnitude of water flowing through the pipe compared to the river. The U.S. Geological Survey reports the median discharge of water from the Schuylkill River in Philadelphia for the last 80 years for any day of the year. For example, for September 13, the median discharge rate was 1.3 million liters per minute; the minimum rate for that day was 39,000 liters per minute (in 1966).[70] The historic flow of water through the cooling water intake pipe averaged over four times more than the river's historic minimum discharge rate for that day. The capacity of the cooling water intake pipe to draw in more water than the river discharges is possible because the river is tidal here, connected to the Atlantic Ocean via Delaware Bay.

Options to protect fish from intake pipes

The Exelon Generation Company considered several alternatives for reducing fish mortality in its cooling water intake structure, but rejected all in favor of the status quo.[71] The U.S. Environmental Protection Agency is currently (as of 2012) considering new standards[72] that may require power plants like Exelon to modify these structures.

Cooling water could be recycled back to the power plants rather than dumped, minimizing the need to pump water out of the river; but such a system would require cooling towers, which in turn would require more space than is available at the existing power plants.[73] Such towers would loom over the neighborhood, and consumers of electric power presumably would bear the costs of construction.

Burdensome accommodations might appear unwarranted on behalf of a species as widespread and generally common as the brown bullhead, but they might be justified on behalf of the river's overall health. Consultants for Exelon estimated that in 2006 the number of fish eggs and larvae "entrained" (i.e., drawn into the intake pipe past the screen mesh to the electrical power plants and discharged back into the river) totaled 1.5 million, encompassing eleven species.[74] In a study of a power plant on the shore of Lake Erie, mortality of juvenile and adult fish pumped into the cooling intake pipe was virtually 100 percent, independent of "entrainment" or "impingement" (i.e., removal and disposal of fish that strike barrier screens).[75]

Charisma

The fate of species in cities is sometimes linked to prejudice either for or against them, as in the case of the common milkweed, gray squirrel, and American robin. Charismatic species may be beneficiaries of special treatment, as exemplified by the presentation of dead rats by the Franklin Institute to red-tailed hawks nesting on its facade.[76]

In the early nineteenth century, Thaddeus Norris observed that the brown bullhead's flavor was better than its reputation.[77] In "A Plea for the Bullhead," the nineteenth-century humorist George Wilbur Peck mused over the brown bullhead's lowly status:

> The same may be said of brook trout. While they will bite a hook, it requires more machinery to catch them than ordinary people can possess without mortgaging a house. A man has got to have a morocco book of expensive flies, a fifteen dollar bamboo jointed rod, a three dollar trout basket with a hole mortised in the top, a corduroy suit made in the latest style, top boots, of the Wellington pattern, with red tassels in the straps, and a flask of Otard brandy in a side pocket. Unless a man is got up in that style, a speckled trout will see him in Chicago, first, and then it won't bite. The brook trout is even more aristocratic than the whitefish, and should not be propagated at public expense.
>
> But there are fish that should be propagated, in the interest of the people. There is a species of fish that never looks at the clothes of the man who throws in the bait, a fish that takes whatever is thrown to it, and when once hold of the hook never tries to shake a friend, but submits to the inevitable, crosses its legs and says "Now I lay me," and comes out on the bank and seems to enjoy being taken. It is a fish that is the friend of the poor, and one that will sacrifice itself in "the interest of humanity." That is the fish that the State should adopt as its trademark, and cultivate friendly relations with, and stand by. We allude to the bullhead.
>
> To catch the bullhead it is not necessary to tempt his appetite with porterhouse steak, or to display an expensive lot of fishing tackle. A pin hook, a piece of liver, and a cistern pole, is all the capital required to catch a bullhead. He lays upon the bottom of a stream or pond, in the mud, thinking. There is no fish that does more thinking, or has a better head for grasping great questions, or chunks of liver, than the bullhead.[78]

In theory, public policy might support intervention to protect brown bullheads if the fish had more charisma, or if they attracted charismatic predators, such as bald eagles. Bald eagles nest in Philadelphia.[79] They occasionally appear in the Schuylkill River outside Bartram's Garden, just two kilometers downstream from the cooling water intake pipe. In a study of bald eagles nesting on the Potomac River, approximately 95 percent of the remains of their prey were catfish, primarily brown bullheads.[80]

The fate of the brown bullhead in the tidal Schuylkill River may depend less on its charisma than on the charisma of its enemies, the channel catfish and flathead catfish. The introduction of these two popular game fish is the most compelling explanation for the brown bullhead's local disappearance, a result that is likely irreversible in Center City. Bald eagles prey on all three species of catfish.[81]

SPOTLIGHT

LEAFHOPPERS AND PLANTHOPPERS

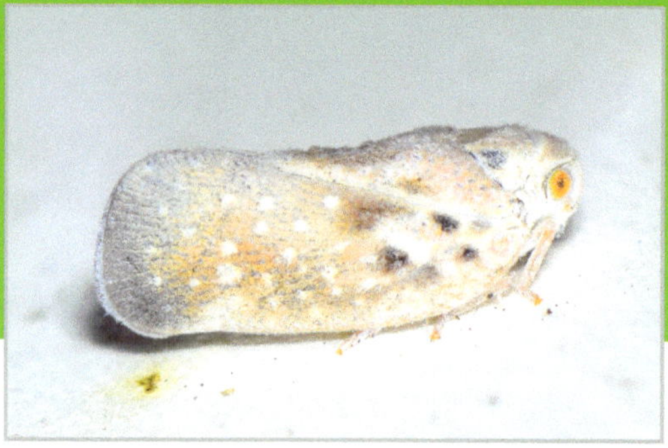

Diversity of leafhoppers and planthoppers (Order Hemiptera) photographed at a lamp at night in our back yard in Center City.

The largest of these insects is 13 mm (half an inch). Starting clockwise from the upper left, they are:

1. Citrus flatid planthopper (*Metcalfa pruinosa*)

2. Green coneheaded planthopper (*Acanalonia conica*)

3. A leafhopper (*Erythroneura calycula*)

4. Broad-headed sharpshooter (*Oncometopia orbona*)

22

RED BACK SALAMANDER

(Plethodon cinereus)

Abundant prey and a strictly terrestrial life cycle helped this non-amphibious amphibian adapt to Center City, Philadelphia.

Figure 22.1 Red back salamander (*Plethodon cinereus*), discovered on a sidewalk in Fitler Square. It was headed away from an enclosed garden, shown in figure 22.3.

At a meeting of the Academy of Natural Sciences of Philadelphia in May 1818, Jacob Green named what he thought was a new species of salamander.[1] Green, then a chemist practicing law in Philadelphia,[2] did not know that two months earlier Constantine Samuel Rafinesque had described and named the same species.[3] Green's naming has endured, even though Rafinesque's has priority.[4] Rafinesque coined the salamander's current vernacular name: *red back salamander.*[5]

Green stated that this salamander is common in New Jersey,[6] while Rafinesque placed it in the highlands of New York.[7] In 1842 John Edwards Holbrook ranked it the most common salamander in the eastern United States from Maryland to Vermont. He was the first to report that it is abundant in Philadelphia.[8] For the next century and a half, a succession of herpetologists noted its presence in Philadelphia.[9] It was recorded in many other cities: lawns in Bethlehem, Pennsylvania, and urban forests in Indianapolis, Montréal, and Cleveland.[10]

Urban dryness hostile to amphibians

Cities typically degrade wetlands, home to many amphibians; and downtown Philadelphia fits this mode. Two rivers border Center City and adjacent neighborhoods, but bulkheading and fill along riverbanks have destroyed all traces of marsh and riparian habitat. Asphalt, concrete, and other impervious material cover most of downtown; runoff from rain drains mostly into sewers. The landscape downtown is dry except after rain or snow. Center City would appear to be among the worst places for amphibians.

Records of amphibians in Philadelphia a century or more ago might be from habitats since destroyed or from remnants of wetland currently protected, as in Fairmount Park or the John Heinz National Wildlife Refuge—far from the urban core. In 1917, Henry Weed Fowler at the Academy of Natural Sciences reported the distribution of red back salamanders within Philadelphia; he listed Fairmount Park and Wissahickon Creek, but also Germantown, Frankford, Holmesburg, Rowlands, La Grange, and "Philadelphia."[11]

Within the barren terrain of downtown are irrigated gardens and parks. In the garden in our backyard, which we water, small animals vulnerable to desiccation thrive. Examples are snails, slugs, earthworms, millipedes, and isopods (such as pillbugs). In experiments conducted in litter from an urban forest near Cleveland, red back salamanders consumed isopods and millipedes.[12] Other studies have shown that this salamander eats introduced earthworms,[13] including *Lumbricus terrestris,*[14] a species common in our garden. Might gardens downtown support amphibians?

Amphibians like the American toad (*Bufo americanus*) require both land and bodies of water. Some salamanders have aquatic and terrestrial stages like those of American toads, whose larvae (tadpoles) have gills, live in water, and metamorphose into adults that live on land.[15] The life cycles of most species of salamanders, in contrast, are strictly terrestrial,[16] even though all salamanders belong to the taxonomic class Amphibia and are called amphibians. The red back salamander exemplifies a non-amphibious amphibian in the sense that its life cycle does not include an aquatic or larval stage. What hatches from its egg is a miniature version of a fully formed, exclusively terrestrial adult, except for vestigial gills that shrivel up and disappear in a day.[17]

Figure 22.2 Bullfrog (*Rana catesbeiana*) in a flooded trench below Boathouse Row. Its dependence on open water limits its distribution in Philadelphia.

Red back salamander's habits adapted to urban gardens

Both Rafinesque and Green noted that the red back salamander is found under stones. In 1961 Frieda B. Taub at Rutgers University discovered that most red back salamanders in a forest lived in soil to a depth of at least 30 centimeters. They preferred moist soil that was not inundated.[18] To nest underground, they burrow, particularly in response to desiccation.[19] Individuals hide under cover during the day and emerge at night.[20]

The red back salamander's secretive behavior makes it easy to overlook. Its habits—predacious, subterranean, and nocturnal—resemble those of firefly larvae (*Photinus pyralis*), which I have never found, even though fireflies flash here every summer. But on October 20, 2012, around 9 a.m., I observed a red back salamander scurrying across a sidewalk in Fitler Square, a neighborhood park occupying less than a square block in Center City. It was headed toward the curb, away from a naturalized garden in the square. This garden is irrigated and has abundant leaf litter. Plantings include old London plane trees, wild ginger, ferns, Virginia bluebells, and oak-leaf hydrangea. A metal fence around it excludes people and dogs.

Figure 22.3 Naturalized habitat surrounded by a fence, located in Fitler Square. The fence, sprinklers, and leaf litter create conditions that resemble the wooded habitat of red back salamanders.

Conceivably this particular red back salamander was searching for a mate. In a study in Rochester, New York, in 1929, the mating period for red back salamanders ranged from October 7 to 16.[21] Red back salamanders are territorial and evict intruders.[22] The salamander on the sidewalk may have been fleeing a rival or a predator, or foraging, driven by hunger and scarcity of prey. Red back salamanders have been found to disperse through open terrain, such as fields.[23]

I found no published reports of red back salamanders colonizing downtown habitats. The naturalized garden in Fitler Square beside the sidewalk where I found the salamander occupies only about 1/100 of a hectare (0.025 acre). What is the smallest habitat that can support a population of red back salamanders? At a meeting of conservation biologists in 2000, Elke Wind considered this question for amphibians generally. In no case could she could find data that provided a definitive answer.[24]

Fragments of habitat may be too small to contain amphibians even though they provide ample food and shelter, as exemplified by species that migrate to ponds and streams for breeding. The red back salamander's nonmigratory life cycle[25] allows it to confine its movement to small areas. One study found that red back salamanders were less likely than migratory amphibians to travel to edges of forest.[26] Another study embedded radioactive tags in red back salamanders and used radiation detectors to monitor their movements. It found that on average red back salamanders traveled less than half a meter a day. Home ranges averaged just over 24 square meters for females and about half that for males.[27] From year to year, the average distance red back salamanders dispersed was less than 2 meters.[28]

Benefits of staying put

Sedentary life cycles can help populations survive in cities. The annual weed *Crepis sancta* produces two kinds of seeds, dispersing and nondispersing. For plants growing in cracks in sidewalks, natural selection favors nondispersing seeds, which are more likely than are dispersing seeds to fall in cracks near their parents and to germinate. Plants in these cracks produce seeds that are predominantly nondispersing, in contrast to dispersing seeds produced by plants growing outside of cracks.[29]

The giant Canada goose (*Branta canadensis maxima*) that populates urban areas along the Atlantic flyway is sedentary compared to long-distance migrant subspecies of Canada geese that overwinter in the same region.[30] In the last half of the twentieth century, the range of this subspecies of Canada goose expanded to cities and suburbs throughout the country.[31] Its urban habitats are safe havens from hunters and predators such as bobcats[32] and wolves,[33] and abandonment of long-distance migration saves the goose energy and avoids the hazards of prolonged flight.

Figure 22.4 Giant Canada geese (*Branta canadensis maxima*) behind the Philadelphia Museum of Art. Their sedentary behavior keeps them within an urban safe haven, protected from hunters and wild predators.

Bedbugs (*Cimex lectularius*) are wingless but belong to a suborder (Heteroptera) of insects whose members, with rare exceptions, are winged, like stinkbugs.[34] Bedbugs likely evolved in caves with bats and later expanded their hosts to include people.[35] The evolutionary loss of wings served the bedbug's need to stay close to its host.

Figure 22.5 Bedbug (*Cimex lectularius*) crawling off a penny. A patient who lives in Philadelphia brought it to me for identification. All bedbugs are wingless—an adaptation that serves their need to remain close to their victims.

Adaptations for an all-terrestrial life cycle

In his description of the red back salamander in 1818, Jacob Green noted its strictly terrestrial life and classified it as a "land salamander." In 1908 W. H. Piersol described its development in eggs. He referred to the embryo as a "larva," due to its resemblance to salamander larvae that develop in water. He noted that mothers care for their offspring for several weeks after they hatch out.[36] More recent observations have shown that mothers brood their eggs before the eggs hatch;[37] and that females take two years to produce eggs with yolks sufficiently large to nurture the salamander's extensive embryonic development.[38] The embryos need the big yokes because they cannot feed on algae and other pond life that nurtures salamander larvae in ponds and streams.

Maternal behavior and big yolks helped liberate red back salamanders from dependence on open water and migratory travel.[39] These traits may have evolved 200 million years ago when the red back salamander's taxonomic family (Plethodontidae) first appeared,[40] but the traits fortuitously prepared this salamander for life in the city.

The six- to eight-month interval between mating in the fall and laying eggs in the spring[41] gives fertilized female red back salamanders an opportunity to disperse and found colonies alone. A tiny isolated urban habitat fragment like the one in Fitler Square might serve as a dispersal site following importation of females in topsoil or root balls.

Urban opportunity and vulnerability

I captured the salamander at the curb and released it into our backyard, which offers an abundance of possible prey, including isopods, millipedes, and earthworms, but also slugs and snails.[42] Among the inhabitants of our garden soil, it could become top predator. Red back salamanders eat arthropods that are predators, including spiders, centipedes, and firefly larvae.[43] If this salamander happened to be a mated female, her release in our backyard may found a new colony, extending the range of this species downtown.

Despite the potential of this species for dissemination downtown, my rescue of this salamander on the sidewalk highlights why in Center City this animal is rare compared to earthworms. Unlike earthworms, red back salamanders cannot disperse independently as eggs in soil; the eggs[44] and hatchlings[45] of red back salamanders require maternal care. Unlike reproduction in earthworms, which are hermaphroditic or parthenogenetic,[46] reproduction in red back salamanders requires mating with members of the opposite sex.[47] And while earthworms on sidewalks are vulnerable to accidental trampling, the rapid crawling of the red back salamander, as on the sidewalk at Fitler Square, attracts attention, inviting attack by birds and people.

SPOTLIGHT

MILKWEED APHID

Milkweed aphids (*Aphis nerii*) on common milkweed (*Asclepias syriaca*) by railroad tracks along the Schuylkill River, Center City. A winged female is shown at the top.

Milkweed aphids, also called oleander aphids, were introduced from southern Europe into the southernmost United States, from which they annually fly north and re-colonize milkweed patches. The aphids do not overwinter in Philadelphia. How do these insects, which measure less than 3 mm, colonize widely scattered, small patches of milkweed and then mate, lay eggs, and re-establish aphid populations in Philadelphia all in one season? They accomplish this feat, at least in part, by elimination of courtship, mating and egg laying: The entire population of aphids consists exclusively of females, which reproduce asexually and bear their young live. A group of giant superclones constitutes the species in North America. Whether populations in Philadelphia make return flights to the south at the end of the season is unknown.[1]

23

LIVERWORT

(*Reboulia hemisphaerica*)

A liverwort known to favor habitats in wild areas has established colonies on brick walkways in Center City.

Figure 23.1 Two colonies of liverwort growing from soil in a brick walkway. *Reboulia hemisphaerica* is on the right, and *Marchantia polymorpha* is on the left. The species on the right is reported to favor "wild" habitats; the species on the left can be weedy. The site of all photos of liverworts illustrated in this chapter is the alley in figure 23.7 unless stated otherwise.

In 1799 the American Philosophical Society of Philadelphia published a list of liverworts found within a mile of the city of Lancaster, 93 kilometers west of Philadelphia. It was the first systematic account of liverworts published in North America. The author, Henrico Muhlenberg, credited his identifications to many authorities, all European. One of the liverworts he found is *Reboulia hemisphaerica*, which has no common name.[1]

Reboulia hemisphaerica in Center City

Reboulia hemisphaerica is shaped like a ribbon about 0.5 centimeter wide and 1–3 centimeters long. In Center City it anchors itself on soil in spaces between brick pavers. The ribbon, or *thallus,* grows flat along the top of the brick and bifurcates once or twice as it grows. If the surface of the soil is below the top of the brick, it grows up the side of the brick. Sometimes many thalli radiate from a sliver of soil between bricks.

For the past five years I have followed two colonies, each occupying less than a square meter on brick walkways, one a residential alley and the other a brick sidewalk along a narrow street (Naudain Street) of two-story row houses. Positioned away from foot traffic, the plants have completed their reproductive cycles, annually sending up spore-filled capsules on slender stalks.

Typical of liverworts and mosses, *R. hemisphaerica* produces male and female structures but no flowers or roots. Fertilization requires that sperm swim in rainwater, dew, or meltwater from male to female organs,[2] both of which are located on each plant.[3] The species is presumably named after the hemispherical shape of its female reproductive organ (archegonium).

Figure 23.2 Reboulia hemisphaerica behind a penny.

Figure 23.3 *Reboulia hemaesphaerica* between brick pavers.

Figure 23.4 Hemispherical structures that presumably inspired the name *hemisphaerica*. Each is a female reproductive organ (archegonium). The dark structure above each hemisphere is the male reproductive organ (antheridium). To fertilize eggs, sperm must swim from male to female organs in rainwater, dew, or meltwater. Photographed January 12, 2008.

Figure 23.5 Spore capsules in heads elevated on stalks of *Reboulia hemisphaerica*. Spores are products of sexual reproduction. Photographed April 26, 2009.

Figure 23.6 Liverwort heads elevated on stalks within a crack between brick pavers. The crack protects them from trampling, but impedes dispersal of spores.

Figure 23.7 Alley with liverworts, mosses, *Mazus,* and *Sagina*—which grow on soil between bricks on the right. The alley is located off Delancey Place near 25th Street. A locked gate has since been installed.

Figure 23.8 Japanese mazus (*Mazus pumilus*) in bloom, with liverwort (*Marchantia polymorpha*), birdseye pearlwort (*Sagina procumbens*), and moss. Here Japanese mazus grows upright just off the beaten path. In areas where it is subjected to more trampling, its flowers stay almost flush with the bricks, as shown in the photo on page 318.

Figure 23.9 Ornately patterned elevated discs from the liverwort *Marchantia polymorpha* in May. They are stalked male organs (antheridea) that produce sperm. Other plants here are *Reboulia hemisphaerica*, birdseye pearlwort (*Sagina procumbens*), Japanese mazus (*Mazus pumilus*), and moss.

Figure 23.10 Star-shaped heads of female reproductive organs containing spore capsules of *Marchantia polymorpha* in July. Like the male sex organs, these female organs are elevated on stalks, but in May when the males are releasing sperm, the female organs are globular and flush with the ground. Photographed on a brick sidewalk near Fitler Square.

R. hemisphaerica is one of an estimated 7,500 species of liverwort worldwide.[4] Centuries ago the shape of some of these species was thought to resemble liver, which led to the vernacular name *liverwort*. *Wort* is from an archaic suffix that means "herb."[5]

R. hemisphaerica is distributed in temperate regions worldwide, including all continents except Antarctica.[6] It has been reported from eighteen of Pennsylvania's sixty-seven counties.[7] It tolerates human disturbance and drought, including several months of desiccation.[8]

Scarcity in "civilized" habitats in the North

R. hemisphaerica is not as common as one might expect considering its global distribution and tolerance of harsh conditions. A monograph on the liverworts and hornworts of eastern North America details its distinctive distribution. Written by Rudolf M. Schuster and published in six volumes from 1966 to 1992, the monograph spans nearly 6,000 pages. Schuster observed that *R. hemisphaerica* inhabits highly disturbed habitats in the North but not in the South:

> In the northern portion of our area one soon gets the impression that *R. hemisphaerica* is a "wild" species, rarely persisting after disturbance. Occasionally it occurs over calcareous cement on old stone walls in long-abandoned areas, but such close associations with "civilization" are rare.
>
> By contrast, in the southeast, it becomes abundant in "civilized," i.e., strongly disturbed, areas. For instance, the old brickwork of Fort Clinch, at Fernandina Fla., is absolutely covered by *Reboulia.* The soil peripheral to the old Biology building at Duke University (Durham, N.C.) supported extensive and luxuriant growths...Similarly, the species is common in lawns and on banks along city streets in Oxford, Miss., and at the edges of old fields in the surrounding country. Southward the tolerance of the species for disturbance is evidently much higher than it is northward. Furthermore, southward (and south-westward) its distribution rarely appears to show any correlation with the occurrence of calcareous soil or rocks.[9]

What might account for the establishment of thriving colonies of *R. hemisphaerica* in downtown Philadelphia—just the kind of "civilized" northern habitat where Schuster found it rarely persisted? One possibility is that global warming and Philadelphia's urban heat island displaced to the north the geographic zone where this liverwort tolerates disturbance. Another possibility is that reduction in air pollution expanded this plant's acceptance of "civilization."

Air pollution a century ago

In the nineteenth and first half of the twentieth century, soot blanketed northern cities, as described in a pamphlet published by the American Civic Association in 1908:

> The dweller in a town burning bituminous coal needs no definition of the smoke nuisance. The great cloud that hangs over the city like a pall can be seen from any neighboring hilltop, and the dweller within is only too well aware of the splotches of soot that settle on every object in the city, bedimming buildings, spoiling curtains, injuring books, and increasing the laundry bill. The direct menace to the public health in fostering tuberculous conditions by loading the air with carbon particles to lodge in the lungs, and by causing housekeepers to keep the windows shut for fear of the soot that floats in when they are open, is equaled only by the mentally and physically depressing effect of the pall which shuts out the life-giving and germ-destroying sunshine. Our city parks have mostly lost their evergreen character,

where it existed, as conifers cannot long endure city smoke. Thus one treatment of the most pleasing variations in landscape is made impossible.[10]

Figure 23.11 B & O (Baltimore and Ohio) passenger train billowing smoke as it chugs south at Spruce Street along the Schuylkill River, 1912. (Photo courtesy of PhillyHistory.org, a project of the Department of Records of the City of Philadelphia)

Impact of air pollution on mosses and liverworts

Severe air pollution eliminates populations of bryophytes, which include mosses and liverworts. Oliver S. Gilbert investigated the impact of air pollution from combustion of coal in an urban area in Britain in the 1960s.[11] He found that, approaching a city center from a distance of 17 kilometers, the number of species of bryophytes progressively declined. The number fell by half to sixteen, of which only four were common in the city center. The four common ones were cosmopolitan mosses, such as silvergreen bryum moss (*Bryum argenteum*),[12] which is abundant in Center City, Philadelphia, and may actually benefit from air pollution.[13] One of the uncommon species present in the center of this coal town was a liverwort, *Marchantia polymorpha,* which is present but rare in Center City; it grows in the alley with *Reboulia hemisphaerica,* a liverwort Gilbert did not encounter. Gilbert determined that the primary cause for the drop in diversity of bryophytes in the city center was air pollution, specifically sulfur dioxide.[14] Many other studies have since confirmed the negative impact of air pollution on the diversity of bryophytes.[15]

Figure 23.12 Silvergreen bryum moss (*Bryum argenteum*). It is abundant in Center City and thrives in air pollution.

Abatement of air pollution in Philadelphia

In 1904 the City of Philadelphia took its first steps to control air pollution. It passed an ordinance regulating emission of smoke, measured using a color scale of darkness. The Bureau of Boiler Inspectors enforced the ordinance, based on standards from the color scale. In 1949 the city established an Air Pollution Control Board with powers of enforcement. Implementation of new regulations shut down almost a thousand incinerators. Philadelphia Electric Company (PECO) reduced sulfur and particulate emissions. The sulfur content of heating oil was reduced, and the city banned the burning of coal for heating and cooking. In three decades starting in 1966, sulfur dioxide in the city fell by 94 percent, particulate pollution by 93 percent, and nitrogen oxides by 61 percent.[16] Pollution from sulfur dioxide in the city is now below levels toxic to bryophytes,[17] and particulate air pollution meets standards of air quality established by the U.S. Environmental Protection Agency under the Clean Air Act.[18]

Industrial melanism in Great Britain

Industrial air pollution in Philadelphia declined to levels believed to be safe, but did the decline make a measurable difference to plants and animals in cities? In 1896 James William Tutt, a British lepidopterist, noted that British moths that rested on tree trunks in industrial regions blackened by soot had evolved black, or *melanic,* forms, which camouflaged them better than their previous pale forms. Here is how Tutt described the transformation of the peppered moth (*Amphidasys betularia,* currently named *Biston betularia*):

> The speckled *A. betularia*, as it rests on a trunk in our southern woods, is not at all conspicuous, and looks like a natural splash or scar, or a piece of lichen, and this is its usual appearance and manner of protecting itself. But, near our large towns where there are factories, and

> where vast quantities of soot are day by day poured out from countless chimneys, falling and polluting the atmosphere with noxious vapours and gases, this Peppered Moth has, during the last fifty years, undergone a remarkable change. The white has entirely disappeared, and the wings have become totally black, so black that it has obtained the cognomen "negro" from naturalists. As the manufacturing centres have spread more and more, so the "negro" form of the Peppered Moth has spread at the same time and in the same districts.[19]

Tutt hypothesized that, near industrial centers, natural selection favored the black form, which concealed the moth and protected it from birds.

After passage of the Clean Air Act in Britain in 1956, air pollution in Britain decreased,[20] populations of lichens recovered,[21] and the frequency of black forms in populations of moths reverted toward levels that had existed in the nineteenth century before industrial pollution.[22] Despite controversy,[23] recent studies confirmed Tutt's hypothesis.[24] Industrial melanism is considered a textbook case of evolution in action.[25]

Industrial melanism in Philadelphia

In 1961 Denis F. Owen at the University of Michigan reported finding industrial melanism in moths around Philadelphia, starting in 1922. He also found it around Detroit, New York City, and Pittsburgh around the same time.[26] In 1963 President Lyndon Baines Johnson signed the Clean Air Act, the first of a series of federal legislative steps to reduce air pollution.[27] In 2002, Bruce S. Grant and Lawrence L. Wiseman at the College of William and Mary in Williamsburg, Virginia, reported that the frequency of melanic forms in populations of American peppered moths in Michigan and Pennsylvania declined from more than 90 percent in 1959 to 6 percent by 2001. In Virginia, melanic forms were practically absent throughout this period.[28]

Recovery of tree moss after abatement of pollution

Industrial melanism in moths surfaced in the nineteenth and early twentieth century and has abated over the last half century. It puts into an ecological context the disappearance and return of urban bryophytes vulnerable to air pollution. Examples are tree mosses, which are exquisitely sensitive to sulfur dioxide in the air.[29] In *The Moss Flora of New York City and Vicinity,* published in 1916, Abel Joel Grout noted the absence of *Orthotrichum* tree mosses downtown.

> As one gets away from the city these mosses begin to appear in normal quantities. For this reason the author is inclined to believe that the gases produced in the city are the cause of this marked absence of arboreal mosses.[30]

A survey of mosses of Philadelphia in 1933 found no *Orthotrichum* tree mosses, and no reports of them since the nineteenth century. It blamed disappearance of mosses in this region on urbanization and smoke.[31] I recently discovered *Orthotrichum pumilum* thriving on tree trunks in Center City five blocks from a municipal power plant. It grows with a second tree moss, *Syntrichia papillosa.* The last published record of these mosses in Philadelphia was by Thomas Potts James,[32] who died in 1882.[33] Except for his specimens, none from Philadelphia is present in collections housed in the Academy of Natural Sciences of Drexel University.

Following reduction of sulfur dioxide pollution, the return of bryophytes on oak trees in London was reported as "spectacular."[34] In Britain, Germany, and Serbia, recent surveys have demonstrated an unprecedented diversity of urban bryophytes, including species that are threatened or endangered.[35] Paradoxically, vehicular traffic may enrich diversity of urban bryophytes by dispersing them[36] and by producing nitrogen and acid pollutants that promote or disrupt their growth, depending on the species and the acid buffering of their substrate.[37]

Figure 23.13 Tree moss (*Orthotrichum pumilum*) on bark of a street tree (Norway maple, *Acer platanoides*) in Center City, 2010. By the early twentieth century, sulfur dioxide air pollution had caused local extinction of *Orthotrichum* tree mosses in New York City and Philadelphia. Levels of sulfur dioxide have since declined, allowing the return of tree mosses like this one.

Figure 23.14 Lichens covering stucco on rear wall of South Square Market, which fronts on South Street. Air pollution suppresses the diversity of lichens, but some lichens thrive in it.

Refuges from air pollution

Tree mosses like *Orthotrichum* may be more sensitive to air pollution than other bryophytes, such as *R. hemisphaerica*, which grows on soil over rocks. In 1952, when industrial melanism was documented in Pennsylvania, *R. hemisphaerica* was collected growing on a rocky, shaded bank in Philadelphia.[38] In his classical studies demonstrating the adverse impact of air pollution on bryophytes in Britain, Gilbert noted that shelter and substrate can protect bryophytes from pollution.[39] *R. hemisphaerica* may have survived air pollution in Philadelphia by colonizing protected sites buffered from sulfur dioxide.

Dispersal of *R. hemisphaerica*

R. hemisphaerica is found in less than half as many Pennsylvania counties as the cosmopolitan liverwort *Marchantia polymorpha*,[40] which is regarded as a weed in the northeastern United States.[41] The two species are most easily distinguished by cup-like structures on their surfaces; only *M. polymorpha* has them. The cups contain asexual propagules (gemmae) that are dispersed by rain.[42] Functioning as simple structures for dispersal and reproduction, gemmae endow *M. polymorpha* with efficient mechanisms for colonization of ephemeral, fragmented habitats such as those present downtown. The moss *Bryum argenteum*, abundant in Center City, disperses by spores, but also by plant fragments transported on the soles of shoes.[43]

Figure 23.15 Liverwort *Marchantia polymorpha* with cups containing vegetative propagules (gemmae), minute asexually generated bodies that can disseminate and develop into new plants. When rainwater strikes the cups, it disperses the propagules. *M. polymorpha* can spread quickly and become weedy. It also reproduces sexually, making spores. Unlike this liverwort, *Reboulia hemisphaerica* produces no vegetative propagules.

Dispersal of spores from refuges buffered or sheltered from pollution is a plausible route by which *R. hemisphaerica* colonized brick walkways in Center City. Center City, with its nineteenth-century landscape largely intact, gave this liverwort time to disperse and colonize despite its lack of gemmae.

Springtails as possible agents of dispersal

Insects may have helped *R. hemisphaerica* to disperse in Center City. I watched garden springtails (*Bourletiella hortensis*) climb up stalks of this liverwort to the heads containing spore capsules. They clambered around the heads and climbed down with liverwort fragments stuck to their backs. The fragments were from sticky, breakable filaments that dangle just under the spore capsules; pieces fall off and adhere to the stems. The long filaments are characteristic of the species, and their function has never been described. Spores released from liverwort capsules would strike these filaments and presumably adhere to them, just as the filaments adhere to the springtail and stems. Carrying the filaments, the springtail would disperse these spores. I observed springtails on *R. hemisphaerica* on different days and on spore capsules in different stages of development, including mature capsules ready to release their spores. This arthropod is common and distributed in all continents.[44] The sticky filaments could disperse spores on other carriers, such as birds or people. Dispersal of spores of *R. hemisphaerica* by animals has not been previously reported, but flies disperse spores of dung moss,[45] and ants disperse propagules (gemmae) of aulacomnium moss.[46] In Center City blow flies (Calliphoridae) disperse spores of the stinkhorn mushroom (*Mutinus caninus*), which attracts them.[47]

Figure 23.16 Garden springtail (*Bourletiella hortensis*) on stalked head of *R. hemisphaerica* bearing spore capsules. Sticky, breakable, white filaments dangle below the spore capsules.

Figure 23.17 Garden springtail (*Bourletiella hortensis*) descending stalk of *R. hemisphaerica*. Fragments of adhesive white filaments from below spore capsules have stuck to its back and head. Transported by the springtail, these fragments could disperse the liverwort's spores.

Figure 23.18 Stinkhorn mushroom (*Mutinus caninus*) sprouting in mulch in landscaped border along the Schuylkill River Trail in Center City. It attracts blow flies (Calliphoridae) that disperse its spores.

Continental drift

What advantage would such complex dispersal offer *R. hemisphaerica*, given that the plant has populated all continents but Antarctica? Marie-Catherine Boisselier-Dubayle and her colleagues at the Muséum National d'Histoire Naturelle in Paris investigated the genetics of *R. hemisphaerica* from five continents. They showed that genetically similar populations are found on continents separated by oceans. Doubting the capacity of this liverwort to disperse long distances, they could not account for this wide distribution.[48]

Continental drift has been invoked to explain species of liverworts with populations separated by oceans,[49] but continental drift does not fit with *R. hemisphaerica's* genetic uniformity. Populations separated from one another by oceans since the continents drifted apart almost 100 million years ago would be expected to have undergone genetic divergence.[50] Presumably, *R. hemisphaerica* has somehow managed to disperse across oceans.

The diameter of this liverwort's spores is 70–80 microns,[51] compared to 5–50 microns for most fungal spores,[52] and 22–32 microns for ragweed pollen grains.[53] A single capsule of *R. hemisphaerica* produces around 3,000 spores,[54] which are viable for at least five months.[55] Considered on a scale that encompasses billions of spores over millions of years, the hypothesis that rare meteorological events blew spores of *R. hemisphaerica* into the stratosphere and across oceans seems possible.[56] Mosses experimentally exposed to the stratosphere by a weather balloon survived despite temperatures of -30°C.[57]

Compared to dispersal across oceans, dispersal of *R. hemisphaerica* within Philadelphia seems prosaic. Possible carriers of its spores include wind, water, vehicles, springtails, birds, rodents, and people.

24

SILVER-HAIRED BAT

(*Lasionycteris noctivagans*)

Silver-haired bats have been turning up in winter in Center City.

Figure 24.1 Female silver-haired bat hanging head down on trunk of Norway maple on Pine Street in Center City, December 27, 2012. Rain has turned the lichens green, revealing the bat. All bat photos in this chapter are of this individual.

Around 1796, the French naturalist Baron Palisot de Beauvois identified one kind of bat as the most common in Philadelphia. He named it after the Latin word for brown, *fuscus,* which designates this species (*Eptesicus fuscus*). His description, published initially in French[1] and later in English,[2] was part of a catalog of Charles Willson Peale's museum in Philadelphia. It stands as the first systematic account of this species.[3]

Today Palisot de Beauvois' bat is known by several common names: big brown bat, barn bat, and house bat.[4] It flies into attics and chimneys.[5] In eighteenth-century Philadelphia, it must also have flown through open windows, which then were unscreened. With a wingspan of 32 centimeters (1 foot), it would have been hard to miss.

Collection of bats in nineteenth-century Philadelphia

Peale's museum housed a collection of bats. One of Peale's sons, Titian Ramsay Peale, described the journey of two red bats (currently named *Lasiurus borealis*) to the museum:

> In June 1823, the son of Mr Gillespie, keeper of the city square, caught a young red bat, (*Vespertilio noveboracensis L.*) which he took home with him. Three hours afterwards, in the evening, as he was conveying it to the Museum in his hand, while passing near the place where it was caught, the mother made her appearance, followed the boy for two squares, flying around him, and finally alighted on his breast, such was her anxiety to save her offspring. Both were brought to the Museum, the young one firmly adhering to its mother's teat. This faithful creature lived two days in the Museum, and then died of injuries received from her captor. The young one, being but half grown, was still too young to take care of itself, and died shortly after.[6]

Peale's museum occupied Independence Hall, known then as the State House, where the Declaration of Independence and the Constitution of the United States were debated and adopted. Despite the prestige attached to this building and the honor attached to describing new species, only three of Pennsylvania's eleven species of bats had been described by 1825 when the museum's curator, Richard Harlan, completed his treatise on North American mammals.[7] In Pennsylvania some of these eleven species remain poorly understood. An example is the silver-haired bat.

Silver-haired bat in Center City

On December 27, 2012, I found a silver-haired bat hanging head down on the trunk of a Norway maple a few doors up from our row house in Center City. The bat did not move for five days, despite freezing temperatures and snow. Passersby did not notice it, even though it was only a meter off the ground, facing the sidewalk. Concerned that a child might find it and suffer a bite, I captured it. It opened its wings and its mouth, baring its teeth, moments after I dislodged it safely into a secure plastic container. I transported it to the Schuylkill Wildlife Rehabilitation Clinic in Philadelphia.

Brenda Malinics, specialist in bat rehabilitation at the clinic, told me that people had brought in six silver-haired bats from Center City in the past month, including two

Figure 24.2 Silver-haired bat hanging by its hind claws.

the past week, one from Rittenhouse Square and another from the former Wanamaker's building, now Macy's. She reported that the bat was female, weighed 11 grams, readily ate mealworms and drank water, and appeared healthy.

Silver-haired bat in North America

The only published record of this bat in Philadelphia dates to before 1864.[8] First described in 1831,[9] it is widely distributed in North America from the Atlantic to the Pacific coast.[10] Unlike communal bats in caves, this bat is usually solitary and difficult to find, although it has been noted to form maternity colonies.[11] The bat is called a "tree bat" because it typically roosts in trees.[12] It breeds in Canada and southern Michigan and migrates south in the fall.[13] In the eastern United States, it overwinters from New York City to Georgia.[14] Whether it breeds in Pennsylvania is unknown.

Center City as thermal refuge

For a bat migrating south or hibernating, Center City's heat island offers a thermal refuge. On clear calm winter nights, Center City is commonly 5.6–11°C (10–20°F) warmer than nearby rural areas.[15] In New York City the bat has been reported hibernating in skyscrapers, churches, wharf houses, and the hulls of ships.[16] It occasionally overwinters in caves.[17] In a forest in Arkansas in winter, almost all silver-haired bats roosted in terrain facing south.[18] The bat I found was hanging on the south side of its tree trunk.

Regulation of body temperature

Despite Center City's heat island, this bat encountered subfreezing temperatures, snow, and no food for at least five days before I took it to the clinic. Bats cope with low temperatures and shortages of food by entering into a state of torpor, or inactivity. Their metabolic rates drop and their body temperatures fall to levels almost matching those of their surroundings. When body temperatures fall sufficiently low, however, metabolism increases and produces body heat, at a cost of energy stored as body fat.[19]

Caves buffer bats from freezing temperatures; roosts in trees offer no such protection. Robert M. R. Barclay at the University of Calgary in Alberta, Canada, and colleagues noted that silver-haired bats roosting in trees during the day in cold weather in Manitoba, Canada, felt cold to the touch and were sluggish and unable to fly. They used flat telethermometers attached to the bats' abdomens to measure their body surface temperatures. They found that body surface temperature matched environmental temperature within 1–2°C. The two measurements coincided over a broad range of environmental temperatures, from 4 to 20°C (39 to 68°F).[20]

Miranda B. Dunbar at the University of Regina in Saskatchewan demonstrated that metabolic rates of torpid silver-haired bats fell with decreasing temperatures in the environment until these temperatures reached 5°C, at which point further decreases in environmental temperature caused metabolic rates to rise. Dunbar concluded that, for the silver-haired bat, 5°C (41°F) is the energetically optimal environmental temperature for hibernation.[21] At this environmental temperature, the bat is drawing the least energy from its reserves of fat, which must last until it can once again draw energy from insect prey.

Comparison with birds

The strategy that the silver-haired bat uses to cope with falling temperatures in winter is opposite that of birds overwintering in our backyard. As winter approaches and outside temperatures begin to fall, the bat lowers its intake of calories, drops its metabolic rate, decreases its body temperature, and enters a state of inactivity, or torpor. In contrast, the white-throated sparrow (*Zonotrichia albicollis*), slate-colored junco (*Junco hyemalis*), and house sparrow (*Passer domesticus*) consume more calories, increase their metabolic rates, and forage for food.[22]

Figure 24.3 White-throated sparrows (*Zonotrichia albicollis*), house finches (*Haemorhous mexicanus*), and a northern cardinal (*Cardinalis cardinalis*) eating sunflower seeds in our backyard, January 26, 2013. Their metabolic approach to cold contrasts with that of the silver-haired bat.

Health of the bat found on the tree

The silver-haired bat I observed on the tree trunk was probably in a normal state of torpor. It was conserving energy in the absence of food. The arousal behavior that it exhibited when I captured it is typical of bats disturbed in hibernation,[23] which is an extended state of torpor. Torpor occurs in the course of both migration[24] and hibernation, but the timing and duration of torpor in this case suggest that the bat was hibernating.

Despite the apparently healthy state of this bat after capture, its choice of an exposed site for roosting raises the possibility that it was sick. Silver-haired bats roosting in trees in forests wedge themselves in narrow crevices, such as splits or forks,[25] or they enter cavities.[26] The scarcity of trees downtown may have prevented this bat from finding a roosting site that was more secure. Acoustic artifacts produced by row houses on either side of the street may have disturbed the bat's echolocation and its search for a better site. Alternatively, an illness such as rabies could have impaired its judgment. An apparently healthy bat may actually be infected with the rabies virus.[27]

Figure 24.4 Snow piled on top of silver-haired bat on tree trunk.

Figure 24.5 Snow on back of silver-haired bat.

To what extent silver-haired bats ordinarily overwinter in Center City is unclear. The monitoring of populations of bats downtown is as primitive today as it was in 1823 when Mr. Gillespie's son brought his red bat to the Peale Museum. Silver-haired bats are well camouflaged against bark. In Center City, a large population of hibernating silver-haired bats could escape detection.

Figure 24.6 Silver-haired bat camouflaged after lichens have dried out and turned gray.

White-nose syndrome

In 2006, a novel disease began killing massive numbers of bats in caves. Called white-nose syndrome, the epidemic rapidly spread from a single cave in New York State to caves throughout much of the eastern United States and adjacent Canada. It has killed over 5 million bats belonging to at least six species, especially the little brown bat (*Myotis lucifugus*),[28] until recently the most common bat in Pennsylvania.[29] An analysis of the high mortality and low rates of reproduction of this bat led to a prediction that this bat would become extinct in the region by the year 2026.[30] The pathogen is a fungus (*Pseudogymnoascus [Geomyces] destructans*) introduced from Europe, where it infects bats but does not cause mass mortality.[31] The fungus grows on the muzzle, wings, and ears of bats hibernating in caves. It arouses bats from torpor and depletes their stores of energy in the form of fat.[32]

So far, silver-haired bats have escaped white-nose syndrome, probably because their usual hibernacula are in trees rather than caves, and they are mostly solitary. Physiological constraints of the fungus have confined white-nose syndrome to caves.[33]

Rabies

Concern that a child might disturb the bat and incur a bite prompted me to remove it. Rabies virus strains tied specifically to the silver-haired bat and one other species of bat (tricolored bat, *Perimyotis subflavus*) account for 70 percent of deaths due to rabies in this country.[34] Most people who die of rabies contracted in the United States have no history of a bite from a rabid animal.[35] Unrecognized exposure leaves victims of rabies clueless about the need to seek timely rabies vaccination and rabies immune globulin, which are lifesaving when administered soon after exposure.[36] High infectivity of rabies virus strains from these particular two species of bat probably contributes to "cryptic" cases of rabies, in which no evidence of exposure can be found.[37]

Despite the importance of the silver-haired bat in fatal human rabies, the risk of death from this bat is low. Only 6.9 percent of silver-haired bats submitted to health departments in the United States tested positive for rabies virus,[38] and only 1 percent of a random sample of silver-haired bats in the wild tested positive.[39] Among all identified species of bat submitted for testing and found to be positive for rabies virus in the United States, the silver-haired bat accounted for only 1 percent.[40] In Pennsylvania in 2011, seven times more raccoons tested positive than did bats, which tested positive less often than did cats, skunks and foxes.[41]

In the United States in the last half century, the incidence of bat rabies in humans was 3.9 cases per billion person-years,[42] about a thousand times less than the incidence of people killed or injured by lightning.[43] The incidence of bat rabies (i.e., rabies with a viral strain specific to bats) in the absence of a history of direct contact with a bat was 0.6 per billion person-years.[44] These figures are based on diagnosed rabies and underestimate the true incidence of rabies. Rabies masquerades as other conditions and can be difficult to diagnose.[45]

Relationship of white-nose syndrome to rabies

Were white-nose syndrome to extirpate populations of Pennsylvania's most common bat (the little brown bat), would Center City experience an increase in other bats, such as the silver-haired bat, increasing the risk of human rabies? Acoustical activity was used to track changes in populations of bats foraging at ponds and streams at Fort Drum in northern New York State before and after the discovery of white-nose syndrome, which first appeared in a cave near Albany. Acoustical activity decreased for little brown bats, but increased for silver-haired bats.[46]

Brenda Malinics told me that the six silver-haired bats that she received in 2012 was a record. A survey of all bat rehabilitation specialists in the state turned up no comparable reports of this bat.

The chance that white-nose syndrome could endanger public health due to a surge in cases of rabies from silver-haired bats is low. Neither Pennsylvania nor New York has ever had a case of diagnosed rabies due to viral strains associated with the silver-haired bat. These strains of rabies virus have caused only eight documented cases of human rabies nationwide, and no clusters of cases.[47] All but one occurred in states that so far have not harbored white-nose syndrome.[48] White-nose syndrome does attack the tricolored bat,[49] the other species of bat that harbors a rabies virus strain linked to human rabies.[50]

Potential harms and benefits of overwintering in Center City

White-nose syndrome is only one of many environmental dangers facing bats.[51] Silver-haired bats migrate along the United States' eastern seaboard,[52] which offers promising sites for offshore energy development, particularly wind farms.[53] Wind turbines have killed large numbers of migrating bats, including silver-haired,[54] due to fatal collisions with rotors.[55] Many of the factors that will determine whether offshore wind farms along the East Coast threaten populations of silver-haired bats remain unknown.[56]

Center City's heat island poses a theoretical hazard to silver-haired bats. Thermoregulation of bats in this country is tuned to the latitude where they overwinter.[57] It controls utilization of energy (fat) stored in the summer and fall and consumed in the winter.[58] Silver-haired bats in Missouri occasionally interrupt hibernation and forage for insects during the winter.[59] The heat island that Center City presents to silver-haired bats could, in theory, desynchronize their thermoregulation from latitude and season.

Would silver-haired bats hibernating in Center City's heat island adjust their metabolic rates appropriately? Would they arouse from torpor and forage when insect prey is on the wing? Depending on the answers to these questions, Center City could serve as either thermal refuge or thermal trap.

I suspect silver-haired bats hibernating in Center City will adapt. During winter they occasionally inhabit other thermal refuges, such as caves and mines.[60]

25

CANADA GOOSE

(Branta canadensis)

Canada geese began nesting in Pennsylvania only after people captured, bred, and conditioned them to change their historic breeding grounds.

Figure 25.1 Canada geese find a patch of open water in ice above Fairmount Dam in early morning, January 24, 2013.

In 1799 Benjamin Smith Barton described the Canada goose as a "passenger-bird" that migrated through Philadelphia from the south as early as March 3 and in the opposite direction in the fall.[1] In 1814 Alexander Wilson wrote that the Canada goose bred far to the north of the United States, but where he could not determine; certainly nowhere near Philadelphia.[2]

How did the Canada goose come to breed in Philadelphia?

Commercial hunting with decoys

Wilson described the bird in Philadelphia:

> The Wild Goose, when in good order, weighs from ten to twelve, and sometimes fourteen pounds. They are sold in the Philadelphia markets at from seventy-five cents to one dollar each; and are estimated to yield half a pound of feathers a piece, which produces twenty-five or thirty cents more.[3]

Gunners shot Canada geese lured with captive geese used as decoys. The shooters pinioned or clipped captives to prevent them from escaping. The captive geese performed their job well:

> They hail every flock that passes overhead, and the salute is sure to be returned by the voyagers, who are only prevented from alighting among them by the presence and habitations of man. The gunners take one or two of these domesticated Geese with them to those parts of the marshes over which the wild ones are accustomed to fly; and concealing themselves within gun-shot, wait for a flight, which is no sooner perceived by the decoy Geese, than they begin calling aloud, until the whole flock approaches so near as to give them an opportunity of discharging two and sometimes three loaded musquets among it, by which great havoc is made.[4]

Migrating geese flew through a gauntlet of carnage from Canada in the north to their overwintering areas in the south and back. Wilson blamed the slaughter for the Canada goose's increasing scarcity.[5]

Breeding of Canada geese in Pennsylvania

In 1935, the use of live decoys for hunting geese became illegal.[6] In the Midwest, hunters transferred captive flocks to wildlife managers, who used them as decoys to attract migratory and overwintering populations into wildlife refuges.[7] In 1936 fifty pinioned (flightless) geese were released in a Pennsylvania state game refuge on Pymatuning Lake, a reservoir on the border between Ohio and Pennsylvania. These birds nested in 1937, and in 1938 produced progeny that were able to fly.[8] This is the first documented instance of Canada geese breeding outside of captivity in Pennsylvania.[9]

Figure 25.2 Canada goose with goslings on the Schuylkill River Trail in downtown Philadelphia, May 3, 2011. When first described in Philadelphia, Canada geese migrated through the region but did not breed here.

Public financing of wildlife restoration

In 1937 President Roosevelt signed the Pittman-Robertson Act, which imposed an 11 percent manufacturer's tax on hunting gear, including shotguns, rifles, ammunition, and archery equipment. The U.S. Treasury collected the money and transferred it to the U.S. Fish and Wildlife Service, which dispersed it to state wildlife agencies; states were required to make matching contributions on a portion of the receipts. In its first fifty years, the Pittman-Robertson Program raised $2 billion for wildlife restoration.[10]

In 1986, Pennsylvania ranked number one in sales of hunting licenses and number three in receipt of Pittman-Robertson funds, compared to all other states. That year alone, the Pittman-Robertson Program transferred to Pennsylvania over $4 million for wildlife restoration.[11] Many other state and federal laws, including the Migratory Bird Treaty Act of 1918 and its amendments, protected Canada geese and their wetland habitats.[12]

Proliferation of urban geese in southeastern Pennsylvania

The flock of resident geese at the state game refuge on Pymatuning Lake thrived, reinforced by introductions of more captive geese.[13] In 1966 the Pennsylvania Game Commission transferred fifteen mating pairs of Canada geese from Pymatuning to the Middle Creek Wildlife Management Area,[14] about 100 kilometers west of Philadelphia. The next year it added more captive geese. Complaints of nuisance geese in Pennsylvania began in the next decade. The Pennsylvania Game Commission then instituted a program to trap and transfer problem geese to destinations within and outside the state. The program ended in 1995;[15] by then Canada geese were distributed in every county in the state.[16] A survey of resident nesting Canada geese in the Atlantic Flyway found that Pennsylvania had 11,819 breeding pairs—more than any other state.[17]

Why Canada geese breed here

Why did these geese nest locally instead of migrating north to ancestral breeding grounds in Canada? In 1970 Dennis C. Surrendi at Montana State University transplanted juvenile Canada geese 100 miles from their place of birth; the following year the transplanted geese homed to the location where they first flew, not to the site where they were born. His findings show that migration to a particular breeding ground is learned, not genetically predetermined.[18] Geese transplanted before they learn to fly nest in their transplant sites.[19] Wildlife managers exploited this phenomenon when they transplanted populations of Canada geese: if they learned to fly near Philadelphia, they would, when mature, return here to breed.

Figure 25.3 Geese grazing in August near the dam at the Fairmount Water Works.

The geese that the Pennsylvania Game Commission released were preadapted to urban habitats. Urban geese in Pennsylvania are derived from a Midwestern subspecies, the giant Canada goose (*Branta canadensis maxima*),[20] which is not native to the Philadelphia area.[21] Compared to other subspecies, this one reaches sexual maturity earlier, nests in more southerly latitudes, and migrates shorter distances; it has large clutch sizes and high rates of nest success and survival;[22] it flies at low altitudes, has a placid disposition, and is readily tamed;[23] its large size conditions it for overwintering in northern latitudes.[24]

Domestication of Canada geese

How did evolution produce a constellation of traits so well suited to human habitats? By the time live decoys were made illegal, wild populations of the giant Canada goose were believed to be extinct.[25] The release of captive flocks of giant Canada geese into parks, wildlife refuges, and game preserves introduced this subspecies into its former breeding grounds and beyond, such as Pymatuning Lake.[26] These geese, although called "wild," were actually feral geese that had been semidomesticated.

The scale of captive breeding must have been big. In 1963, almost three decades after the outlawing of live decoys for hunting, the number of permits for breeding Canada geese by game breeders under the U.S. Bureau of Sports Fisheries and Wildlife numbered 1,242, and the number of Canada geese held was 14,581.[27] Once released, these geese could incorporate their semidomesticated behavior into free-flying flocks.

Alexander Wilson's essay on the Canada goose reveals how self-selection could lead to domestication of captive geese. It describes a clipped-wing captive goose that wandered several miles away on foot during migration season.[28] Were such birds to escape, they would cease contributing to the captive flock's gene pool. Natural selection in the flock would favor geese that did not wander off.

In short, the Canada geese that nest in Philadelphia are descendants of captive geese conditioned to breed locally rather than in breeding grounds far to the north. Wildlife managers disseminated these geese in programs funded by state and federal revenue raised by taxes on guns and ammunition. Philadelphia offered these geese sanctuary from gunners and predators. It provided mowed grass (a favorite food)[29] and shelter along the Schuylkill River. In winter, when the Schuylkill froze, the Fairmount Dam afforded geese a haven of open water.

Mixed populations of Canada geese in the winter

The geese that overwinter in southeastern Pennsylvania include year-round residents as well as long-distance migrants from the north from as far away as Canada and Greenland.[30] Some of the geese that breed here in the summer overwinter in the Chesapeake Bay and Delmarva Peninsula.[31] Morphometric and molecular studies have distinguished Canada geese populations that migrate to breeding grounds that are geographically separated.[32]

Figure 25.4 The Schuylkill River, frozen, January 26, 2013. View is from Spring Garden Street Bridge looking south. Canada geese congregate on ice only when it abuts open water.

Figure 25.5 The Fairmount Dam, providing the only open water in the vicinity. The birds on the edge of the ice are Canada geese.

Canada geese that nest in urban and suburban areas are probably continuing to adapt. On an evolutionary time scale, the short time that Canada geese have inhabited suburbs and cities suggests that adaptations, both genetic and learned, may take more time. The tendency of Canada geese to evolve genetically distinct geographic races[33] encourages the evolution of local adaptations.

The story of Canada geese in Philadelphia parallels stories of other wildlife, including mallards (*Anas platyrhynchos*), wood ducks (*Aix sponsa*), wild turkeys (*Meleagris gallopavo*), and white-tailed deer (*Odocoileus virginianus*). All of these species have been beneficiaries of the Pittman-Robertson Act.[34] The mallard and Canada goose have both established semidomesticated populations that breed in Philadelphia, and the Canada goose and white-tailed deer have later become targets of programs to reduce overpopulation.[35]

Figure 25.6 Canada geese grazing while sitting on their feet in the snow on the same lawn as in figure 25.3.

Retention of adaptations to cold

During a cold snap in 2013, the Schuylkill River froze, except around the Fairmount Dam. I observed Canada geese grazing in snow-covered grass near the dam. These geese allowed me to observe them from a distance of only a meter. This tame behavior is characteristic of giant Canada geese that breed here. The geese grazed sitting rather than standing, their usual posture when feeding. They sat on their feet in the snow and browsed the grass by craning their necks. This behavior protected their feet and breasts from exposure to cold air. Despite their ancestral origin as semidomesticated captive geese, their instincts for thermal protection under frigid conditions have endured.

SPOTLIGHT

PAVEMENT ANT

Pavement ant (*Tetramorium caespitum*; also named *Tetramorium* species e) collected as it was emerging from a hole in the sand between granite pavers in Logan Circle, Center City. Head is on the left.

The length of this ant is about 3 mm (a tenth of an inch). Pavement ants were introduced into North America, perhaps in ship's ballast or in horticultural produce. Their natural history in the New World was first described in 1878, when two colonies of this species were observed engaged in a battle in Penn Square, current location of City Hall. The battle lasted over two weeks.[1]

26

STINK BUG HUNTER

(Sand wasp; *Bicyrtes quadrifasciatus*)

The native sand wasp, *Bicyrtes quadrifasciatus*, specializes in preying on nymphs of stink bugs, including the brown marmorated stink bug (*Halyomorpha halys*), an introduced pest first found in North America in Allentown, Pennsylvania, in 1996.

Figure 26.1 Logan Circle, where stink bug hunters (*Bicyrtes quadrifasciatus*) dig nests in sand between granite pavers.

In 2001 Karen M. Bernhard, a Pennsylvania State University entomologist, collected two specimens of a strange species of stink bug in Allentown, 79 kilometers northwest of Philadelphia. She submitted them to Cornell University's Insect Diagnostic Lab, where E. Richard Hoebeke identified them as the brown marmorated stink bug (*Halyomorpha halys*; family Pentatomidae), an agricultural and domestic pest in China, Korea, Taiwan, and Japan. Published records had never before documented the species in North America, although Hoebeke and Maureen E. Carter later determined the species had been found in Allentown as early as 1996.[1]

Two days after Hoebeke identified the insect, he went to Allentown:

> On that unseasonably warm day, adults of *H. halys* were extremely numerous on the foundations, outer wall surfaces, eaves, and window and door frames of homes and sidewalks.[2]

Spread of the brown marmorated stink bug to downtown Philadelphia

By the time Hoebeke's observations were published in 2003, the outbreak had spread to five counties in Pennsylvania.[3] By 2007, the *Philadelphia Inquirer* reported the stink bug invading people's homes in suburban Philadelphia.[4] By January 2012, the stink bug had been reported in thirty-seven counties in Pennsylvania and in thirty-six states, including California and Oregon.[5] The insect has since turned up in Switzerland[6] and New Zealand.[7] I first noted adults of the insect attracted to my black light on the rear wall of our house on Pine Street in Center City on July 4, 2010. The nymphs (immature stages) arrived twelve days later.

Figure 26.2 Nymph (*immature stage*) of brown marmorated stink bug (*Halyomorpha halys*) climbing up the outside doorframe of our house in July 2010.

Figure 26.3 Head of adult brown marmorated stink bug crawling up stucco rear wall of our house, February 2011.

Agricultural pest

A trade organization for apple growers in the mid-Atlantic region estimated that in 2010 the pest caused crop losses valued at $37 million.[8] The insect attacks peaches as well, and many kinds of garden vegetables and ornamentals.[9]

To control infestations, apple growers in Maryland quadrupled the number of applications of insecticides and reduced intervals between applications.[10] In Japan pyrethroid and neonicotinoid insecticides administered together, plus three additional insecticides, reduced fruit injury to one-fifth the level achieved with conventional insecticidal regimens;[11] however, neonicotinoid insecticides used against pests have unintentionally harmed bees.[12] Several species of fungi pathogenic against this stink bug[13] have not been shown effective as weapons against outbreaks. Parasitoids of the stink bug's eggs in China and Japan[14] have yet to be released as biological control agents in the United States, and whether they will succeed in establishing populations here and safely controlling outbreaks of the stink bug is unknown.

Native wasps (*Trissolcus* sp.) parasitize eggs of this stink bug but have not substantially increased its mortality rate.[15] Feather-legged flies (*Trichopoda pennipes*), native to North America, were found parasitizing brown marmorated stink bugs in Allentown, Pennsylvania.[16] Although used for biological control of other stink bugs,[17] feather-legged flies have not controlled infestations of brown marmorated stink bugs.

Outbreaks of this agricultural pest in peach and apple orchards have challenged entomologists in Pennsylvania. The insect is wary and feeds on fruit at night, so visual counts in an orchard during the day may underestimate the potential for damage. Monitoring populations with traps based on pheromones, bait, or light has proved disappointing. Stink bugs that attack fruit in orchards spend most of their reproductive time outside orchards, so targeting them in orchards may fail to destroy them where they propagate. The few classes of insecticides effective against stink bugs are also toxic to beneficial species useful in pest management.[18]

Comparison with other outbreaks

The outbreak of the brown marmorated stink bug in Pennsylvania is reminiscent of nineteenth-century outbreaks of pests of shade trees in Philadelphia. Parasitoids then had not had sufficient time to establish themselves within the city, so even native species like bagworms proliferated out of control, causing epidemics of defoliation. The city responded by importing house sparrows (*Passer domesticus*), with dubious benefit.[19] Broad-spectrum insecticides such as arsenic were manufactured in the city and sprayed on trees.[20] Eventually parasites dispersed into the city, and massive defoliation of shade trees abated. David J. Biddinger and his colleagues from Pennsylvania State University have noted that the use of broad-spectrum insecticides against the brown marmorated stink bug in Pennsylvania threatens to turn back the clock on integrated pest management, which attempts to minimize use of such agents.[21]

Theoretical ideal agent for control of stink bugs

The best insect for biological control of the brown marmorated stink bug might not be a parasitoid; insecticides that kill the host also kill parasitoids inside the host, and parasites like the feather-legged fly allow their stink bug victims to reproduce before succumbing.[22] The optimal insect for biocontrol might be a predator that carries the stink bug away—distancing itself from insecticides targeting its prey's habitat—and kills or incapacitates its victims before they have a chance to reproduce. Such a predator ideally would be locally native, obviating the risks associated with importation of foreign agents for biocontrol. It would also be adaptable—capable of living in disturbed habitats, including cities and farms—and its geographic distribution would be broad.

Figure 26.4 Feather-legged fly (*Trichopoda pennipes*) in Fairmount Park just outside Center City. Although native to North America, it lays its eggs on the brown marmorated stink bug (in addition to native stink bugs), which its larvae eat.[23]

Discovery of *Bicyrtes quadrifasciatus*

In 1823, John C. Calhoun, Secretary of War under James Monroe, ordered Major Stephen H. Long to lead an expedition to the headwaters of the St. Peters River, now known as the Minnesota River. He appointed Thomas Say of Philadelphia to serve as official naturalist. Among Say's accomplishments on this trip were discoveries of many new species of insects, including a wasp with four conspicuous cream-colored abdominal stripes. Say described its morphology and documented its location in Pennsylvania, but he offered no clues about its habits or habitat.[24] It is *Bicyrtes quadrifasciatus*, named in reference to the four abdominal stripes. It has no specific common name.[25]

The life history of this wasp was not examined until the twentieth century. In 1966 Howard E. Evans reviewed all that had been learned about it in the chapter "*Bicyrtes:* A genus of stink bug hunters."[26] The best-known member of this genus is the species Say discovered. It is found throughout temperate North America east of the Rockies. Females prey exclusively on nymphs (immature stages) of true bugs (Heteroptera), predominantly stink bugs (Pentatomidae) and leaf-footed bugs (Coreidae). A female seizes a nymph, paralyzes it with her stinger, and clasping it beneath her, flies to her nest, a chamber she has excavated underground in sand. She uncovers a hidden entrance hole leading to a tunnel connecting to the chamber, and, clutching her prey, disappears down the hole to her nest. She lays an egg on the first nymph she brings to her nest, and then returns to stock the nest with additional prey. Her offspring feed on her prey, while she feeds on nectar at flowers. All other members of the genus *Bicyrtes* are sand wasps that specialize in hunting true bugs.[27] Biddinger et al. reported that prey of *Bicyrtes* wasps include brown marmorated stink bugs.[28]

Bicyrtes quadrifasciatus at work in Logan Circle

I observed Say's stink bug hunter at work in sand between granite pavers in Logan Circle. Jon Gelhaus, entomologist at the Academy of Natural Sciences of Drexel University, called my attention to the sand wasp during Bug Fest, an occasion when entomologists at the Academy guide visitors on tours of Logan Circle. Male sand wasps flew around close to the ground while females landed with prey or emerged from their holes. At any one moment, two or three males would be cruising over the pavers. They generally ignored me, except for avoiding me when I blocked their way. I photographed a female carrying into her hole a nymph of a leaf-footed bug, probably the western conifer seed bug, *Leptoglossus occidentalis*, another home invader recently introduced into Pennsylvania. It feeds on seeds of cones of conifers.[29]

Figure 26.5 Sand wasp (*Bicyrtes quadrifasciatus*) emerging from her hole in sand at Logan Circle.

Figure 26.6 Preparing to plug her hole and take off in search of prey, which she will paralyze with her stinger.

Figure 26.7 Dragging prey down the hole leading to her nest. The hole is too small.

Figure 26.8 Backing out of the hole and positioning herself for a change in strategy.

Figure 26.9 Shoving prey down the hole. The fit is still too tight.

Figure 26.10 Enlarging the hole.

Figure 26.11 Portrait of her prey on a granite paver. While she was distracted enlarging her hole, I seized the opportunity to snatch her prey and place it on a paver for this diagnostic photo. Her prey is a nymph (immature stage) of a leaf-footed bug (family Coreidae), in the same suborder as stink bugs (family Pentatomidae)

Figure 26.12 The hole after her descent into her nest. While I was photographing her prey a couple of pavers away, she discovered my theft and in an instant snatched it back and disappeared with it down this hole. At this moment she may be laying an egg on it in her nest.

Figure 26.13 Stink bug hunter feeding on late boneset (*Eupatorium serotinum*) along the Schuylkill River Trail in Grays Ferry Crescent immediately south of Center City. Stink bug hunters do not eat their prey; they dedicate their prey exclusively to their offspring, which as larvae consume their parents' victims.

Use of specialized predators for biocontrol

To what extent this wasp might succeed as a biological control agent against brown marmorated stink bugs has not been studied. The impact of specialized predators on populations of their prey may be less than one might suppose. Cicadas can be heard singing throughout residential areas of Center City despite the presence of cicada killers (*Sphecius speciosus*), large wasps that prey exclusively on them. Cicada killers have not eradicated cicadas in Center City, but they may suppress their populations sufficiently to prevent destruction of the trees upon which immature cicadas feed. Biddinger et al. suggest that goals for biological control agents need to be scaled back. They recommend that a biological control agent be introduced not as a single solution, but as an incremental step, one of many safeguards that, in combination, provide protection against outbreaks.[30]

Figure 26.14 Annual cicada (*Tibicen tibicen*) attracted to light behind the screen door in our backyard. It can be heard singing in summer in parks and residential areas throughout Center City, despite the presence of cicada killers.

Figure 26.15 Cicada killer (*Sphecius speciosus*) found dead in Fitler Square. I have posed it for scale. Cicada killers are the largest wasps in North America. Females grab and paralyze cicadas and drag them into their burrows, where their larvae consume them.

Controlled trials of *Bicyrtes* wasps deployed against the brown marmorated stink bug would be technically challenging. Methods to cultivate or attract the wasps have yet to be investigated. The wasps' enemies, particularly flies (*Senotainia trilineata* and *S. rubriventris* [Miltogramminae]), may suppress populations of the wasps. These flies have been described hovering around *Bicyertes* and depositing their larvae on *Bicyrtes'* prey as the wasps carry their prey into their holes or after the wasps have deposited their prey inside their nests.[31] I saw no such flies around the colony of wasps at Logan Circle.

The *Bicyrtes* colony at Logan Circle suggests how an empiric trial using *Bicyrtes* wasps for biocontrol might proceed: Construct a spacious sunny patio or broad promenade with pavers embedded in sand, and add beds of flowers that bloom all summer long. Evans noted that the wasps visit many kinds of flowers, especially white sweet clover (*Melilotus alba*) and Queen Anne's lace (*Daucus carota*).[32]

The question of how to control infestations of the brown marmorated stink bug may become moot if its enemies like *Bicyrtes* wasps and feather-legged flies achieve this control independent of human intervention. Farmers selling apples in Rittenhouse Square told me that stink bugs were less numerous in 2012. Until longer periods of observation confirm such improvement, the plague of brown marmorated stink bugs is likely to continue.

Figure 26.16 Frecon Farms' stand in Rittenhouse Square's farmer's market, September 2012. Brown marmorated stink bugs, which attack apple orchards, were less numerous this year, according to farmers at this stand.

SPOTLIGHT

RED-EARED SLIDER

Red-eared sliders (*Trachemys scripta elegans*) sunning in the Schuylkill River in Center City just downstream from the Spring Garden Street Bridge.

This turtle first appeared in Philadelphia in the twentieth century after its young became popular as pets, which were purchased in pet shops. It is now the turtle most commonly seen in the Schuylkill River in Center City. It may be displacing native species, such as the eastern redbelly turtle (*Pseudemys rubriventris*), classified as threatened in Pennsylvania. Red-eared sliders are native to the Mississippi valley.[1]

27

MULTICOLORED ASIAN LADY BEETLE

(*Harmonia axyridis*)

The multicolored Asian lady beetle first appeared in Pennsylvania two decades ago. It now is Center City's most abundant lady beetle.

Figure 27.1 Larva of multicolored Asian lady beetle (*Harmonia axyridis*) eating an aphid (*Aphis nerii*) on common milkweed (*Asclepias syriaca*) in Center City.

In 1806 Frederick Valentine Melsheimer produced a catalogue of the beetles of Pennsylvania. It was the first systematic entomological work published in North America. It recorded and classified 1,363 different kinds of beetles.[1] Because of this achievement Melsheimer is recognized as "The Father of American Entomology."[2]

Despite its renown, Melsheimer's catalogue has been dismissed as worthless to science because it presents names of species without descriptions, illustrations or references.[3] Some of these names were newly coined, appearing in print for the first time. Taxonomic references to this catalogue are currently rare.

Even with its inscrutable names, however, Melsheimer's catalogue offers a useful perspective, dating back over two centuries. For example, it enumerates twenty-three species under the heading *Coccinella*, the current name of a genus in the beetle family Coccinellidae, commonly known as lady beetles or ladybugs (which are beetles). By comparison, today in Center City, only two species in this family are generally common, and both were introduced into North America in the twentieth century.

Why are all of Melsheimer's lady beetles rare or absent here? Lady beetles are, with few exceptions, predaceous on small arthropods such as aphids, mealy bugs, scale insects and mites. In contrast to plant-eating insects, predatory lady beetles are not physiologically tied to particular kinds of host plants. One might have expected that at least some of the species of lady beetle that Melsheimer documented in Pennsylvania would be common in Center City today. Other groups of predatory insects living in Center City include common native species. Examples mentioned earlier in this book are wasps such as bald-faced hornets, stink bug hunters, cicada killers, mud daubers and, until recently, native yellowjackets.

Frederick Valentine Melsheimer

The story of Melsheimer's life puts his entomological findings into context. He was born in 1749 in the Duchy of Brunswick, Germany. He studied at the University of Helmstedt and served as a chaplain in a regiment of Hessian dragoons under the British during the Revolutionary War. Captured and imprisoned by American troops in the battle of Bennington, he and his regiment were incarcerated in Massachusetts for 14 months, released on parole, and re-incarcerated in Bethlehem, Pennsylvania, where he resigned his military commission. He settled in Lancaster and York Counties, Pennsylvania and pursued a career of teaching and preaching. Here he indulged in entomology as a recreational diversion, collaborating with August Wilhelm Knoch, an entomologist in Germany and a childhood friend. Melsheimer died in 1814.[4]

The Pennsylvania towns where Melsheimer resided are all located within 200 kilometers (120 miles) of Philadelphia, and within Pennsylvania's Piedmont, which encompasses much of southeastern Pennsylvania, including Center City's northern edge at Fairmount.

Twenty-spotted lady beetle

A black light I operate in our garden at night occasionally attracts the twenty-spotted lady beetle, a rare finding in Center City. Melsheimer's catalogue lists this beetle as *Coccinella 20 maculata*, a name it attributes to Knoch. A paper published in 1824, ten years after Melsheimer's death, provides the first description of the species.[5] The author is Philadelphia entomologist Thomas Say, also recognized as "The Father of American Entomology."[6] Say designates the species by the same name, *Coccinella 20 maculata*, and credits the name to Knoch in Melsheimer's catalogue. Say collected his specimen in Missouri.

Figure 27.2 Twenty-spotted lady beetle attracted at night to a pillowcase illuminated by a black light facing our garden. It is about a 2.5 mm (a tenth of an inch) in length.

How could Say recognize Melsheimer's *Coccinella 20 maculata* without a description or illustration? Say probably examined specimens labeled with this name in Melsheimer's collection of beetles. The Melsheimer collection contained 14,075 beetles representing 4,674 species housed in 41 homemade wooden boxes when Harvard's Museum of Comparative Zoology accessioned it decades later. Melsheimer's oldest son, who shared his father's interest in beetles, collaborated with Say.[7]

Melsheimer's catalogue, reinforced with Say's description, documents that the twenty-spotted lady beetle (now named *Psyllobora vigintimaculata*, a synonym for the original name) inhabited Pennsylvania at the beginning of the beetle's recorded history more than two centuries ago.

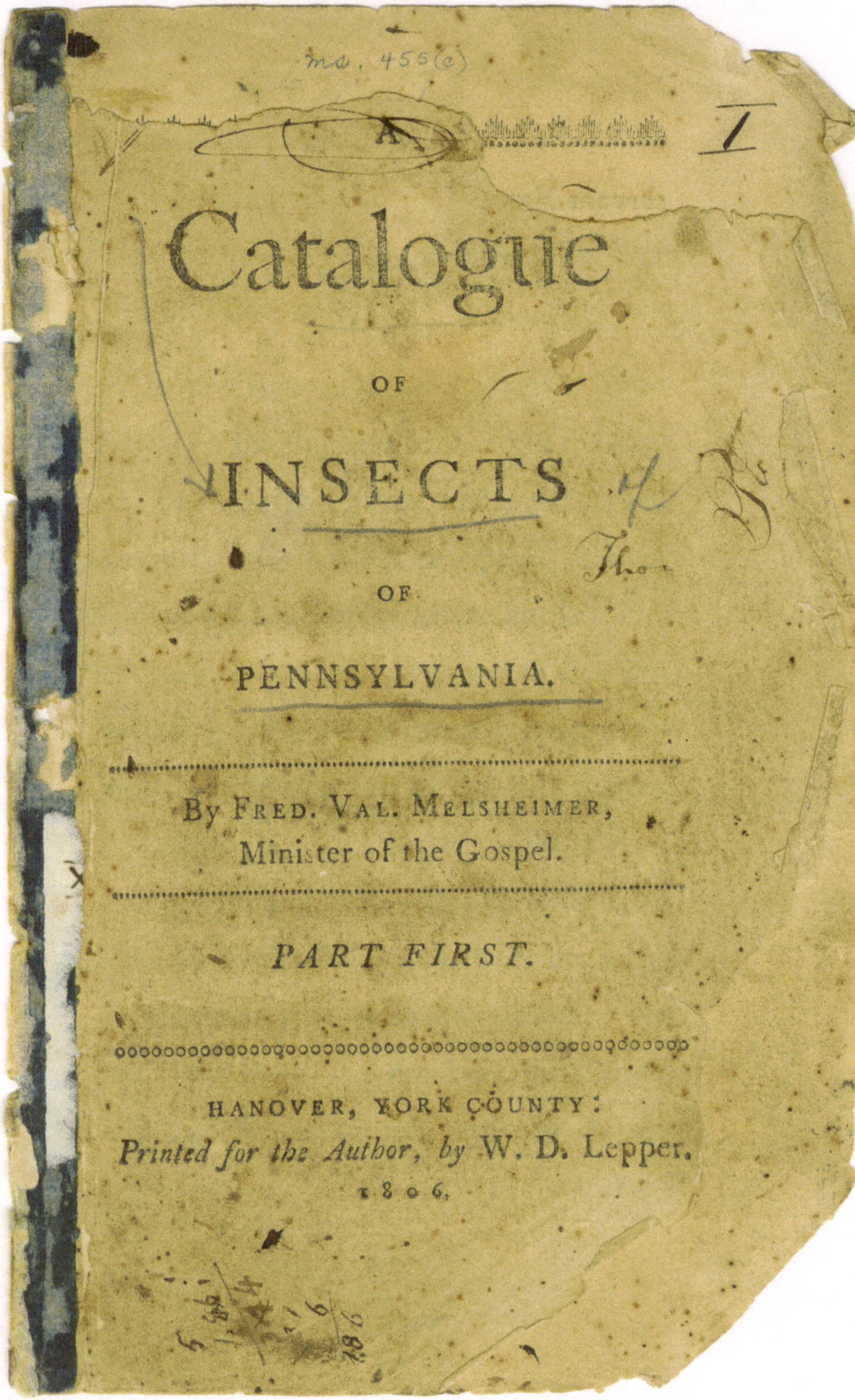

Figure 27.3 Cover of *A Catalogue of Insects of Pennsylvania* by Frederick Valentine Melsheimer, published in 1806. The work is devoted exclusively to beetles. This copy belonged to Thomas Say. (From folder 6, Archive Collection 455, Library of the Academy of Natural Sciences of Drexel University. Digitally scanned by the Academy. Reproduced courtesy of the Academy.)

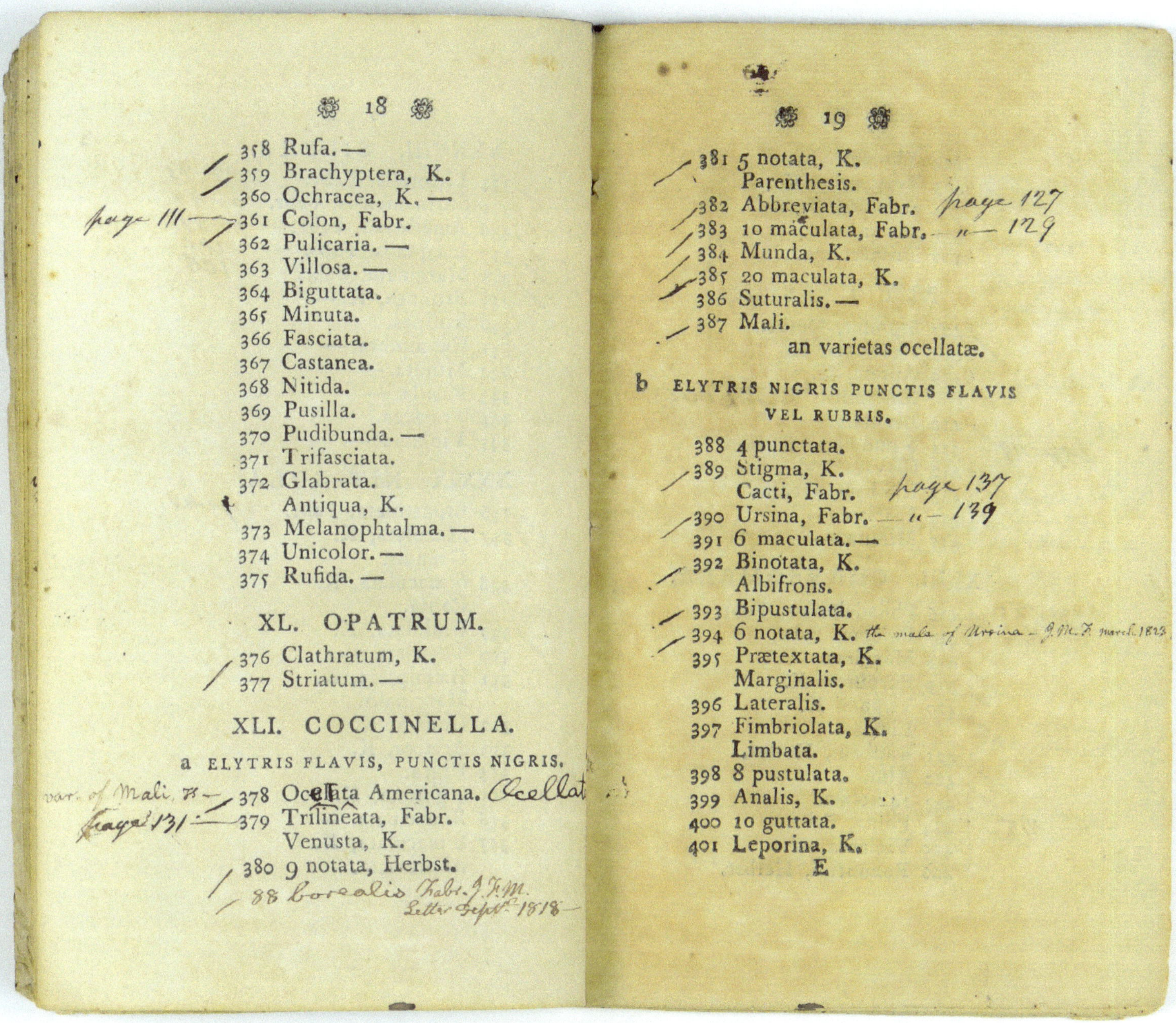

18

358 Rufa.—
359 Brachyptera, K.
360 Ochracea, K.—
361 Colon, Fabr.
362 Pulicaria.—
363 Villosa.—
364 Biguttata.
365 Minuta.
366 Fasciata.
367 Castanea.
368 Nitida.
369 Pusilla.
370 Pudibunda.—
371 Trifasciata.
372 Glabrata.
Antiqua, K.
373 Melanophtalma.—
374 Unicolor.—
375 Rufida.—

XL. OPATRUM.

376 Clathratum, K.
377 Striatum.—

XLI. COCCINELLA.

a ELYTRIS FLAVIS, PUNCTIS NIGRIS.

378 Ocellata Americana.
379 Trilineata, Fabr.
Venusta, K.
380 9 notata, Herbst.

19

381 5 notata, K.
Parenthesis.
382 Abbreviata, Fabr.
383 10 maculata, Fabr.
384 Munda, K.
385 20 maculata, K.
386 Suturalis.—
387 Mali.
an varietas ocellatæ.

b ELYTRIS NIGRIS PUNCTIS FLAVIS VEL RUBRIS.

388 4 punctata.
389 Stigma, K.
Cacti, Fabr.
390 Ursina, Fabr.
391 6 maculata.—
392 Binotata, K.
Albifrons.
393 Bipustulata.
394 6 notata, K.
395 Prætextata, K.
Marginalis.
396 Lateralis.
397 Fimbriolata, K.
Limbata.
398 8 pustulata.
399 Analis, K.
400 10 guttata.
401 Leporina, K.
E

Figure 27.4. Species enumerated on pages 18 and 19 in *A Catalogue of Insects of Pennsylvania.* Lady beetles are covered under the heading Coccinella. Handwritten annotations are by Thomas Say. (Source and credits as in Figure 27.3)

Figure 27.5 Seven-spotted lady beetle (*Coccinella septempunctata*) foraging on a leaf of common milkweed in late October. Parasitic larvae of wasps infected all the aphids here and transformed them into spherical "mummies," which have already hatched into wasps, leaving empty shells and no food for the beetle.

Seven-spotted lady beetle

Importation of lady beetles for biological control of insect pests in the United States began toward the end of the nineteenth century.[8] It includes Center City's two common lady beetles. The seven-spotted lady beetle (*Coccinella septempunctata*) was the first of these two to be established in Pennsylvania. It appeared in this state in 1979 just outside of Harrisburg[9] as it disseminated across North America. It is native to Europe and Asia.[10]

Explosive growth in populations of this lady beetle[11] has been blamed for increasing scarcity or local extinction of native lady beetles, especially the nine-spotted lady beetle (*Coccinella novemnotata*),[12] which Melsheimer's catalogue lists as *Coccinella 9 notata*. It may be impossible to tease out the extent that any one factor has reduced populations of native lady beetles. Multiple factors may be responsible, including exotic lady beetles but also habitat loss, pesticides, pollution, pathogens, predators and parasites.[13]

Ivo Hodek at the Czech Academy of Sciences, and J.P. Michaud at Kansas State University, tried to explain the tendency of populations of the seven-spotted lady beetle to predominate over those of native lady beetles in Europe and Asia and also in North America. They pointed out that this species preys on a wide variety of aphids. It is able to detect trails of larvae of seven-spotted lady beetles and, in response, to reduce its egg laying. It is able to adjust the number of its broods per season, and to reduce activity (that is, enter diapause) according to local conditions. It mates in the fall before hibernation. Its reproductive rate is high, as is its tolerance of environmental disturbance.[14]

Multicolored Asian lady beetle

Despite the seven-spotted lady beetle's adaptability, it is not the most abundant lady beetle in Center City. This distinction belongs to the multicolored Asian lady beetle (*Harmonia axyridis*), which first appeared in Pennsylvania in 1993.[15] Introduced to control agricultural pests, it is native to China, Japan, Korea, Mongolia and eastern Russia. It has spread rapidly in North America, South America, Europe, and Africa.[16]

Figure 27.6 Multicolored Asian lady beetle (*Harmonia axyridis*) foraging on common milkweed a meter from the seven-spotted lady beetle shown in Figure 27.5. Both lady beetles are finding only vestiges of prey.

The range of prey that can sustain multicolored Asian lady beetles is broader than that for seven-spotted lady beetles and extends to prey other than aphids.[17] The multicolored Asian lady beetle eats eggs, larvae, pupae and adults of other species of lady beetle. Its prey includes the seven-spotted lady beetle. Its large size and larval spines deter other lady beetles from attacking it.[18] When disturbed it reflexly exudes toxic hemolymph, which repels seven-spotted lady beetles.[19] During periods of scarcity of food, its larvae engage in cannibalism, which increases their survival and accelerates their development.[20] Cannibalistic larvae use chemical cues to avoid eating relatives.[21] Multicolored Asian lady beetles track populations of aphids in space and time.[22] Exploiting urban habitats, multicolored Asian lady beetles overwinter in houses and in crevices of buildings—to the point that people in Center City occasionally view them as a nuisance.

Biological arsenal

Andreas Vilcinskas and colleagues in Germany reported that multicolored Asian lady beetles use "biological weapons against native competitors." They discovered that multicolored Asian lady beetles carry spores of parasitic microsporidia harmless to them but potentially lethal to other lady beetles, including seven-spotted lady beetles.[23] Vilcinskas and colleagues also found that multicolored Asian lady beetles have unusually potent immunological defenses against pathogenic fungi and gram-negative bacteria.[24] They raised the possibility that microsporidia in eggs of multicolored Asian lady beetles fatally infect competing lady beetles that eat these eggs.[25]

In addition to harboring microbial "weapons," multicolored Asian lady beetles produce potent chemical defenses. They synthesize poisonous alkaloids (harmonine) that protect them from predators, including other lady beetles. They also produce methoxyprazines that endow them with a deterrent odor. Other lady beetles produce similar defensive chemicals; but multicolored Asian lady beetles nevertheless attack them and their immature stages,[26] despite hidden costs.[27]

Diverse Enemies

Even though the multicolored Asian lady beetle is fortified with protective chemicals and microbes, it is vulnerable to many enemies, including flies, wasps, mites, nematodes and pathogens.[28] Rates of parasitism of the multicolored Asian lady beetle are higher where it is native than where it is introduced.[29]

In winter in Center City multicolored Asian lady beetles frequently host a fungus on their wing covers (elytra), mouthparts, and legs. A parasitic fungus, *Hesperomyces virescens,* infected more than half of a winter population of this lady beetle in Pennsylvania.[30] This fungus has been found to decrease survival of multicolored Asian lady beetles.[31] Ted E. Cottrell and Eric W. Riddick of the United States Department of Agriculture Research Service noted that the fungus attacks six other species of lady beetle. In the laboratory they found that forced bodily contact between different species of lady beetle rarely if ever transmitted the fungal infection.[32]

Figure 27.7 Fungus studding the sides of a multicolored Asian lady beetle hibernating indoors in Philadelphia.

Three non-predaceous lady beetles

Gardens in Center City host two kinds of lady beetle (family Coccinellidae) that feed on plants instead of prey. They are the squash lady beetle (*Epilachna borealis*) and the Mexican bean beetle (*Epilachna varivestis*). The twenty-spotted lady beetle feeds on mildew,[33] which sprinklers in our garden promote. The diets of these three lady beetles spare them from competition with the multicolored Asian lady beetle and, theoretically, distance them from fatal encounters with this voracious predator.

Seaports

William H. Day at the U.S. Department of Agriculture's Beneficial Insects Research Laboratory in Newark, Delaware, and his colleagues, evaluated the introduction of six species of exotic lady beetle, including the seven-spotted lady beetle and the multicolored Asian lady beetle. They noted that entomologists had made numerous unsuccessful attempts to establish all six as biological control agents in agricultural settings in the eastern United States. Day and colleagues concluded that all six of these lady beetles established themselves independently through accidental importation, most likely through inland seaports.[34]

Shipping or other modes of commerce may introduce new enemies of the multicolored Asian lady beetle. While such introductions theoretically could undermine populations of this species here, the multicolored Asian lady beetle has been resilient, thriving in thirty-eight countries on five continents, and in every state in the United States except Alaska and Wyoming.[35]

Figure 27.8 Brown lacewing (*Micromus posticus*) attracted to a porch light facing our garden. It is a competitor and predator of lady beetles: Adults and larvae feed on aphids,[36] and larvae also prey on eggs of lady beetles.[37] It is native to North America.[38]

Preying on other lady beetles as it disseminates globally, the multicolored Asian lady beetle in Center City is likely, at least in the short term, to remain more abundant than all other predaceous lady beetles—including those that Melsheimer recorded in Pennsylvania over 200 years ago; however, in the long term, it presents a growing target for evolution of ever more formidable enemies: competitors, predators, parasites and pathogens.

SPOTLIGHT

COMMON RAGWEED

Male flowers of common ragweed (*Ambrosia artemisiifolia*) along Schuylkill River Trail, Center City.

In the last century ragweed pollen was blamed for causing more hay fever than all other plants combined.[1] Over the past five years, flowering ragweed has been absent in Center City except near railroad tracks by the river, where it has been scarce. Ragweed pollen counts have been falling in the United States in the East and Midwest since the mid-twentieth century.[2] In Center City mugwort (*Artemisia vulgaris*) may have contributed to ragweed's scarcity. Mugwort and ragweed colonize similar habitats. Mugwort is abundant here, having evolved competitive strains and expanded its range starting in the nineteenth century, as described in Chapter 19.

28

OVERVIEW

Populations of wild plants and animals in Center City have had to cope with conditions that were never part of their evolutionary past. Those that adapted, or that were preadapted, survived; and those that benefited multiplied. Others dwindled or went locally extinct, sometimes reappearing a century or more later, after hardships had abated.

Figure 28.1 Triangulate cobweb spider (*Steatoda triangulosa*) wrapping silk around a multicolored Asian lady beetle (*Harmonia axyridis*) snared in the spider's web, which was anchored to a door lamp outside our house at night. The beetle had flown and crawled toward the lamp's light. Both species were introduced into North America.[1]

The preceding chapters examined Center City's plants and animals from diverse points of view: geologic time; pollution; prejudice; persecution; populations; habitats; evolution; and conservation. This chapter will explore each of these viewpoints separately, integrating observations from earlier chapters. It will omit references previously cited. The goal is to present an overview of the ecology of Center City.

Geologic time

The oldest tree on record in Center City grew when the area was a cypress swamp. The time was a warm interglacial period in the Pleistocene more than 36,000 years ago—the age of the Subway Tree, a bald cypress unearthed here in 1931 during excavation for a tunnel. Glacial meltwater flooded the swamp and buried it in sediment. Twenty thousand years ago the last glaciation in this region reached its maximum. It stopped 100 kilometers north of what is now Philadelphia.

Permafrost extended south of the glacier along the coastal plain to southern Delaware and Maryland. It wiped out all terrestrial isopods (pillbugs and sow bugs) outside of caves and seashores. Isopods did not repopulate the area until Europeans introduced them, probably incidentally in garden soil.

Around 6,000 years ago, Native Americans built houses upon which the black and yellow mud dauber constructed its nest. In 1745 John Bartram described this wasp making nests under the eaves of his house in Philadelphia. In Philadelphia the black and yellow mud dauber continues to produce nests—exclusively on man-made structures, including John Bartram's house and buildings and bridges in Center City.

Most preadaptations to Center City can only be inferred. An example is the predilection of the American robin to forage near houses and gardens. The affinity of this bird for human habitation dates to our earliest records of its behavior in Pennsylvania. The robin displayed this trait even when people hunted it and sold it as a delicacy in markets. It probably exhibited the trait around gardens of Native Americans before European settlement of North America. The red-tailed hawk, in contrast to the American robin, avoided people until persecution abated late in the twentieth century.

Preadaptation to downtown habitats typically evolved long before people or human disturbance. The red back salamander was able to inhabit Fitler Square in Center City in part because it is a non-amphibious amphibian. Evolution to an all-terrestrial life cycle liberated it from dependence on open water such as ponds and streams. Its ancestors evolved the capacity to live exclusively on land early—perhaps 200 million years ago, around the beginning of the Jurassic period.

Air pollution

The end of the eighteenth century marked the beginning of what has been dubbed the *Anthropocene*, a term for a geologic epic Paul J. Crutzen popularized in a paper called "The Geology of Mankind," published in Nature in 2002. *Anthropocene* is derived from *anthros*, Greek for "human," and *cene*, meaning "new." Crutzen dated the onset of his Anthropocene to the late eighteenth century, when concentrations of carbon dioxide and methane began increasing in air trapped in polar ice.[2]

In Philadelphia, the beginning of Crutzen's Anthropocene coincided with the Industrial Revolution, which exposed plants and animals to conditions that they had never before encountered and that were never part of their evolutionary history. Industrial air pollution is one example.

Tree mosses in the genus *Orthotrichum* are exquisitely vulnerable to sulfur dioxide in air. Published studies of mosses in the early twentieth century reported no tree mosses in New York City and Philadelphia. At that time, air pollution caused by combustion of coal contained sulfur dioxide in high concentrations. Levels of sulfur dioxide in Philadelphia have since reverted to levels not toxic to mosses. I recently discovered the tree moss *Orthotrichum pumilum* growing on trunks of street trees in Center City. Published records indicate that Thomas Potts James, who was born just outside Philadelphia in 1803 and died at age 79,[3] was the last person before me to have found this moss in Philadelphia.[4]

Figure 28.2 Cemetery at Old Pine Street Church. Burials began here in 1764. Acid rain accelerates natural weathering of marble and limestone. The acidity is due to sulfuric acid and nitric acid from sulfur dioxide and nitrogen oxides in air pollution,[5] which improved in Philadelphia in the 20th century after reduction in the burning of coal.

The rise of industrial pollution of air in Philadelphia preceded introduction of methodical monitoring of air quality, but historic collections of moths provide objective evidence of it. In Britain, the peppered moth (*Biston betularia*) has melanic (black) forms that increased in frequency when soot in air pollution turned tree bark black. When soot subsided, the abundance of melanic forms decreased. The phenomenon, called industrial melanism, offered a textbook case of evolution in action. As in Britain, industrial melanism showed up in populations of the peppered moth in Philadelphia, and subsequently abated in Pennsylvania as the quality of air improved.

Water pollution

Shad, initially abundant in the tidal Schuylkill River in Philadelphia, disappeared by 1914, the year Henry Fowler at the Academy of Natural Sciences in Philadelphia produced an inventory of fish collected in the polluted Schuylkill River.[6] Like tree mosses and light-colored peppered moths, shad returned when pollution abated. Construction of a fish ladder at the Fairmount Dam in Philadelphia and, later, release of shad fry upstream supported the recovery.

In the nineteenth and twentieth centuries the native, pollution-tolerant brown bullhead catfish inhabited the Schuylkill River, whose pollution probably protected the fish from pollution-sensitive enemies such as tapeworms. The brown bullhead disappeared from inventories of fish in the Schuylkill River below the Fairmount Dam in the twenty-first century after pollution declined and nonnative competitors and predators, such as flathead and channel catfishes, established abundant local populations.

In nineteenth-century Philadelphia pollution-tolerant freshwater sponges such as *Spongilla* (*Eunapius*) *fragilis* were abundant in the polluted Schuylkill River. They tolerate pollution better than do specialized predators such as the spongillafly, *Climacea areolaris* (order Neuroptera), a flying insect whose aquatic larvae feed exclusively on freshwater sponges. In Center City spongillaflies fly to light in our backyard, a few blocks from the river. Their presence implies that the river here continues to support freshwater sponges.

Thermal pollution

Temperatures typically run from 1.7 to 3.3°C (3 to 6°F) higher in Center City than in surrounding suburbs, and on clear, calm winter nights they often exceed those in nearby rural areas by 5.6 to 11°C (10 to 20°F).[7] Philadelphia's urban heat island is due to passive trapping of solar energy plus active production of heat, such as that released into the environment by gasoline engines, electric motors, computers, and furnaces. In the winter, man-made sources of heat are estimated to account for 2–3°C of Philadelphia's nighttime heat island.[8]

Philadelphia's heat island advances the onset of leafing out in the spring. This shift is superimposed on advances due to global warming. Satellite imaging of the Washington-Philadelphia-New York corridor in 2001 showed that leafing out occurred almost nine days earlier inside urban zones compared to 8–10 kilometers outside these zones.[9]

Thermal mapping graphically displays geographic differences in temperature within Center City's heat island. Shaded by trees, Rittenhouse Square stands out as a "cool island" during the day within the heat island.[10] Studies performed in Raleigh, North Carolina, demonstrated the ecological importance of temperature differences within an urban heat island. In Raleigh, higher temperatures drove the abundance of a scale insect (order Hemiptera: *Parthenolecanium quercifex*) that infests willow oak trees. The study controlled for confounding variables such as parasitism and habitat. Abundance of the scale insect in Raleigh was thirteen times greater in hot areas compared to cooler areas.[11]

In Center City, animals vulnerable to desiccation from heat limit their activities to the night; during the day they find shelter underground or beneath leaf litter or stones. Examples include pillbugs, slugs, millipedes, and the red back salamander. Pill-bugs and slugs avoid desiccation by bunching together. The silvergreen bryum moss (*Bryum argenteum*), a common moss in Center City, tolerates desiccation.[12]

In theory, Center City's heat island might attract birds such as American robins that overwinter here but also migrate south. Against this theory are William Bartram's detailed records of robins around his home in Philadelphia during frigid spells in January in the early nineteenth century, long before one would expect development of a significant urban heat island effect. In the winter of 2012, the number of silver-haired bats people discovered in Center City and transported to a local animal rehabilitation clinic totaled six. Theoretically Center City's heat island offered these bats thermal protection during winter hibernation.

Pumps serving power plants at Christian Street draw cooling water from the Schuylkill River, circulate it around the plants, and discharge heated water into the river. The average volume of water they draw from the river has exceeded the river's historic minimum rate of flow by more than fourfold. The river here is tidal, so that the Atlantic Ocean maintains water levels in the river close to sea level despite the withdrawals. The thermal plume of the power plants' heated discharge could point upstream toward Center City, depending on the plant's discharge rate, river discharge rate, and tidal flow.

In 1962 at the annual conference of the American Water Works Association in Philadelphia, Gerald E. Arnold, superintendent of the Philadelphia Water Department, presented a paper on industrial thermal pollution. It reported results of studies on the Allegheny, Monongahela, and Ohio Rivers in the vicinity of Pittsburgh. Among the findings: as the temperature of water rises, water holds less dissolved oxygen, and dissolved oxygen required by aquatic organisms increases. A reduction in dissolved oxygen reduces the upper temperatures that fish can tolerate. As temperatures rise, bacteria deplete dissolved oxygen. Some fish swimming into hot water are killed, while fish acclimated to warm water are rapidly killed when they swim into cold water. Elevated temperature disturbs activity, feeding, growth, and spawning of fish; and it acts as a force repelling or attracting them. Lethal high temperatures vary according to species, developmental stage, and rate of temperature change.[13] Arnold added a worrisome caveat:

> Water temperatures do not have to reach lethal levels in order to wipe out a species. Temperatures which favor competitors, predators, parasites, and diseases can destroy a species at levels far below those which are lethal.[14]

I have found no ecological reports on thermal discharges into the Schuylkill River from the power plants just below Center City.

One of these power plants cogenerates electricity and steam. Customers in Center City and University City, located across the river, purchase the steam for heating. The power plant's capacity is 170 MW of electricity plus 1.5 million lb/hr of steam, which it distributes through steam pipes.[15] In theory, heat in the form of steam piped to customers reduces heat in the form of hot water discharged into the Schuylkill River.

Light pollution

At the beginning of the twentieth century, powerful arc lamps mounted on top of City Hall Tower disoriented nocturnal migrating birds, which collided with the building. After storms, hundreds of dead and wounded birds accumulated below City Hall Tower, which has since stopped producing the bright light.

Later in the twentieth century and into the twenty-first, windows and glass facades of tall buildings in Philadelphia compounded the danger of collisions due to artificial light. Birds fail to recognize transparent or reflective glass as a barrier, and fatally collide with it. Annually, the number of birds killed in collisions with buildings in just two locations in Philadelphia has been estimated at over 2,000.

When municipal electric lighting was first introduced to big cities at the end of the nineteenth and the beginning of the twentieth century, it attracted entomologists because of the rich assortment of insects that flew to the lamps. Today, in comparison, urban lamps are among the worst places to look for insects. Paradoxically, light pollution downtown can reduce attraction of insects to light. By increasing background light, it decreases contrast between light sources and their backgrounds. Insects downtown still fly to lamps surrounded by relative darkness, as in our backyard garden, which is shielded from streetlights. From the perspective of moths, outdoor electric lighting in Center City may be viewed as a double-edged sword, disturbing or protecting them depending on circumstances.

Bridge spiders (*Larinioides sclopetarius*) have colonized lamps along the east bank of the Schuylkill River in Center City. In the nineteenth century this spider was observed capturing prey in its web primarily during the day, but now in Center City the spider captures prey at electric lamps at night. Lamp fixtures along the river offer the bridge spider shelter and aquatic insects, and they distance the spider from a predator, the black and yellow mud dauber.

The common eastern firefly (*Photinus pyralis*) appears in June and July in Center City. It flies at dusk despite exposure to streetlamps. Center City endows this firefly with a safe haven from its usual predator, femme fatale fireflies (*Photuris*). Femme fatales emit flashes that lure in male common eastern fireflies, which the femme fatales overpower and devour. The reason these predatory fireflies are absent from Center City is unknown; they are present in Philadelphia's suburbs. One hypothesis is that femme fatale fireflies are more vulnerable than their victims to light pollution.

Intense light pollution was found to advance the onset of early-morning singing by American robins (*Turdus migratorius*) in Pennsylvania; the shift averaged 107 minutes. In urban areas in Britain, the European robin (*Erithacus rubecula*) has shifted its song from day to night; but noise pollution, rather than light pollution, caused the shift. In Center City, noise and light pollution theoretically could act together to advance the time in the morning when American robins begin singing.

Sound (noise) pollution

Noise from interstate highways surrounds Center City on three sides. Within Center City's commercial core, vehicular noise peaks at rush hour. Intermittent sources of noise downtown are geographically scattered, and include car alarms, sirens, air conditioners, street cleaners, helicopters, freight trains, and amplified music. A city ordinance defines limits of lawful noise.[16]

Noise pollution may protect house finches (*Haemorhous mexicanus*), which are common at our backyard feeder. In an experimental study in New Mexico, house finches preferred to nest in a noisy habitat compared to a control habitat that was similar but quiet. Its enemies, jays and cowbirds (nest predators and parasites, respectively), avoided the noisy habitat.[17]

Urban noise can interfere with communication among songbirds. House finches studied in Mexico City increased the minimal acoustic frequency (pitch) of their songs according to the level of background urban noise.[18] This capability has been postulated to explain the house finch's success in colonizing cities.[19] American robins and cardinals (*Cardinalis cardinalis*) also modulate the pitch of their songs according background noise.[20]

Mourning doves (*Zenaida macroura*), like house finches, are frequent visitors to our backyard, where they forage for seed on the ground. In contrast to house finches, mourning doves in the experiments in New Mexico preferentially nested in the habitat that was less noisy.[21]

In Center City, common insects that sing include cicadas, crickets, and katydids. The response of these insects to man-made noise has not been studied, but in Germany, the maximum pitch of songs of grasshoppers was tested under standard background noise in the laboratory. The maximum pitch of grasshoppers from noisy roadside habitats was higher than that of grasshoppers of the same species from relatively quiet habitats.[22]

Mouse-eared bats, which find food by passively listening for rustling sounds of prey on vegetation, avoid foraging near highway noise.[23] Using an ultrasonic monitor, I observed bats emitting ultrasonic sounds as they foraged for insects flying in Schuylkill River Park, well within human earshot of the noisy Schuylkill Expressway across the river.

Persecution

In Center City green space is scarce and manicured. Wild plants colonize vulnerable sanctuaries like pavement cracks, strips along curbs, and untended borders of parking lots or playgrounds.

Wild plants successful at escaping persecution are those that colonize small spaces, mature quickly, keep a low profile, and are hard to pull up. Tall, broad-leaved perennials that occupy more than a few square centimeters of ground are attractive targets for removal; an example is common milkweed (*Asclepias syriaca*), recently extirpated from Center City.

Persecution downtown takes aim against habitat as well as particular plants. Outside Center City, a fern called purple-stemmed cliffbrake (*Pellaea atropurpurea*) thrives in cracks in old mortar that has weathered. Downtown, the fern and its habitat are both absent, casualties of property maintenance such as painting and repairing of masonry. An old granite masonry pier supporting the Chestnut Street Bridge is an exception. In an inaccessible refuge near the top of this stone structure is a colony of ebony spleenwort (*Asplenium platyneuron*), another fern with an affinity for old masonry. Ferns have populated old inaccessible masonry retaining walls around the Fairmount Water Works, but they have disappeared from nearby rocky cliffs scoured by maintenance crews with weed whackers.

Figure 28.3 Community of wild ferns on an old stone retaining wall near Fairmount Water Works. It contains ebony spleenwort (*Asplenium platyneuron*), purple-stemmed cliffbrake (*Pellaea atropurpurea*), and bladder fern (*Cystopteris*). So far, inaccessibility has protected this wall from weeding.

Persecution in Center City targets animals as well as plants. Property owners try to rid facades of unsightly nests of black and yellow mud dauber wasps. Positioned in inaccessible nooks high on exterior walls below eaves, these nests often defy removal. The nest of the organ pipe mud dauber, in contrast, is more vulnerable, because this mud dauber positions its nest lower, and the male hangs around loudly buzzing—although it cannot sting. I have yet to find the organ pipe mud dauber downtown, although I have found it under suburban railroad trestles. Mud daubers are solitary and less aggressive than paper wasps, which also build nests under eaves.

Prejudice

In 1808 Alexander Wilson puzzled over why people persecute the American robin less than other birds. He suggested that people like this bird because they associate its song with spring and its name with the European robin, a traditional English favorite. Popular affection for the American robin endures even though the bird strips ornamental berries off garden shrubbery; harbors deer ticks and the pathogen for Lyme disease; and is a reservoir for West Nile virus, an introduced cause of human encephalitis.

In Center City animus toward the Norway rat has been constant since earliest records, while historical attitudes toward the gray squirrel and house sparrow have shifted according to circumstances. The status of the Canada goose reversed as the bird changed from migratory icon to perennial nuisance. Love for the ailanthus tree turned to hate as people discovered its invasiveness. When first described in Philadelphia in 1819, the brown bullhead was esteemed as dinner fare, but dismissed as a game fish; in Center City it has disappeared, and so is no longer a game fish, and in Philadelphia's Darby Creek, where it can still be found, PCB contamination has spoiled its status as fine fare.

Downtown, rats, geese, sparrows, and squirrels have all survived bad reputations, which may, paradoxically, serve the animals better than do good reputations. For example, in the winter, well-meaning people pick egg cases of Chinese mantids and bring them indoors, where they hatch prematurely, before the hatchlings can survive outdoors. Chinese mantids, which have reputations as "beneficial," have become scarce downtown compared to Carolina mantids, whose egg cases are smaller and better camouflaged.

Populations

People

Center City's high density of buildings and streets fosters an illusion that the city is insulated from disruptive forces of nature except for geophysical disturbances such as hurricanes, floods, and heat. In 1793 an epidemic of yellow fever swept through Philadelphia and killed a tenth of the population. Nearly half of the people in the city fled. Yellow fever epidemics in Philadelphia recurred six times until the final one in 1805.[24]

The yellow fever virus and its mosquito vector (*Aedes aegypti*) have not re-established themselves in Philadelphia since the last epidemic, but a close relative, the Asian tiger mosquito (*Aedes albopictus*), was discovered here in 2000 for the first time.[25] In other parts of the world, this mosquito has been a vector for viruses causing dengue and Chikungunya fever,[26] while in the Philadelphia region it carries West Nile virus. [27]

Public surveillance programs in Philadelphia are finding West Nile virus most often in two other mosquitoes, *Culex pipiens* and *Culex restuans*.[28] From 1999 to 2012, West Nile virus caused over 37,000 cases of encephalitis nationwide[29] and over a thousand deaths.[30] A state-sponsored program to control this disease in Philadelphia uses trucks to spray insecticide against populations of mosquitoes found to be infected with the virus.[31]

Figure 28.4 Asian tiger mosquito (*Aedes albopictus*) on the stucco rear wall of our row house. A recently introduced vector for viral pathogens, it is spreading globally, and elsewhere has caused outbreaks of dengue and Chikungunya fever.

Bagworms

Diverse forces have shaped populations of plants and animals downtown over the past 200 years. In public squares in mid-nineteenth-century Philadelphia, plagues of insects attacked municipal shade trees. One of the pests was the bagworm. Its populations are controlled primarily by parasitic wasps, which take time to establish numbers sufficient to suppress infestations. In disturbed habitats, this grace period offers bagworms windows of opportunity, which bagworms likely exploited in the nineteenth century as Philadelphia expanded its plantings of trees in public spaces. At that time, the absence of flowerbeds and flowering shrubs in public squares may have contributed to population explosions of bagworms: parasitic wasps whose larvae eat bagworms feed on nectar and pollen of flowers. Use of insecticides against bagworms may have disrupted parasitic control and paradoxically prolonged infestations.

House sparrows

Populations of some introduced species exploded initially, only to taper off with the passage of time and loss of favorable conditions. In 1869 the City of Philadelphia released 1,000 house sparrows imported from England to prey on insects infesting municipal shade trees. Numbers of house sparrows at first multiplied exponentially, but lately their numbers have been declining. Hypotheses to explain the decline have cited many factors: exotic ornamental plants resistant to insects have reduced the house sparrow's supply of insect prey; destruction of weeds has reduced availability of both seeds and insects; parasites of insect pests reduced the house sparrow's prey; predation by raptors and cats increased house sparrow mortality; introduction of house finches has increased competition; the design of new buildings denies house sparrows suitable nesting sites. Since 1966 numbers of house sparrows have declined in Pennsylvania by 62 percent, and nationally by 85 percent.

Yellowjackets

Changes in abundance of some species downtown are shrouded in mystery. Populations of yellowjackets in Center City exploded in the late 1990s. During this period a series of papers in medical and veterinary journals documented yellowjacket stings in Philadelphia, but the outbreak was neither formally monitored nor evaluated, and the causes for its onset and remission remain unknown. Its geographic distribution was never mapped. During the outbreak, which peaked annually in late August and September, yellowjackets would swarm around food and drink of people dining outdoors; meat, fruit, and sweet beverages would attract yellowjackets within minutes. In recent years, yellowjackets downtown have been rare or absent. Spikes in abundance of yellowjackets have recurred for reasons unknown in England and the Pacific Northwest..

Geese

Canada geese outside of captivity never bred in Pennsylvania until the twentieth century. In 1814 Canada geese were selling in Philadelphia markets for seventy-five cents to a dollar per bird. Hunters kept captive flocks of decoy geese to lure in wild Canada geese as they migrated overhead. Over generations, these captive geese lost their migratory behavior, which is learned rather than innate. With their migratory habits broken, captive Canada geese nested and reproduced locally, even when freed. After use of decoy flocks for hunting geese was outlawed, hunters released their semidomesticated Canada geese or turned them over to state game managers, who in turn transplanted them throughout Pennsylvania, unintentionally unleashing the current proliferation of urban Canada geese. The Canada geese game managers distributed in Pennsylvania were giant Canada geese of Midwestern stock.

Stink bugs

In Center City accidental introduction of pests continues to occur. In 1996 the brown marmorated stink bug showed up for the first time in North America in Allentown, Pennsylvania, 79 kilometers outside of Philadelphia. This bug had been known as an agricultural pest in Asia. By 2010 the stink bug had reached our house in Center City. In Pennsylvania it has proved damaging to crops and difficult to control with insecticides and parasites. A sand wasp discovered in Pennsylvania in 1823 is a predator of stink bug nymphs, including those of the brown marmorated stink bug. In Logan Circle it drags prey down holes to its underground nests. This wasp has so far not controlled the outbreak of stink bugs.

Evolution

Industrial melanism demonstrated the power of big cities like Philadelphia to generate selective forces strong enough to drive evolution, as in the case of the melanic (black) form of the peppered moth. Cities also drive evolution by importing foreign strains; hosting genetic mixing; and subjecting hybrids to new selective pressure.

Mugwort naturalized in New England and Canada after Jesuits introduced it into North America from Europe in the sixteenth century. Around Philadelphia, populations of mugwort stayed quiescent or disappeared until the latter half of the nineteenth century, when sailing ships deposited it in ballast dumps in the city. These dumps allowed strains of mugwort imported from around the world to hybridize. The plant evolved competitive strains whose populations exploded out of port cities. Today this perennial populates pavement cracks and edges of gardens and parking lots throughout Center City. It forms dense monocultures that exclude other plants.

Habitats

Row house courtyards

A habitat characteristic of Center City and surrounding neighborhoods is the courtyard behind a row house. Courtyards have graced the rear of Center City's row houses since they were built in the nineteenth century. Even when not cultivated, rear courtyards function as refuges for wild plants and animals, especially invertebrates, in a downtown otherwise saturated with buildings and pavement.

The longevity of Center City's courtyards allowed time for introduction and establishment of populations of earthworms, which may in part explain the abundance here of fireflies, whose larvae prey on earthworms. Longevity also contributed to the diversity of isopods such as pillbugs, which support pillbug hunters, spiders that specialize in consuming them. The abundance of soil invertebrates attracts American robins, which in turn harbor parasitic worms that infect pillbugs, which transmit the worms back to robins.

The community of animals on the ground in our garden decomposes leaf litter and aerates the soil. Although most of them are introduced, some are native, such as fireflies, or support species that are native, such as American robins. In a pristine habitat, exotic animals such as those in Center City courtyards might be condemned as invasive or destructive, but here they inhabit territory long ago stripped of endemic flora and fauna.

Sidewalk cracks

In Center City sidewalk cracks are the only places where I have found some plants, like lawn pennywort (*Hydrocotyle sibthorpioides*), which leafs out and produces flowers and seed entirely within cracks between the pavers of brick walkways. Sidewalk cracks are the predominant habitat for Japanese mazus (*Mazus pumilus*), whose flowers barely poke above the pavement surface in zones that are heavily trampled. Cracks at the juncture of sidewalks and buildings are the most common location for Pennsylvania pellitory (*Parietaria pensylvanica*), a tall, upright native annual. On the paved island in the middle of Broad Street, fine cracks between snugly fitted pavers accommodate plants with stems that are thin, as exemplified by lovegrasses (*Eragrostis pectinacea* and *E. spectabilis*). Between the pavers of brick sidewalks on the south side of east-west streets, mosses flourish—beneficiaries of shade cast by row houses.

Plants in sidewalk cracks in Center City contribute not only to botanical diversity, but also to the city's hydrology. Center City's sewage system, a legacy of infrastructure dating back 200 years, drains storm water and sewage through a single pipe rather than through two separate pipes, the standard in modern municipal drainage systems. Ordinarily Center City's sewers direct their flow to treatment plants. During heavy storms, however, surface runoff overwhelms the system, and the overflow containing raw sewage mixed with storm water spills into the Schuylkill River. By impeding storm runoff and directing it into cracks, plants growing in pavement cracks retard the flow of storm water into the sewers. They also soak up storm water and divert it into the ground. Wide, deep cracks with thick vegetation offer the most effective barriers.

A common inhabitant of sidewalk cracks is purslane (*Portulaca oleracea*). Leroy G. Holm and his colleagues highlighted it in *The World's Worst Weeds*. They rank weeds according to type and number of published reports in agricultural literature. Among the world's seventy-six worst weeds, purslane ranked number nine. It infests the earth's major crops: corn, wheat, rice, potatoes, sugarcane, linseed, safflower, sugar beets, sorghum, bananas, citrus, millet, peanuts, vineyards, cotton, vegetables, and coffee. It poisons and kills livestock and is an alternative host of harmful agricultural viruses, nematodes, and insects.[32] In Center City, it is an attractive, maintenance-free, compact flowering succulent that tolerates harsh conditions on pavement.

Figure 28.5 Purslane (*Portulaca oleracea*) in bloom at curb on Spruce Street. Here it impedes storm runoff that pollutes the Schuylkill River, although wider cracks with thicker vegetation offer more protection. Agronomists have ranked purslane number nine among the world's worst weeds.

Five of the world's ten worst weeds as cited by Holm et al. inhabit Center City's sidewalk cracks. They are Bermuda grass (*Cynodon dactylon*), barnyard grass (*Echinochloa crus-galli*), goose grass (*Eleusine indica*), lambsquarters (*Chenopodium album*), and purslane. House sparrows, which are declining in Pennsylvania, feed on the seeds of common wild plants such as these. Flora in sidewalk cracks may degrade pavement and impede foot traffic even as it supports wildlife and enriches urban vegetation.

Tidal alluvial mudflat

Specialized habitats downtown include an artificially created tidal alluvial mudflat in an 8-meter gap in the bulkheading along the Schuylkill River just below the Walnut Street Bridge. Here wild purple loosestrife (*Lythrum salicaria*) quickly overran a garden of carefully selected plants native to Pennsylvania wetlands. The Pennsylvania Department of Agriculture has posted this notice online:[33]

Purple Loosestrife Alert

An Attractive but Deadly Threat to Pennsylvania's Wetlands and Waterways

Purple loosestrife is an aggressive plant that is invading our wetlands, replacing valuable wetland plants; eliminating food and shelter for wildlife; and choking waterways.

Figure 28.6 Purple loosestrife (*Lythrum salicaria*) an exotic species growing wild in tidal alluvial mud below the Walnut Street Bridge along the east bank of the Schuylkill River. It attracts native bees, as shown in Figure 29.1. It has replaced native wetland flora that had been planted in a garden here several years before.

In this site along the Schuylkill River, purple loosestrife has colonized former wharves and industrial property, now a recreational park. It attracts honeybees and native pollinators such as bumblebees, to the admiration of visitors in the park. It deters children from wandering into the river. It is care free. Occupying but a small patch of ground, it belongs to a lush community of wild plants in fierce competition. Such a botanical spectacle flourishing unfettered in Center City is unique.

Just outside Center City, north of Chinatown and the Vine Street Expressway, are more extensive communities of wild plants. They inhabit the Reading Viaduct, an abandoned elevated commuter railway operational from 1893 to 1984. When trains ran here, the railroad used herbicide and maintenance crews to suppress vegetation, but now trees, shrubs, and herbaceous annuals and perennials thrive along the tracks high above city streets. Common milkweed (*Asclepias syriaca*), recently extirpated from Center City, thrives here. Conversion of the Reading Viaduct into a public park would inevitably transform these botanical communities.

Figure 28.7 Reading Viaduct, December 11, 2011. Elevated above city streets, it carried trains into Reading Terminal Station from to 1893 to 1984. In 1894, 290 trains on thirteen sets of tracks were scheduled to run daily from this station onto tracks such as those that once ran here.[34]

Center City as beneficiary of environmental protection

In Center City, the tree moss *Orthotrichum pumilum* returned to Philadelphia after disappearing over a century earlier. It embodies the triumph of campaigns for clean air. Its counterparts among animals include the American shad, which returned with restoration of clean water, and the red-tailed hawk, once a victim of DDT and persecution. Center City is no Wilderness Area, but environmental activism has nevertheless rewarded it.

Downtown the fate of populations of plants and animals attracts interest either when it defies human control and comprehension, or when it appears, directly or indirectly, to affect people. The attraction may also come from what E. O. Wilson calls *biophilia*—the human bond with other species.[35] In Center City, accounts of the journeys of populations of plants and animals out of the past—documented in publications and museums and linked to familiar streets and buildings—illuminate the present and introduce the future.

SPOTLIGHT

BOXELDER

Seedlings of boxelder (*Acer negundo*), also known as ash-leaved maple, colonizing stone stairs rising from Bonsall to Walnut Streets in Center City.

In Philadelphia two hundred years ago boxelder was described as a "very large tree," occurring "on the Schuylkill, near the falls, east side, and elsewhere."[1] In 1831 it was offered for sale in Bartram's Garden in Philadelphia for 25 cents each.[2] Currently it grows wild in Center City where it typically occurs as a shrub or small tree, except along the Schuylkill River bank where it still attains a large size.

29
CONCLUSION

The ecology of Center City is dynamic and resilient.

Figure 29.1 A digger bee (*Melissodes bimaculata*) visiting purple loosestrife (*Lythrum salicaria*), the same plant shown in figure 28.6. The bee is native to North America; the plant is native to Europe and Asia.

At first glance, downtown Philadelphia seems to embody what Bill McKibben projected in his book *The End of Nature*: complete subordination of Earth to people.[1] Here every parcel of land is zoned, and each zone is identified with a code specifying permissible development and use.[2] Michael L. McKinney, at the University of Tennessee, pointed out that plants and animals in central core districts of big cities tend to be similar—mostly cosmopolitan species, products of ecological homogenization. He blamed this uniformity on the dedication of downtowns everywhere to serving just one species: human beings.[3]

People do dominate the landscape of Center City, and exotic organisms are common here, especially in soil; but the ecology of Center City is largely hidden. While buildings and pavement cover the landscape, wild plants and animals, including native species, thrive—albeit in small fragments such as in pavement cracks and courtyards, or underwater or in darkness; or as tiny, taxonomically obscure organisms. While Center City's mass of concrete and asphalt epitomizes habitat destruction, it also exemplifies habitat creation, represented by dry vertical walls—ideal nesting sites for black and yellow mud daubers, for example.

Figure 29.2 Lawn pennywort (*Hydrocotyle sibthorpioides*) in sidewalk cracks in front of our row house in Center City. The plant leafs out, blooms, makes seed, and spreads entirely within these cracks, which protect it from trampling. A creeping perennial introduced as an ornamental from Asia, it was first reported naturalized in Philadelphia in 1909.[4]

Ecological disturbance in Center City has benefited some urban populations at the expense of others. Water pollution benefited brown bullheads at the expense of its enemies, who were intolerant of it. Municipal streetlamps nurtured bridge spiders at the expense of insects attracted to artificial light. Some populations reaped benefits at no cost to others. Disruption of migratory behavior in Canada geese opened urban and suburban territory, including Center City, as breeding grounds. *Creative destruction*, a process Joseph A. Schumpeter described in *Capitalism, Socialism and Democracy*,[5] shaped Center City's ecology much as described for ecosystems generally.[6]

Center City is ecologically dynamic. Populations of the ailanthus silkmoth exploded in the nineteenth century, only to go locally extinct in the twentieth. Yellowjackets that just over a decade ago swarmed around outdoor food and drink have mysteriously vanished as table pests. Red-tailed hawks once absent here have proliferated, while cries of nighthawks, once heralding summer nights, have gone silent. Numbers of house sparrows and starlings in Pennsylvania are declining after a century of superabundance. In the tidal Schuylkill River, populations of channel catfish and flathead catfish have surged, while brown bullhead catfish have disappeared; American shad have returned, while northern snakeheads have just moved in. Mugwort, absent in early nineteenth-century Philadelphia, is now one of its most common wild herbaceous plants. Japanese mazus, a denizen of sidewalk cracks in old residential sections of Center City, is also relatively new. Ailanthus, imported here at the end of the eighteenth century and naturalized in the nineteenth, faces an uncertain future in the twenty-first: a fungal epidemic is destroying stands of ailanthus in a Pennsylvania state forest 210 kilometers to the west.

While communities of plants and animals in Center City have been transformed, so has its human population, increasing in number and wealth. New construction is increasing the height and density of buildings. These physical and demographic changes present Center City's wild inhabitants with new demands and opportunities, maintaining pressures for ecological change.

Political and cultural shifts have driven ecological change in Center City. Action to protect the environment has brought back to Center City plants and animals once locally extirpated. Timothy Beatley's *Biophilic Cities* details programs that cities, including Philadelphia, have implemented to promote environmental health.[7]

Figure 29.3 Green roof on top of PECO (Philadelphia Electric Company) building in Center City.

Commerce and transportation will continue to introduce exotic strains and species into Center City, enabling genetic mixing among strains once geographically isolated. Center City will continue to apply selective pressure and catalyze evolution. Center City does not epitomize the end of nature; on the contrary, it exemplifies nature's resilience.

SPOTLIGHT

JAPANESE MAZUS

Japanese mazus (*Mazus pumilus*) filling in most of the cracks on sidewalk, Pine Street.

Japanese mazus was first reported naturalized in Philadelphia in 1935.[1] In residential neighborhoods, it colonizes sidewalk cracks, where it keeps its top close to the surface of the bricks. If not trampled, it grows upright with its top several centimeters above the bricks (as shown in Figure 23.8).

Figure 29.4: Paved islands in the middle of South Broad Street, terminating at City Hall. I have found 26 species of plants growing in these islands.

ADDITIONAL INFORMATION

Urban Ecology

Alberti M (2009) *Advances in Urban Ecology: Integrating Humans and Ecological Processes in Urban Ecosystems* (Springer Science + Business Media, New York).

Forman, RTT (2014) *Urban Ecology: Science of Cities* (Cambridge University Press, Cambridge, UK).

Gilbert OL (1989) *The Ecology of Urban Habitats* (Chapman and Hall, London).

Marzluff JM, Shulenberger E, Endlicher W, Alberti M, et al., eds. (2008) *Urban Ecology: An International Perspective on the Interaction between Humans and Nature* (Springer Science + Business Media, New York).

Müller N, Werner P, & Kelcey JG, eds. (2010) *Urban Biodiversity and Design* (Wiley-Blackwell, West Sussex, UK).

Niemelä J, Breuste JH, Guntenspergen G, McIntyre NE, et al., eds. (2011) *Urban Ecology: Patterns, Processes and Applications* (Oxford University Press, Oxford).

Urban Natural History

Feinstein J (2011) *Field Guide to Urban Wildlife* (Stackpole Books, Mechanicsburg, PA).

Kieran J (1959) *A Natural History of New York City* (Houghton Mifflin, Boston).

McCully B (2007) *City at the Water's Edge: A Natural History of New York* (Rutger's University Press, New Brunswick, NJ).

Vessel MF & Wong HH (1987) *Natural History of Vacant Lots.* California Natural History Guides No. 50 (University of California Press, Berkeley).

Guides to Plants

Brown L (1979) *Grasses: An Identification Guide.* Peterson Nature Library (Houghton Mifflin, Boston).

Cobb B, Farnsworthy E, & Lowe C (2005) *Ferns of Northeastern and Central North America.* Peterson Field Guide (Houghton Mifflin, New York).

Crum HA & Anderson LE (1981) *Mosses of Eastern North America* (Columbia University Press, New York).

Del Tredici P (2010) *Wild Urban Plants of the Northeast: A Field Guide* (Comstock Publishing/Cornell University Press, Ithaca, NY).

Gleason HA & Cronquist A (1991) *Manual of Vascular Plants of Northeastern United States and Adjacent Canada,* 2nd ed. (New York Botanical Garden Press, New York).

Hinds JW & Hinds PL (2007) *The Macrolichens of New England* (New York Botanical Garden Press, New York).

Holmgren NH (1998) *Illustrated Companion to Gleason and Cronquist's Manual: Illustrations of the Vascular Plants of Northeastern United States and Adjacent Canada* (New York Botanical Garden Press, New York).

Lord TR & Travis HJ (2006) *The Ferns and Fern Allies of Pennsylvania* (Pinelands Press, Pemberton, NJ).

Munch S (2006) *Outstanding Mosses & Liverworts of Pennsylvania and Nearby States* (Sunbury Press, Mechanicsburg, PA).

Newcomb L (1977) *Newcomb's Wildflower Guide* (Little, Brown & Co, New York).

Plotnik A (2000) *The Urban Tree Book* (Three Rivers Press, New York).

Rhoads AF & Block TA (2005) *Trees of Pennsylvania: A Complete Reference Guide* (University of Pennsylvania Press, Philadelphia).
Rhoads AF & Block TA (2007) *The Plants of Pennsylvania: An Illustrated Manual,* 2nd ed. (University of Pennsylvania Press, Philadelphia).
Russell B (2009) *Field Guide to Mushrooms of Pennsylvania and the Mid-Atlantic* (Pennsylvania State University Press, University Park).
Uva RH, Neal JC, & DiTomaso JM (1997) *Weeds of the Northeast* (Comstock Publishing/ Cornell University Press, Ithaca, NY).

Guides to Invertebrates

Cech R & Tudor G (2005) *Butterflies of the East Coast: An Observer's Guide* (Princeton University Press, Princeton, NJ).
Dindal DL, ed. (1990) *Soil Biology Guide* (John Wiley & Sons, New York).
Grimm FW, Forsyth RG, Schueler FW, & Karstad A (2009) *Identifying Land Snails and Slugs in Canada: Introduced Species and Native Genera* (Canadian Food Inspection Agency, Ottawa, ON).
Hopkin S (1991) A key to the woodlice of Britain and Ireland. *Field Studies* 7(4):599–650, reprinted by the Field Studies Council, Unit C591, Stafford Park 515, Telford, TF593 593BB, UK. email: publications@field-studies-council.org.
Howell WM & Jenkins RL (2004) *Spiders of the Eastern United States* (Pearson Education, Boston).
Marshall SA (2006) *Insects: Their Natural History and Diversity: With a Photographic Guide to the Insects of Eastern North America* (Firefly Books, Buffalo, NY).
Nardi JB (2007) *Life in the Soil: A Guide for Naturalists and Gardeners* (University of Chicago Press, Chicago).
Oliver PG & Meechan CJ (1993) *Woodlice: Keys and Notes for the Identification of Species.* Synopses of the British Fauna (New Series) No. 49 (Field Studies Council, Shrewsbury, UK).
Shapiro AM (1966) *Butterflies of the Delaware Valley.* Special Publication of the American Entomological Society (American Entomological Society, Philadelphia).
Thorp JH & Rogers DC (2011) *Field Guide to Freshwater Invertebrates of North America* (Academic Press, Burlington, MA).
Tietz HM (1952) *The Lepidoptera of Pennsylvania: A Manual* (School of Agriculture, Agricultural Experiment Station, Pennsylvania State College, State College).

Guides to Vertebrates

Cooper EL (1983) *Fishes of Pennsylvania and the Northeastern United States* (Pennsylvania State University Press, University Park).
Fergus C (2000) *Wildlife of Pennsylvania and the Northeast* (Stackpole Books, Mechanicsburg, PA).
Hulse AC, McCoy CJ, & Censky EJ (2001) *Amphibians and Reptiles of Pennsylvania and the Northeast* (Cornell University Press, Ithaca, NY).
McWilliams GM & Brauning DW (2000) *The Birds of Pennsylvania* (Comstock Publishing/ Cornell University Press, Ithaca, NY).

Brun C (2009) Biodiversity changes in highly anthropogenic environments (cultivated and ruderal) since the Neolithic in eastern France. *The Holocene* 19(6):861–871.

De Schweinitz LD (1828) Remarks on the plants of Europe which have become naturalized in a more or less degree, in the United States. *Annals of the Lyceum of Natural History of New York* 3(1):148–155.

Larson D, Matthes U, Kelly PE, Lundholm J, et al. (2004) *The Urban Cliff Revolution: New Findings on the Origins and Evolution of Human Habitats* (Fitzhenry and Whiteside, Markham, ON).

La Sorte FA & Pyšek P (2009) Extra-regional residence time as a correlate of plant invasiveness: European archaeophytes in North America. *Ecology* 90:2589–2597.

Mack RN (2003) Plant naturalizations and invasions in the eastern United States: 1634–1860. *Annals of the Missouri Botanical Garden* 90(1):77–90.

Müller N (2010) Most frequently occurring vascular plants and the role of non-native species in urban areas—A comparison of selected cities in the Old and the New Worlds. In *Urban Biodiversity and Design*, eds. PW Müller & JG Kelcey (Wiley-Blackwell, Oxford), 227–242.

Wittig R (2004) The origin and development of the urban flora of Central Europe. *Urban Ecosystems* 7(4):323–329.

Community Resources for the Study of Urban Ecology in Philadelphia

Institutions

Academy of Natural Sciences of Drexel University, 1900 Benjamin Franklin Parkway, Philadelphia, PA 19103

Bartram's Garden, 54th Street and Lindbergh Boulevard, Philadelphia, PA 19143

Cusano Environmental Education Center at the John Heinz National Wildlife Refuge and Nature Center, 8601 Lindbergh Boulevard, Philadelphia, PA 19153

Fairmount Water Works Interpretive Center, 640 Waterworks Drive, Philadelphia, PA 19130

McLean Library of the Pennsylvania Horticultural Society, 100 N 20th Street # 500, Philadelphia, PA 19103

Schuylkill Center for Environmental Education, 8480 Hagys Mill Road, Philadelphia, PA 19128

Wagner Free Institute of Science of Philadelphia, 1700 W Montgomery Avenue, Philadelphia, PA 19121

Organizations

American Entomological Society, http://darwin.ansp.org/hosted/aes/

Delaware Valley Ornithological Club, http://www.dvoc.org/Main.htm

Pennsylvania Horticultural Society, http://phsonline.org/

Philadelphia Botanical Club, http://darwin.ansp.org/hosted/botany_club/

CHAPTER ENDNOTES

Preface

1 Peck RM & Stroud PT (2012) *A Glorious Enterprise: The Academy of Natural Sciences of Philadelphia and the Making of American Science* (University of Pennsylvania Press, Philadelphia).
2 Meyers ARW ed. (2011) *Knowing Nature: Art and Science in Philadelphia, 1740–1840* (Yale University Press, New Haven, CT).
3 http://www.ansp.org/research/systematics-evolution/collections.
4 http://phillyhistory.org/PhotoArchive/.

Introduction

1 Center City District & Central Philadelphia Development Corporation (2013) State of Center City 2013 (Philadelphia, PA), http://www.centercityphila.org/docs/SOCC2013.pdf.
2 United States Census Bureau (2012) Top 20 Cities, 1790–2010, http://www.census.gov/dataviz/visualizations/007/508.php.
3 United States Census Bureau (2013) State & County QuickFacts. Philadelphia County, Pennsylvania, http://quickfacts.census.gov/qfd/states/42/42101.html; United States Census Bureau (2013) State & County QuickFacts. Boston (city), Massachusetts, http://quickfacts.census.gov/qfd/states/25/2507000.html; United States Census Bureau (2013) State & County QuickFacts. New York (city,) New York, http://quickfacts.census.gov/qfd/states/36/3651000.html; United States Census Bureau (2013) State & County QuickFacts. Baltimore City, Maryland, http://quickfacts.census.gov/qfd/states/24/24510.html; United States Census Bureau (2013) State and County QuickFacts. District of Columbia, http://quickfacts.census.gov/qfd/states/11000.html.
4 Forman RTT (2008) *Urban Regions: Ecology and Planning beyond the City* (Cambridge University Press, Cambridge, UK).
5 Clemants SE & Moore G (2003) Patterns of species richness in eight northeastern United States cities. *Urban Habitats* 1(1):4–16, http://www.urbanhabitats.org/v01n01/speciesdiversity_pdf.pdf.

Front Matter

Illustration on page iv: Philadelphia fleabane
1 Linnæus C (1753) *Species Plantarum. Exhibentes Plantas Rite Cognitas, ad Genera Relatas, cum Differentiis Specificis, Nominibus Trivialibus, Synonymis Selectis, Locis Natalibus, Secundum Systema Sexuale Digestas* (Laurentii Salvii, Stockholm, Sweden), 863.

Chapter 1: The Subway Tree

[1] Richards HG (1931) The subway tree—A record of a Pleistocene cypress swamp in Philadelphia. *Bartonia* 13:1–6.
[2] Ibid.
[3] Ibid.
[4] Richards HG (1960) The date of the "subway tree" of Philadelphia. *Proceedings of the Pennsylvania Academy of Science* 34:107–108.
[5] Sevon WD, Fleeger GM, & Shepps VC (1999) *Pennsylvania and the Ice Age,* 2nd ed. Pennsylvania Geological Survey, 4th series, Educational Series 6. (Commonwealth of Pennsylvania, Department of Conservation and Natural Resources, Office of Conservation and Engineering Services, Bureau of Topographic and Geologic Survey, Harrisburg, PA).
[6] Richards, The date of the "subway tree" of Philadelphia.
[7] Lewis HC (1880) The surface geology of Philadelphia and vicinity. *Proceedings of the Academy of Natural Sciences of Philadelphia* 32:65.
[8] Lewis HC (1883) Geology of Philadelphia. *Journal of the Franklin Institute* (May), 359.
[9] Sevon WD & Braun DD (1997) *Glacial Deposits of Pennsylvania,* 2nd ed. (map). (Commonwealth of Pennsylvania, Department of Conservation and Natural Resources, Bureau of Topographic and Geologic Survey), http://www.dcnr.state.pa.us/cs/groups/public/documents/document/dcnr_016200.pdf.
[10] Lewis, Geology of Philadelphia, 366.
[11] Hopkins TC (1898–99) Clays and clay industries of Pennsylvania: II. Clays of southeastern Pennsylvania (in part), Appendix to the *Annual Report of Pennsylvania State College, for 1898–1899,* Official Document No. 22 (Pennsylvania State College, Centre County), 1–76.
[12] Gillingham HE (1929) Some early brickmakers of Philadelphia. *The Pennsylvania Magazine of History and Biography* 53(1):1–27.
[13] Hopkins, Clays and clay industries of Pennsylvania.
[14] Gillingham, Some early brickmakers of Philadelphia.
[15] Weiss JL, Overpeck JT, & Strauss B (2011) Implications of recent sea level rise science for low-elevation areas in coastal cities of the conterminous USA: A letter. *Climatic Change* 105(3–4):635–645.
[16] Lambeck K, Esa TM, & Potter E-K (2002) Links between climate and sea levels for the past three million years. *Nature* 419(12):199–206.
[17] Schaeffer M, Hare W, Rahmstorf S, & Vermeer M (2012) Long-term sea-level rise implied by 1.5° C and 2° C warming levels. *Nature Climate Change* 2(12):867–870.
[18] Rhoads AF & Block TA (2005) *Trees of Pennsylvania: A Complete Reference Guide* (University of Pennsylvania Press, Philadelphia).

Chapter 2: Eastern Gray Squirrel

[1] Kalm P (1771) *Travels into North America: Containing Its Natural History, and a Circumstantial Account of Its Plantations and Agriculture in General, with the Civil, Ecclesiastical and Commercial State of the Country, the Manners of the Inhabitants, and Several Curious and Important Remarks on Various Subjects,* vol 2. trans. JR Forster (T. Lowndes, London), 243–245.
[2] Ibid., 250.
[3] Ibid., 246–247.
[4] Ibid., 245–246.
[5] Benson E (2013) The urbanization of the eastern gray squirrel in the United States. *Journal of American History* 100(3):691–710.
[6] Wilson A & Bonaparte CL (1831) Falco borealis, Wilson. Red-tailed hawk. In *American Ornithology: Or the Natural History of the Birds of the United States*, ed. R Jameson, vol. 1 (Constable & Co and Hurst, Chance & Co., Edinburgh and London), 82.
[7] Warren BH (1888) *Report on the Birds of Pennsylvania: With Special Reference to the Food Habits, Based on Over Three Thousand Stomach Examinations* (E. K. Meyers, State Printer, Harrisburg, PA).
[8] Pennsylvania Department of Agriculture (1897) *Second Annual Report of the Pennsylvania Department of Agriculture*, 820.
[9] Warren, *Report on the Birds of Pennsylvania,* 84.
[10] Gillespie JA (1944) Birds of Rittenhouse Square. *Cassinia* 33:24–26.
[11] Richards KC (1975) Some declining bird species of southeastern Pennsylvania. *Cassinia* 55:33–36.
[12] Carson R (1962) *Silent Spring* (Fawcett, New York).
[13] Environmental Protection Agency (1972) Press release. DDT ban takes effect, http://www.epa.gov/history/topics/ddt/01.html.
[14] Kosak J (1995) *The Pennsylvania Game Commission 1895–1995* (Pennsylvania Game Commission, Harrisburg, PA).

[15] Kerlinger P & Brett J (1995) Hawk Mountain Sanctuary: A case study of birder visitation and birding economics. In *Wildlife and Recreationists: Coexistence through Management and Research*, eds. RL Knight & KJ Gutzwiller (Island Press, New York), 271–280.

[16] Bildstein K (2001) Raptors as vermin: A history of human attitudes towards Pennsylvania's birds of prey. *Endangered Species Update* 18(July–August):124–128.

[17] Sauer JR, Hines JE, Fallon JE, Pardieck KL, et al. (2011) *The North American Breeding Bird Survey Results and Analysis 1966–2009. Version 3.23.2011* (USGS Patuxent Wildlife Research Center, Laurel, MD), http://www.mbr-pwrc.usgs.gov/bbs/.

[18] Saffron I (2011) City's new pastime: Talon shows. *Philadelphia Inquirer*, February 20, 2011, http://www.philly.com/philly/news/20110220_City_s_new_pastime__Talon_shows.html?viewAll=y&c=y.

[19] Winn M (1998) *Red-tails in Love: A Wildlife Drama in Central Park* (Pantheon Books, New York).

[20] The Franklin Institute (2012) The Franklin Institute Hawk Nest, http://www.fi.edu/hawks/.

[21] Micah D (2009–2012) Hawkwatch at the Franklin Institute (blog), http://sunnydixie.blogspot.com/.

[22] Luttich S, Rusch DH, Meslow EC, & Keith LB (1970) Ecology of red-tailed hawk predation in Alberta. *Ecology* 51(2):190–203.

[23] Zanette LY, White AF, Allen MC, & Clinchy M (2011) Perceived predation risk reduces the number of offspring songbirds produced per year. *Science* 334(6061):1398–1401.

[24] Mosby HS (1969) The influence of hunting on the population dynamics of a woodlot gray squirrel population. *The Journal of Wildlife Management* 33(1):59–73.

[25] Lawton C & Rochford J (2007) The recovery of grey squirrel (*Sciurus carolinensis*) populations after intensive control programmes. *Biology & Environment: Proceedings of the Royal Irish Academy* 107(1):19–29.

[26] Barkalow FS Jr. & Soots RF Jr. (1975) Life span and reproductive longevity of the gray squirrel, *Sciurus c. carolinensis* Gmelin. *Journal of Mammalogy* 56(2):522–524.

[27] Thompson DC (1978) Regulation of a Northern grey squirrel (*Sciurus carolinensis*) population. *Ecology* 59(4):708–715.

[28] Barkalow & Soots, Life span and reproductive longevity of the gray squirrel.

[29] Fergus C (2000) *Wildlife of Pennsylvania and the Northeast* (Stackpole Books, Mechanicsburg, PA).

[30] Tester J (1987) Changes in daily activity rhythms of some free-ranging animals in Minnesota. *The Canadian Field-Naturalist* 101(1):13–21.

[31] Godman JD (1831) *American Natural History*, vol. II, pt. I, *Mastology*. 2nd ed. (Stoddard and Atherton, Philadelphia), 79–82.

[32] Grant-Hoffman MN & Barboza PS (2010) Herbivory in invasive rats: Criteria for food selection. *Biological Invasions* 4(12):805–825.

[33] Koprowski JL (1991) Response of fox squirrels and gray squirrels to a late spring–early summer food shortage. *Journal of Mammalogy* 72(2):367–372.

[34] Stone W, Okoniewski J, & Stedelin J (1999) Poisoning of wildlife with anticoagulant rodenticides in New York. *Journal of Wildlife Diseases* 35:187–193.

[35] Cook HB (July 21–28, 2004) The use of bromethalin in Rittenhouse Square, *The Weekly Press*, http://208.109.172.241/pesticides/bromethalin.rittenhouse.sq.htm.

[36] Beitler AM (1892) Annual report of the Board of Health for the year ending December 31, 1891. In *First Annual Message of Edwin S. Stuart, Mayor of the City of Philadelphia, with Annual Report of Abraham M. Beitler, Director of the Department of Public Safety, and Annual Report of the Board of Health for the Year Ending December 31, 1891, Issued by the City of Philadelphia* (Dunlap and Clarke, Philadelphia), 58; Abbott SW (1889) Some historical and statistical facts pertaining to the use of arsenic as a poison. *Boston Medical and Surgical Journal* 120:477–480.

[37] Vogleson JA (1913) Annual report of the chief of the Bureau of Health for the year 1912. In *Second Annual Message of Rudolph Blankenburg, Mayor of the City of Philadelphia, with the Annual Reports of the Departments of Public Health and Charities, Supplies, Law, City Controller, City Treasurer, Commissioners of the Sinking Funds, Receiver of Taxes, and Board of Revision of Taxes for the Year Ending December 31, 1912, Issued by the City of Philadelphia*, vol. III (Dunlap Printing Company, Philadelphia), 67–516.

[38] Vogel C & Cadwallader C (1935) Rat-flea survey of the port of Philadelphia, Pa. *Public Health Reports* 50:952–957.

[39] Dyer R, Rumreich A, & Badger L (1931) Typhus fever: A virus of the typhus type derived from fleas collected from wild rats. *Public Health Reports* 46:334–338.

[40] Vogel & Cadwallader, Rat-flea survey of the port of Philadelphia, Pa.

[41] Gerhard W (1837) On the typhus fever which occurred at Philadelphia in the spring and summer of 1836. *American Journal of the Medical Sciences* 38:289–322.

[42] Whitmore JT, Cerda JJ, & Offutt RG (1971) Urban leptospirosis presenting as afebrile jaundice. *Digestive Diseases and Sciences* 16(5):455–459.

[43] Lewis M (1942) The incidence of *Leptospira icterohaemorrhagiae* in trapped rats in Philadelphia. *American Journal of Tropical Medicine* 22:571–576.

[44] Hirschhorn RB & Hodge RR (1999) Identification of risk factors in rat bite incidents involving humans. *Pediatrics* 104(3):e35.

[45] Meerburg B, Singleton G, & Kijlstra A (2009) Rodent-borne diseases and their risks for public health. *Critical Reviews in Microbiology* 35(3):221–270.

[46] McDavid CG & Mood EW (1972) Biological aspects of urban rat control. *HSMHA Health Reports* 87(1):17–24.
[47] Gates J (1972) Red-tailed hawk populations and ecology in east-central Wisconsin. *The Wilson Bulletin* 84(4):421–433.
[48] Tibbetts J (2005) Combined sewer systems: Down, dirty, and out of date. *Environmental Health Perspectives* 113(7):A464–A467.
[49] Beck JR & Rodeheffer PW (1965) Cause and control of sewer rats. *Public Works* 96(April):116–118.
[50] Clinton JM (1969) Rats in urban America. *Public Health Reports* 84(1):1-7.
[51] Ibid.
[52] Ibid.
[53] Ferritti, Thomas, Telephone conversation with the author, October 17, 2012.
[54] Harris LK (May 5, 2004) Effort launched to end rat poisoning in city park, Animal activists say that squirrels and other life are being harmed in the effort to control pests, http://articles.philly.com/2004-05-05/news/25381809_1_rat-population-rat-poisoning-animal-advocacy-group.
[55] Gómez A, Kramer LD, Dupuis AP II, Kilpatrick AM, et al. (2008) Experimental infection of eastern gray squirrels (*Sciurus carolinensis*) with West Nile virus. *American Journal of Tropical Medicine and Hygiene* 79:447–451.
[56] Rushton SP, Lurz PW, Gurnell J, Nettleton P, et al. (2006) Disease threats posed by alien species: The role of a poxvirus in the decline of the native red squirrel in Britain. *Epidemiology & Infection* 134(3):521–533.
[57] Don BAC (1983) Home range characteristics and correlates in tree squirrels. *Mammal Review* 13(2–4):123–132.
[58] Hadidian J, Manski D, Flyger V, Cox C, et al. (1987) Urban gray squirrel damage and population management: A case history. *Proceedings of the Eastern Wildlife Damage Control Conference* (1987), Paper 19, http://digitalcommons.unl.edu/ewdcc3/19.
[59] Durden LA, Ellis BA, Banks CW, Crowe JD, et al. (2004) Ectoparasites of gray squirrels in two different habitats and screening of selected ectoparasites for Bartonellae. *Journal of Parasitology* 90(3):485–489.

Chapter 3: House sparrow

[1] Leidy J (1862) *Report to the Councils of Philadelphia on Some of the Insects Injurious to Our Shade Trees* (s.n., Philadelphia).
[2] Harris TW (1862) *Treatise on Some of the Insects Injurious to Vegetation* (Crosby and Nichols, Boston).
[3] Anderson TR (2006) *The Biology of the Ubiquitous House Sparrow. From Genes to Populations* (Oxford University Press, Oxford, UK).
[4] Barrows WB (1889) *The English Sparrow* (Passer domesticus) *in North America: Especially in Its Relations to Agriculture* (U.S. Government Printing Office, Washington, DC).
[5] Gentry TG (1878) *The House Sparrow at Home and Abroad, with Some Concluding Remarks upon Its Usefulness, and Copious References to the Literature of the Subject* (Claxton, Remsen, and Haffelfinger, Philadelphia).
[6] *The Evening Telegraph*, An ordinance, May 18, 1869, 4th ed., 6.
[7] Barrows, *The English sparrow in North America.*
[8] Gentry TG (1874) English sparrows. *The American Naturalist* 8(11):667–672.
[9] Brewer TM (1874) The European house sparrow. *The American Naturalist* 8(9):556–557.
[10] Coues E (1878) The ineligibility of the European house sparrow in America. *The American Naturalist* 12(8):499–505.
[11] Gentry, *The House Sparrow at Home and Abroad,* 100–101.
[12] Ibid., 98.
[13] Coates P (2006) *American Perceptions of Immigrant and Invasive Species—Strangers on the Land* (University of California Press, Berkeley).
[14] Brodhead MJ (1971) Elliott Coues and the sparrow war. *The New England Quarterly* 44(3):420–432.
[15] Fine GA & Christoforides L (1991) Dirty birds, filthy immigrants, and the English sparrow war: Metaphorical linkage in constructing social problems. *Symbolic Interaction* 14(4):375–393.
[16] McCook H (1889) *Tenants of an Old Farm. Leaves from the Note Book of a Naturalist. Illustrations from Nature.* (Fords, Howard and Hulbert, New York), 109.
[17] Folsom E (1983) The mystical ornithologist and the Iowa tufthunter: Two unpublished Whitman letters and some identifications. *Walt Whitman Quarterly Review* 1(1):18–29.
[18] Barrows, *The English Sparrow in North America,* 164.
[19] Ibid., 150.
[20] Forbush EH & May JB (1955) *A Natural History of American Birds of Eastern and Central North America* (Bramhall House, New York).
[21] Sauer JR, Hines JE, Fallon JE, Pardieck KL, et al. (2011) *The North American Breeding Bird Survey Results and Analysis 1966–2009. Version 3.23.2011* (USGS Patuxent Wildlife Research Center, Laurel, MD), http://www.mbr-pwrc.usgs.gov/bbs/.

[22] Erskine AJ (2006) Recent declines of house sparrows, *Passer domesticus*, in Canada's Maritime Provinces. *The Canadian Field-Naturalist* 120(1):43–49.

[23] De Laet J & Summers-Smith JD (2007) The status of the urban house sparrow *Passer domesticus* in north-western Europe: A review. *Journal of Ornithology* 148(Supplement 2):S275–S278.

[24] Dandapat A, Banerjee D, & Chakraborty D (2010) The case of the disappearing house sparrow (*Passer domesticus indicus*). *Veterinary World* 3(2):97–100.

[25] De Laet & Summers-Smith, The status of the urban house sparrow *Passer domesticus* in north-western Europe; Anderson, *The Biology of the Ubiquitous House Sparrow*; Balmori A & Hallberg Ö (2007) The urban decline of the house sparrow (*Passer domesticus*): A possible link with electromagnetic radiation. *Electromagnetic Biology and Medicine* 26(2):141–151; Beckerman AP, Boots M, & Gaston KJ (2007) Urban bird declines and the fear of cats. *Animal Conservation* 10(3):320–325; Bell CP, Baker SW, Parkes NG, Brooke MDL, et al. (2010) The role of the Eurasian sparrowhawk (*Accipiter nisus*) in the decline of the house sparrow (*Passer domesticus*) in Britain. *Auk* 127:411–420; Cooper CB, Hochachka WM, & Dhondt AA (2007) Contrasting natural experiments confirm competition between house finches and house sparrows. *Ecology* 88(4):864–870; Erskine, Recent declines of house sparrows in Canada's Maritime Provinces; Everaert J & Bauwens D (2007) A possible effect of electromagnetic radiation from mobile phone base stations on the number of breeding house sparrows (*Passer domesticus*). *Electromagnetic Biology and Medicine* 26(1):63–72; Summers-Smith J (2003) The decline of the house sparrow: A review. *British Birds* 96(9):439–446; Summers-Smith JD (2005) Changes in the house sparrow population in Britain. *International Studies on Sparrows* 30:23–37; Zanette LY, White AF, Allen MC, & Clinchy M (2011) Perceived predation risk reduces the number of offspring songbirds produce per year. *Science* 334(6061):1398–1401.

[26] Tinbergen JM (1981) Foraging decisions in starlings (*Sturnus vulgaris*). *Ardea* 69:1–67; Feare C (1984) *The Starling* (Oxford University Press, Oxford, UK).

[27] European Bird Census Council (2010) Trends in common birds in Europe, 2010 update, http://www.ebcc.info/index.php?ID=387.

[28] Sauer et al., *The North American Breeding Bird Survey.*

[29] Downes C, Blancher P, & Collins B (2011) *Landbird Trends in Canada, 1968–2006. Canadian Biodiversity: Ecosystem Status and Trends 2010, Technical Thematic Report No. 12.* (Canadian Councils of Resource Ministers, Ottawa, ON).

[30] Gentry, *The House Sparrow at Home and Abroad.*

[31] Port A, Hein J, Wolff A, & Bielory L (2006) Aeroallergen prevalence in the northern New Jersey–New York City metropolitan area: A 15-year summary. *Annals of Allergy, Asthma and Immunology* 96(5):687–691.

[32] Peach WJ, Vincent KE, Fowler JA, & Grice PV (2008) Reproductive success of house sparrows along an urban gradient. *Animal Conservation* 11(6):493–503.

[33] Barrows, *The English Sparrow in North America.*

[34] Tallamy DW (2009) *Bringing Nature Home: How You Can Sustain Wildlife with Native Plants (Updated and Expanded)* (Timber Press, Portland, OR).

[35] Morris Arboretum (2011) The Pennsylvania Flora Project. Philadelphia County Species, http://www.paflora.org/. (This database presents Pennsylvania's flora according to county, species, and nativity.)

[36] Rhoads AF & Klein WM (1983) *The Vascular Flora of Pennsylvania: Annotated Checklist and Atlas* (American Philosophical Society, Philadelphia).

[37] Schuyler AE (2011) *Trees of Center City, Philadelphia* (Wagner Free Institute of Science, Philadelphia).

[38] Downing AJ (1841) *A Treatise on the Theory and Practice of Landscape Gardening, Adapted to North America; with a View to the Improvement of Country Residences. Comprising Historical Notices and General Principles of the Art, Directions for Laying Out Grounds and Arranging Plantations, the Description and Cultivation of Hardy Trees, Decorative Accompaniments to the House and Grounds, the Formation of Pieces of Artificial Water, Flower Gardens, etc.; with Remarks on Rural Architecture* (Wiley and Putnam, Boston).

[39] Wheeler A Jr. (1975) Insect associates of *Ginkgo biloba*. *Entomological News* 86:37–44.

[40] Shapiro AM (2002) The Californian urban butterfly fauna is dependent on alien plants. *Diversity and Distributions* 8:31–40.

[41] Matteson KC & Langellotto GA (2011) Small scale additions of native plants fail to increase beneficial insect richness in urban gardens. *Insect Conservation and Diversity* 4(2):89–98.

[1] Leidy J (1862) *Report to the Councils of Philadelphia on Some of the Insects Injurious to Our Shade Trees* (s.n., Philadelphia), 9–10.
[2] Davis DR (1964) Bagworm moths of the Western Hemisphere (Lepidoptera: Psychidae). In *Smithsonian Institution, United States National Museum Bulletin 244* (U.S. Government Printing Office, Washington, DC).
[3] Rhainds M, Davis DR, & Price PW (2009) Bionomics of bagworms (Lepidoptera: Psychidae). *Annual Review of Entomology* 54:209–226.
[4] Covell CV Jr. (1984) *A Field Guide to the Moths of Eastern North America* (Houghton Mifflin, Boston).
[5] Jones FM (1927) The mating of Psychidae. *Transactions of the American Entomological Society* 53(4):293–312, 314; Haseman L (1912) The evergreen bagworm. In *Bulletin 104* (University of Missouri Agricultural Experiment Station, Columbia, MO), 308–330.
[6] Jones, The mating of Psychidae; Kaufmann T (1968) Observations on the biology and behavior of the evergreen bagworm moth, *Thyridopteryx ephemeraeformis* (Lepidoptera: Psychidae). *Annals of the Entomological Society of America*. 61(1):38–44.
[7] Haseman L (1920) Bagworms destructive in Missouri. In *Circular 92* (University of Missouri Agricultural Experiment Station, Columbia, MO).
[8] Ibid.
[9] Moore RG *& Hanks LM (2004) Aerial dispersal and host plant selection by neonate* Thyridopteryx ephemeraeformis (Lepidoptera: Psychidae). *Ecological Entomology* 29:327–335.
[10] Davis, Bagworm moths of the Western Hemisphere.
[11] Horticultural Society of Pennsylvania (1831) *Report of the Committee Appointed by the Horticultural Society of Pennsylvania for Visiting the Nurseries and Gardens in the Vicinity of Philadelphia, July 13, 1830* (Wm. F. Geddes, printer, Philadelphia).
[12] Barrows E (1974) Some factors affecting population size of the bagworm, Thyridopteryx ephemeraeformis (Lepidoptera: Psychidae). *Environmental Entomolology* 3(6):929–932.
[13] Leidy, Report to the Councils of Philadelphia.
[14] Riley CV (1887) Our shade trees and their insect defoliators, being a consideration of the four most injurious species which affect the trees of the capital; with means of destroying them. In *United States Department of Agriculture, Division of Entomology, Bulletin Number 10* (U.S. Government Printing Office, Washington, DC); Howard L & Chittenden F (1916) The bagworm, an injurious shade-tree insect. In *Farmer's Bulletin 701* (U.S. Department of Agriculture, Washington, DC).
[15] Menke W & Kelly S, Historic Restoration: Fairmount Water Works, Philadelphia, http://www.landscapeonline.com/research/article/11737 (accessed April 10, 2013).
[16] Kemble FA (1891) Letter from 1812 Rittenhouse Square, Thursday, February 5, 1874. In *Further Records, 1848–1883: A Series of Letters by Frances Ann Kemble Forming a Sequel to Records of a Girlhood and Records of Later Life* (Henry Holt and Company, Philadelphia), 6–7.
[17] Ellis JA, Walker AD, Tooker JF, et al. (2005) Conservation biological control in urban landscapes: Manipulating parasitoids of bagworm (Lepidoptera: Psychidae) with flowering forbs. *Biological Control* 34:99–107.
[18] Tallamy DW, Ballard M, & D'Amico V (2010) Can alien plants support generalist insect herbivores? *Biological Invasions* 12:2285–2292.
[19] Barrows, Some factors affecting population size of the bagworm.
[20] Frost SW (1964) Insects taken in light traps at the Archbold Biological Station, Highlands County, Florida. *The Florida Entomologist* 47(2):129–161.
[21] Jones, The mating of Psychidae.
[22] Frank KD (1988) Impact of outdoor lighting on moths: An assessment. *Journal of the Lepidopterists' Society* 42(2):63–93.
[23] Jones, The mating of Psychidae.
[24] Frank, Impact of outdoor lighting on moths.
[25] Schaefer PW, Fuester RW, Chianese RJ, Rhoads LD, & Tichenor RB (1989) Introduction and North American establishment of *Coccygomimus disparis* (Hymenoptera: Ichneumonidae), a polyphagous pupal parasite of Lepidoptera, including gypsy moth. *Environmental Entomology* 18(6):1117-1125.
[26] Ellis et al., Conservation biological control in urban landscapes.
[27] Berisford YC & Tsao CH (1975) Distribution and pathogenicity of fungi associated with the bagworm, *Thyridopteryx ephemeraeformis* (Haworth). *Environmental Entomology* 4(2):257–261; Berisford & Tsao (1975) Parasitism, predation and disease in the bagworm. *Environ Entomol* 4(4)549-554; Ellis et al., Conservation biological control in urban landscapes; Lackey JA, Huckaby DG, & Ormiston BG (1985) *Peromyscus leucopus.* Mammalian Species (247):1–10; Lord RD, Lord VR, Humphreys JG, & McLean RG (1994) Distribution of *Borrelia burgdorferi* in host mice in Pennsylvania. *Journal of Clinical Microbiology* 32(10):2501–2504; Moore RG & Hanks LM (2000) Avian predation of the evergreen bagworm (Lepidoptera: Psychidae). *Proceedings of the Entomological Society of Washington* 102(2):350–352.

[1] Leidy J (1862) *Report to the Councils of Philadelphia on Some of the Insects Injurious to Our Shade Trees* (s.n., Philadelphia).

[2] Downing AJ (1847) Trees in towns and villages. *The Horticulturist and Journal of Rural Art and Rural Taste* 1(9):393–397.

[3] Downing AJ (1852) Shade trees in cities. *The Horticulturist and Journal of Rural Art and Rural Taste* 7(8):345–349.

[4] Stewardson T (1861) Remarks by Dr. Stewardson, on the ailanthus silk worm. *Proceedings of the Academy of Natural Sciences of Philadelphia* 13:525.

[5] Covell CV Jr. (1984) *A Field Guide to the Moths of Eastern North America* (Houghton Mifflin, Boston).

[6] Stewardson, Remarks by Dr. Stewardson, on the ailanthus silk worm.

[7] Tuskes PM, Tuttle JP, & Collins MM (1996) *The Wild Silk Moths of North America: A Natural History of the Saturniidae of the United States and Canada* (Cornell University Press, Ithaca, NY).

[8] Clarke J (1839) *A Treatise on the Mulberry Tree and Silkworm. And on the Production and Manufacture of Silk*, 2nd ed. (Thomas, Cowperthwait, Philadelphia).

[9] d'Homergue J (1839) *The Silk Culturist's Manual: Or a Popular Treatise on the Planting and Cultivation of Mulberry Trees, the Rearing and Propagating of Silk Worms, and the Preparation of the Raw Material for Exportation: Addressed to the Farmers and Planters of the United States* (Hogan & Thompson, Philadelphia).

[10] Barbour IR & Blydenburgh S (1844) *The Silk Culture in the United States: Embracing Complete Accounts of the Latest and Most Approved Modes of Hatching, Rearing, Feeding the Silk-Worm, Managing a Cocoonery, Reeling, Spinning, and Manufacturing the Silk, &c. &c. &c. With Brief Historical Sketches of the Silk Business, Natural History of the Silk-Worm, the Mulberry, &c. Compiled from the Most Approved and Reliable Works, Illustrated by Numerous Engravings of Machinery and Processes. To Which Is Added Blydenburgh's Manual of the Silk Culture; Statistics of Silk Imports, &c. &c.* (Greeley & McElrath, New York). [Note: Authorship is ambiguous, as this work is a compilation with sparse attribution.]

[11] Morris JG (1963) Additional observations on the ailanthus silkworm of China. In *Report of the Commissioner of Agriculture for the Year 1862* (U.S. Government Printing Office, Washington, DC), 390–394.

[12] Holland WJ (1905) *The Moth Book: A Popular Guide to a Knowledge of the Moths of North America* (Doubleday, Page & Company, New York).

[13] Nolan EJ (1892) Introduction of the ailanthus silk worm moth. *Entomological News* 3(8):193–195.

[14] Hu SY (1979) Ailanthus. *Arnoldia* 39:29–50.

[15] Ibid; Pyle RM (1975) Silk moth of the railroad yards. *Natural History* 84:45–51.

[16] Ferguson DC (1972) *Bombycoidea: Saturniidae, Comprising Subfamilies Hemileucinae (Conclusion) Saturniinae. The Moths of America North of Mexico. Fascicle 20.2B* (E.W. Classey Ltd and R. B. D. Publications Inc., London).

[17] Packard AS (1914) *Monograph of the Bombycine Moths of North America, Including Their Transformations and Origin of the Larval Markings and Armature, pt. III, Families Ceratocampidae (Exclusive of Ceratocampinae), Saturniidae, Hemileucidae and Brahmaeidae.* Memoirs of the National Academy of Sciences, vol. XII, pt. I, First Memoir, ed. TDA Cockerell (U.S. Government Printing Office, Washington, DC).

[18] Denton SF (1900) *As Nature Shows Them. Moths and Butterflies of the United States East of the Rocky Mountains, vol. 1: Moths* (J. B. Millet Company, Boston), 105.

[19] Shapiro A (1986) Quotation on page 44 in Frank KD (1986) History of the ailanthus silk moth (Lepidoptera: Saturniidae) in Philadelphia: A case study in urban ecology. *Entomological News* 97:41–51.

[20] Weast RD (1989) *Saturniidae. Ecological and Behavioral Observations of Select Attacini* (Self-published, Johnston, IA).

[21] Pyle, Silk moth of the railroad yards.

[22] Weast, *Saturniidae.*

[23] Ferguson, *Bombycoidea Saturniidae, Comprising Subfamilies Hemileucinae (Conclusion) Saturniinae.*

[24] Weast, *Saturniidae.*

[25] Frank KD (1986) History of the ailanthus silkmoth (Lepidoptera: Saturniidae) in Philadelphia: A case study in urban ecology. *Entomological News* 97:41–51.

[26] Howard LO (1885) *U.S. Department of Agriculture, Bureau of Entomology, Bulletin No. 5. Descriptions of North American Chalcididae from the Collections of the U.S. Department of Agriculture and of Dr. C. V. Riley: with Biological Notes. First paper* (U.S. Government Printing Office, Washington, DC).

[27] Weast, *Saturniidae*; Schaffner JV & Griswold CL (1934) *U.S. Department of Agriculture Miscellaneous Publication No. 188. Macrolepidoptera and Their Parasites Reared from Field Collections in the Northeastern Part of the United States.* (U.S. Government Printing Office, Washington, DC).

[28] Schaffner & Griswold, *U.S. Department of Agriculture Miscellaneous Publication No. 188.*

[29] Arnaud PH (1978) *A Host-Parasite Catalog of North American Tachinidae (Diptera)* (U.S. Department of Agriculture, Science and Education Administration, Washington, DC).

[30] Shapiro AM (1966) *Butterflies of the Delaware Valley.* Special Publications of the American Entomological Society. (The American Entomological Society, Philadelphia).

[31] Scudder SH (1887) The introduction and spread of *Pieris rapae* in North America 1860–1885. *Memoirs of the Boston Society of Natural History* 4(3):53–69.

[32] Paxon OS (1908) Numerical distribution of some insects. *Entomological News* 9:324–337.

[33] Tietz HM (1952) *The Lepidoptera of Pennsylvania: A Manual* (Pennsylvania State College, School of Agriculture, Agricultural Experiment Station, State College).
[34] Parry D (2009) Beyond Pandora's box: Quantitatively evaluating nontarget effects of parasitoids in classical biological control. *Biological Invasions* 11:47–58.
[35] Elkinton JS & Boettner GH (2004) The effects of *Compsilura concinnata*, an introduced generalist tachinid, on non-target species in North America. In *Assessing Host Ranges for Parasitoids and Predators Used for Classical Biological Control: A Guide to Best Practice*, eds. RG Van Driesche & R Reardon (U.S. Department of Agriculture, United States Forest Service, Morgantown, WV), 4–14.
[36] Kellogg SK, Fink LS, & Brower LP (2003) Parasitism of native luna moths, *Actias luna* (L.) (Lepidoptera: Saturniidae), by the introduced *Compsilura concinnata* (Meigen) (Diptera: Tachinidae) in central Virginia, and their hyperparasitism by trigonalid wasps (Hymenoptera: Trigonalidae). *Environmental Entomology* 32(5):1019–1027.
[37] Kellogg S (2002) *Parasitism of Silk Moths at Sweet Briar College* (honors thesis, Sweetbriar College), http://www2.sbc.edu/honors/HJ_2002/kellogg.htm.

Chapter 6: Ailanthus Webworm Moth

[1] Ilg C (1911) The life history of *Atteva aurea* Fitch. *Entomological News, and Proceedings of the Entomological Section of the Academy of Natural Sciences of Philadelphia* 22(5):229.
[2] Fitch A (1857) Third report on the noxious and other insects of the state of New York. *Transactions of the New York State Agricultural Society* 16:315–490.
[3] Wilson JJ, Landry JF, Janzen DH, Hallwachs W, et al. (2010) Identity of the ailanthus webworm moth (Lepidoptera, Yponomeutidae), a complex of two species: Evidence from DNA barcoding, morphology and ecology. *ZooKeys* 46:41–60.
[4] Ibid.
[5] Dyar HG (1897) *Oeta floridana* Neumoegen. *Journal of the New York Entomological Society* 5:48.
[6] Wilson et al., Identity of the ailanthus webworm moth.
[7] Becker VO (2009) A review of the New World *Atteva* Walker moths (Yponomeutidae: Attevinae). *Revista Brasileira de Entomologia* 53(3):349–355.
[8] Powell J, Comestock J, & Harbison C (1973) Biology, geographical distribution and status of *Atteva exquisita* (Lepidoptera: Yponomeutidae). *Transactions of the San Diego Society of Natural History* 17:175–186.
[9] Peigler RS & Naumann S (2003) *A Revision of the Silkmoth Genus* Samia (University of the Incarnate World, San Antonio, TX).
[10] Duffey S & Scudder GGE (1972) Cardiac glycosides in North American Asclepiadaceae, a basis for unpalatability in brightly coloured Hemiptera and Coleoptera. *Journal of Insect Physiology* 18:63–78; Malcolm SB, Cockrell BJ, & Brower LP (1989) Cardenolide fingerprint of monarch butterflies reared on common milkweed, *Asclepias syriaca* L. *Journal of Chemical Ecology* 15(3):819–853.
[11] Brower LP, Brower JVZ, & Corvino JM (1967) Plant poisons in a terrestrial food chain. *Proceedings of the National Academy of Science* 57:893–898.
[12] Kundu P & Laskar S (2010) A brief resume on the genus *Ailanthus*: Chemical and pharmacological aspects. *Phytochemistry Reviews* 9(3):379–412; De Feo V, De Martino LD, Quaranta E, & Pizza C (2003) Isolation of phytotoxic compounds from tree-of-heaven (*Ailanthus altissima* Swingle). *Journal of Agricultural and Food Chemistry* 51(5):1177–1180; De Feo V, De Martino L, Santoro A, Leone A, et al. (2005) Antiproliferative effects of tree-of-heaven (*Ailanthus altissima* Swingle). *Phytotherapy Research* 19(3):226–230.
[13] Jones FM (1932) Insect coloration and the selective acceptability of insects to birds. *Transactions of the Royal Entomological Society of London* 80:345–385; Evans DL (1983) Relative defensive behavior of some moths and the implications to predator-prey interactions. *Entomologia Experimentalis et Applicata* 33(1):103–111.
[14] Peigler R, Personal communication by email, March 2, 2012.
[15] Marples NM, Kelly DJ, & Thomas RJ (2005) Perspective: The evolution of warning coloration is not paradoxical. *Evolutionary Biology* 59(5):933–940.
[16] Coppinger RP (1970) The effect of experience and novelty on avian feeding behavior with reference to the evolution of warning coloration in butterflies. II. Reactions of naive birds to novel insects. *The American Naturalist* 104(938):323–335.
[17] Prysby MD (2004) Natural enemies and survival of monarch eggs and larvae. In *The Monarch Butterfly: Biology and Conservation*, eds. KS Oberhauser & MJ Solensky (Cornell University Press, Ithaca, NY), 27-37; Rayor LS (2004) Effects of monarch larval host plant chemistry and body size on *Polistes* wasp predation. In *The Monarch Butterfly: Biology and Conservation*, eds. KS Oberhauser & KS Solensky MJ (Cornell University Press, Ithaca, NY), 39-46.
[18] Ilg, The life history of *Atteva aurea* Fitch.
[19] Schall MJ (2008) *Verticillium Wilt of* Ailanthus altissima (dissertation in plant pathology submitted in partial fulfillment of the requirements for the degree of doctor of philosophy, The Pennsylvania State University, University Park).
[20] Ibid.

[21] Davis DD, Kasson M, & Schall M (2010) Potential for using *Verticillium albo-atrum* as a biocontrol agent for tree-of-heaven (*Ailanthus altissima*). In *Proceedings. 21st U.S. Department of Agriculture Interagency Research Forum on Invasive Species 2010*, eds. KA McManus & KW Gottschalk (U.S. Department of Agriculture, Forest Service, Northern Research Station, Newtown Square, PA), 11.

[22] Schall, *Verticillium wilt of* Ailanthus altissima.

[23] Inderbitzin P, Bostock RM, Davis RM, Usami T, et al. (2011) Phylogenetics and taxonomy of the fungal vascular wilt pathogen *Verticillium*, with the descriptions of five new species. *PLoS ONE* 6(12):e28341.

[24] Radišek S, Jakše J, & Javornik B (2006) Genetic variability and virulence among *Verticillium albo-atrum* isolates from hop. *European Journal of Plant Pathology* 116(4):301–314.

[25] Fradin EF & Thomma BPHJ (2006) Physiology and molecular aspects of verticillium wilt diseases caused by *V. dahliae* and *V. albo-atrum*. *Molecular Plant Pathology* 7(2):71–86.

[26] Davis et al., Potential for using *Verticillium albo-atrum* as a biocontrol agent for tree-of-heaven.

[27] Kasson MT, Davis MD, & Davis DD (2013) The invasive *Ailanthus altissima* in Pennsylvania: A case study elucidating species introduction, migration, invasion, and growth patterns in the northeastern U.S. *Northeastern Naturalist* 20(10):1–60.

[28] These plants include maple, hackberry, hawthorn, beech, hazel, walnut, ash, magnolia, mock orange, sumac, willow, linden, cypress, and viburnum, as reported by Corbett SL & Manchester SR (2004) Phytogeography and fossil history of *Ailanthus* (Simaroubaceae). *International Journal of Plant Sciences* 165(4):671–690.

[29] Clayton JW (2008) *Evolutionary History of Simaroubaceae (Sapindales): Systematics, Biogeography and Diversification* (dissertation submitted in partial fulfillment of the requirements for the degree of doctor of philosophy, University of Florida, Gainesville).

[30] Sohn J & Wu C (2013) A taxonomic review of Attevidae (Lepidoptera: Yponomeutoidea) from China with descriptions of two new species and a revised identity of the ailanthus webworm moth, *Atteva fabriciella*, from the Asian tropics. *Journal of Insect Science (Madison)* 13:66, http://www.insectscience.org/13.66/i1536-2442-13-66.pdf.

Chapter 7: Northern Parula

[1] Culver DE (1916) Mortality among birds at Philadelphia, May 21–22, 1915. *Cassinia* 19:33–36.

[2] Wilson A (1828) Species 29, *Sylvia pusilla,* blue yellow-back warbler. In *American Ornithology; or The Natural History of the Birds of the United States,* vol. 2 (Collins & Co. and Harrison Hall, New York and Philadelphia), 381–382.

[3] City of Philadelphia (1901) *Official Handbook, City Hall, Philadelphia* (City Publishing Company, Philadelphia).

[4] Ibid., 11.

[5] Ibid.

[6] Baily WL (1900) Migration data on City Hall Tower. *Cassinia* 3:15.

[7] Ibid., 15–19.

[8] Cooke WW (1904) *Distribution and Migration of North American Warblers. United States Department of Agriculture, Division of Biological Survey Bulletin No. 18* (U.S. Government Printing Office, Washington, DC), 17–18.

[9] Ibid., 49.

[10] Sauer EGF & Sauer EM (1960) Star navigation of nocturnal migrating birds. The 1958 planetarium experiments. *Cold Spring Harbor Symposia on Quantitative Biology* 25:463-473; Moore FR & Phillips JB (1988) Sunset, skylight polarization and the migratory orientation of yellow-rumped warblers, *Dendroica coronata*. *Animal Behaviour* 36:1770–1778; Moore FR (1986) Sunrise, skylight polarization, and the early morning orientation of night-migrating warblers. *The Condor* 88(4):493–498; Mettke-Hofmann C & Gwinner E (2003) Long-term memory for a life on the move. *Proceedings of the National Academy of Sciences* 100(10):5863–5866; Wiltschko W & Gwinner E (1974) Evidence for an innate magnetic compass in garden warblers. *Naturwissenschaften* 61(9):406–406.

[11] Wiltschko R, Stapput K, Bischof H-J, & Wiltschko W (2007) Light-dependent magnetoreception in birds: Increasing intensity of monochromatic light changes the nature of the response. *Frontiers in Zoology* 4:5. doi:10.1186/1742-9994-4-5, http://www.frontiersinzoology.com/content/pdf/1742-9994-4-5.pdf; Gauthreaux Jr S & Belse C (2006) Effects of artificial night lighting on migrating birds. In *Ecological Consequences of Artificial Night Lighting*, eds. C Rich & T Longcore (Island Press, Covelo, CA), 67–93; Wiltschko R, Stapput K, Thalau P, & Wiltschko W (2010) Directional orientation of birds by the magnetic field under different light conditions. *Journal of the Royal Society Interface* 7(Suppl 2):S163–S177; Poot H, Ens BJ, de Vries H, Donners MAH, et al. (2008) Green light for nocturnally migrating birds. *Ecology and Society* 13(2):47, http://www.ecologyandsociety.org/vol13/iss2/art47/.

[12] Emporis Data Committee (2012) Philadelphia's Tallest Buildings—Top 20. (Emporis), http://www.emporis.com/statistics/tallest-buildings-philadelphia-pa-usa.

[13] Klem D Jr. (2009) Preventing bird-window collisions. *The Wilson Journal of Ornithology* 121(2):314–321.

[14] Klem D Jr. (1989) Bird-window collisions. *The Wilson Bulletin* 101(4):606–620.

[15] Dunn EH (1993) Bird mortality from striking residential windows in winter. *Journal of Field Ornithology* 64(3):302–309.

[16] Klem D Jr., Farmer CJ, Delacretaz N, Gelb Y, et al. (2009) Architectural and landscape risk factors associated with bird–glass collisions in an urban environment *The Wilson Journal of Ornithology* 121(1):126–134.

[17] Evans Ogden LJ (2002) Effect of light reduction on collision of migratory birds. *Special Report of the Fatal Light Awareness Program (FLAP),* Paper 5, http://digitalcommons.unl.edu/flap/5.

[18] Kousky C (2004) A building less bright. Chicago skyscrapers go dark for migratory birds. *Terrain.org: A Journal of the Built and Natural Environments* No. 15 (Fall/Winter), http://www.terrain.org/articles/15/kousky.htm.

[19] Hertzler L (2011) Bird deaths on campus prompt window-design project, art exhibit. (*The Temple News*), http://temple-news.com/news/2011/11/14/bird-deaths-on-campus-prompt-window-design-project-art-exhibit/; Johnson K (2012) Winning design would save birds through music. (*The Temple News*), http://temple-news.com/tag/keith-russell/.

[20] Bauers S (2012) Birds blinded by the light of Philly's glass-clad buildings. Philly.com. http://www.philly.com/philly/news/science/152412745.html?page=3&c=y.

[21] Drewitt AL & Langston RHW (2008) Collision effects of wind-power generators and other obstacles on birds. *Annals of the New York Academy of Sciences* 1134:233–266.

[22] Longcore T, Rich C, Mineau P, MacDonald B, et al. (2012) An estimate of avian mortality at communication towers in the United States and Canada. *PLoS ONE* 7(4):e34025. doi:34010.31371/journal.pone.0034025.

[23] Johnston DW & Haines TP (1957) Analysis of mass bird mortality in October 1954. *The Auk* 74(4):447–458; Johnston DW (1955) Mass bird mortality in Georgia, October 1954. *Oriole* 20:17–26.

[24] Larkin RP & Frase BA (1988) Circular paths of birds flying near a broadcasting tower in cloud. *Journal of Comparative Psychology* 102(1):90–93.

[25] Kriska G, Malik P, Szivák I, & Horváth G (2008) Glass buildings on river banks as "polarized light traps" for mass-swarming polarotactic caddis flies. *Naturwissenschaften* 95(5):461–467.

[26] Málnás K, Polyák L, Prill E, Hegedüs R, et al. (2011) Bridges as optical barriers and population disruptors for the mayfly *Palingenia longicauda*: An overlooked threat to freshwater biodiversity? *Journal of Insect Conservation* 15:823–832.

[27] Horváth G, Blahó M, Egri A, et al. (2010) Reducing the maladaptive attractiveness of solar panels to polarotactic insects. *Conservation Biology* 24(6):1644–1653.

[28] Kriska G, Horváth G, & Andrikovics S (1998) Why do mayflies lay their eggs en masse on dry asphalt roads? Water-imitating polarized light reflected from asphalt attracts Ephemeroptera. *Journal of Experimental Biology* 201(15):2273–2286.

[29] Horváth G, Kriska G, Malik P, & Robertson B (2009) Polarized light pollution: A new kind of ecological photopollution. *Frontiers in Ecology and the Environment* 7(6):317–325.

[30] Ibid.

[31] Moldenhauer RR & Regelski DJ (2012) Northern parula (*Setophaga americana*). In *The Birds of North America Online*, ed. A Poole (Cornell Lab of Ornithology, Ithaca, NY), http://bna.birds.cornell.edu/bna/species/215.

[32] Schwalbe PW (1992) Northern parula, *Parula americana.* In *Atlas of Breeding Birds in Pennsylvania,* ed. DW Brauning (University of Pittsburgh Press, Pittsburgh, PA), 306–307.

[33] Devlin J (1954) Effects of weather on nocturnal migration as seen from one observation point at Philadelphia. *The Wilson Bulletin* 6(2):93–101.

[34] Sauer JR, Hines JE, Fallon JE, Pardieck KL, et al. (2011) The *North American Breeding Bird Survey, Results and Analysis 1966–2010. Version 12.07.2011* (USGS Patuxent Wildlife Research Center, Laurel, MD), http://www.mbr-pwrc.usgs.gov/cgi-bin/atlasa10.pl?06480&1&10.

[35] Moldenhauer & Regelski, Northern parula.

[36] Robbins CS, Sauer JR, Greenberg RS, & Droege S (1989) Population declines in North American birds that migrate to the neotropics. *Proceedings of the National Academy of Sciences* 86(19):7658–7662.

[37] Graveland J (1990) Effects of acid precipitation on reproduction in birds. *Cellular and Molecular Life Sciences* 46(9):962–970.

[38] Böhning Gaese K, Taper M, & Brown J (2002) Are declines in North American insectivorous songbirds due to causes on the breeding range? *Conservation Biology* 7(1):76–86.

[39] Nebel S, Mills A, McCracken JD, & Taylor PD (2010) Declines of aerial insectivores in North America follow a geographic gradient. *Avian Conservation and Ecology* 5(2):1, http://www.ace-eco.org/vol5/iss2/art1/.

[40] Both C, Van Turnhout CAM, Bijlsma RG, Siepel H, et al. (2010) Avian population consequences of climate change are most severe for long-distance migrants in seasonal habitats. *Proceedings of the Royal Society B* 277:1259–1266.

[41] Faaborg J, Holmes RT, Anders AD, Bildstein KL, et al. (2010) Recent advances in understanding migration systems of New World land birds. *Ecological Monographs* 80(1):3–48.

[42] Bent AC (1963) *Life Histories of North American Wood Warblers.* Part one of two parts. [Reprint of Smithsonian Institution, United States National Museum Bulletin 203, published in 1953 by the U.S. Government Printing Office, Washington, DC] (Dover Publications, New York).

[43] Böhning Gaese, Taper, & Brown, Are declines in North American insectivorous songbirds due to causes on the breeding range?

[44] Arnold T & Zink R (2011) Collision mortality has no discernible effect on population trends of North American birds. *PLoS ONE* 6(9): e24708. doi:10.1371/journal.pone.0024708.

[45] Loss SR, Will T, & Marra PP (2012) Direct human-caused mortality of birds: Improving quantification of magnitude and assessment of population impact. *Frontiers in Ecology and the Environment* 10:357–364.

[46] Bauers S (2012) Concerns over bird migration provokes change in Philadelphia light show. (Philly.com). http://articles.philly.com/2012-09-09/news/33697443_1_bird-migration-birds-use-stars-dead-birds.

[47] Wilson E O (1984) *Biophilia: The Human Bond with Other Species* (Harvard University Press, Cambridge, MA).

Chapter 8: Polyphemus Moth

[1] Peale TR (1833) *The Titian Peale Butterfly and Moth Collection in the Academy of Natural Sciences.* Box 7, cork 5 and 6, curated by Jason Weintraub, http://clade.ansp.org/entomology/collections/peale/peale.php?-mode=specimen&box=7.

[2] Tuskes PM, Tuttle JP, & Collins MM (1996) *The Wild Silk Moths of North America. A Natural History of the Saturniidae of the United States and Canada* (Cornell University Press, Ithaca, NY).

[3] Cotter JL, Roberts DG, & Parrington M (1992) *The Buried Past. An Archeological History of Philadelphia* (University of Pennsylvania Press. Philadelphia), 524.

[4] Frost CA (1915) Remarks on collecting at light, with a list of the Coleoptera taken. *Psyche* 22:207–210.

[5] Engraving by anonymous artist in Bowen D (1839) *A History of Philadelphia, with a Notice of Villages in the Vicinity Embellished with Engravings Designed as a Guide to Citizens and Strangers, Containing a Correct Account of City Improvements, up to the Year 1839; Also, the State of Society, in Relation to Science, Religion and Morals; with an Historical Account of the Military Operations of the Late War, Including the Names of Over Two Thousand Patriotic Officers, and Citizen Soldiers, Who Volunteered Their Services in Defence of This City, When Threatened by an Hostile Army in 1812–13 & 14* (Daniel Bowen, Philadelphia), 116.

[6] Strickland W, Gill EH, & Campbell HR, eds. (1841) *The Philadelphia Gas Works* (J. Weale, London).

[7] Huston RM (1839) *Third Annual Report Made to the Select and Common Councils of the City of Philadelphia, In Behalf of the Trustees of the Philadelphia Gas Works. Reports, Specifications, and Estimates of Public Works in the United States of America: Comprising the Philadelphia Gas Works; Reservoir Dam across the Swatara; Twin Locks on the Schuylkill Canal; Delaware Breakwater; Philadelphia Water Works; Dam and Lock on the Sandy and Beaver Canal; Dam on the James River and Kanawha Canal, Virginia; Locks of Eight Feet Lift, On the Same; Aqueducts across Rivanna River and Byrd Creek, On the Same; Superstructure, Etc., of Farm Bridges, On the Same; Lock Gates and Mitre Sills,* eds. W Strickland, EH Gill, & HR Campbell (J. Weale, London), 15–27.

[8] Bowen D (1839) *A History of Philadelphia, with Notice of Villages, in the Vicinity, Embellished with Engravings, Designed as a Guide to Citizens and Strangers, Containing a Correct Account of the City Improvements up to the Year 1839…* (Daniel Bowen, Philadelphia), 116–117.

[9] Wainwright NB (1961) *History of the Philadelphia Electric Company 1881–1961* (Philadelphia Electric Company, Philadelphia).

[10] Riley CV (1892) *Directions for Collecting and Preserving Insects. Part F of Bulletin of the United States National Museum No. 39.* Smithsonian Institution, United States National Museum (U.S. Government Printing Office, Washington, DC), 51.

[11] Denton SF (1900) *As Nature Shows Them. Moths and Butterflies of the United States East of the Rocky Mountains, vol 1: Moths* (J. B. Millet Company, Boston), 35.

[12] Frank KD (1988) Impact of outdoor lighting on moths: An assessment. *Journal of the Lepidopterists' Society* 42(2):63–93.

[13] Ibid.

[14] Langevelde Fv, Ettema JA, Donners M, Wallis de Vries MF, et al. (2011) Effect of spectral composition of artificial light on the attraction of moths. *Biological Conservation* 144:2274–2281.

[15] Eisenbeis G (2006) Artificial night lighting and insects: Attraction of insects to streetlamps in a rural setting in Germany. In *Ecological Consequences of Artificial Night Lighting,* eds. C Rich & T Longcore (Island Press, Washington, DC), 281–304.

[16] Frank, Impact of outdoor lighting on moths.

[17] Yela JL & Holyoak M (1997) Effects of moonlight and meteorological factors on light and bait trap catches of noctuid moths (Lepidoptera: Noctuidae). *Environmental Entomology* 26(6):1283–1290.

[18] Robinson HS & Robinson PJM (1950) Some notes on the observed behavior of Lepidoptera in flight in the vicinity of light sources together with a description of a light trap designed to take entomological samples. *Entomologists' Gazette* 1:3–20.

[19] Yela & Holyoak, Effects of moonlight and meteorological factors on light and bait trap catches of noctuid moths.

[20] Kyba C, Ruhtz T, Fische J, & Hölker F (2011) Cloud coverage acts as an amplifier for ecological light pollution in urban ecosystems. *PLoS ONE* 6(3):e17307. doi: 10.1371/journal.pone.0017307

[21] Tietz HM (1952) *The Lepidoptera of Pennsylvania: A Manual* (Pennsylvania State College, School of Agriculture, Agricultural Experiment Station, State College, PA).

[22] Capinera JL (2001) *Handbook of Vegetable Pests* (Academic Press, San Diego, CA).

[23] Covell CV Jr. (1984) *A Field Guide to the Moths of Eastern North America* (Houghton Mifflin, Boston).
[24] Capinera, *Handbook of Vegetable Pests.*
[25] Covell Jr., *A Field Guide to the Moths of Eastern North America.*
[26] Adamski D, Hevel GF, & Pultyniewicz A (2009) Redescription and immature stages of *Promalactis suzukiella* (Matsumura) (Gelechioidea: Oecophoridae), a new introduction into the United States. *Proceedings of the Entomological Society of Washington* 111(1):204–214.
[27] Covell Jr., *A Field Guide to the Moths of Eastern North America.*
[28] Tietz, *The Lepidoptera of Pennsylvania: A Manual.*
[29] Ibid.
[30] Tuskes et al., *The Wild Silk Moths of North America.*
[31] Forstner M, Breer H, & Krieger J (2009) A receptor and binding protein interplay in the detection of a distinct pheromone component in the silkmoth *Antheraea polyphemus. International Journal of Biological Sciences* 5(7):745–757.
[32] Vickers NJ (2006) Winging it: Moth flight behavior and responses of olfactory neurons are shaped by pheromone plume dynamics. *Chemical Senses* 31:155–166.
[33] Kochansky JP, Cardé RT, Taschenberg EF, & Roelofs WL (1977) Rhythms of male *Antheraea polyphemus* attraction and female attractiveness, and an improved pheromone synthesis. *Journal of Chemical Ecology* 3(4):419–427.
[34] Janzen DH (1983) Introduction [Insects]. In *Costa Rican Natural History*, ed. DH Janzen (University of Chicago Press, Chicago), 619–645.
[35] Kolankiewicz L & Beck R (2001) *Weighing Sprawl Factors in Large U.S. Cities. A Report on the Nearly Equal Roles Played by Population Growth and Land Use Choices in the Loss of Farmland and Natural Habitat to Urbanization. Analysis of the Census Data on the 100 Largest Urbanized Areas of the United States* (NumbersUSA, Arlington, VA).
[36] Wagner WH & Mayfield MR (1980) Foodplants and cocoon construction in *Antheraea polyphemus* (Lepidoptera: Saturniidae) in southern Michigan. *Great Lakes Entomologist* 13:131–138.
[37] Young A (1982) Predation on the pupae of Saturniidae (Lepidoptera) by gray squirrels in Wisconsin. *The Great Lakes Entomologist* 15(2):145.
[38] Tan M (2009) *The Effect of Alien Plants on the Survival of Larval Lepidoptera* (honors thesis submitted in partial fulfillment of the requirements for the degree of bachelor of science, University of Delaware, Newark), http://dspace.udel.edu:8080/dspace/bitstream/handle/19716/4438/Tan,%20 Milton.pdf?sequence=1; Tietz, *The Lepidoptera of Pennsylvania: A Manual.*
[39] Wagner D (2012) Moth decline in the northeastern United States. *News of the Lepidopterists' Society* 54(2):52–56.
[40] Frick T & Tallamy D (1996) Density and diversity of nontarget insects killed by suburban electric insect traps. *Entomological News* 107(2):77–82; Frick WF, Pollock JF, Hicks AC, Langwig KE, et al. (2010) An emerging disease causes regional population collapse of a common North American bat species. *Science* 329(5992):679–682.

Chapter 9: Bridge Spider

[1] McCook HC (1890) *American Spiders and Their Spinning Work,* vol. 2 (self-published, Philadelphia), 265.
[2] Ibid., 294.
[3] Ibid., 289.
[4] Dale WE & Axtell RC (1975) Flight of the salt marsh Tabanidae (Diptera), *Tabanus nigrovittatus, Chrysops atlanticus* and *C. fuliginosus*: Correlation with temperature, light, moisture and wind velocity. *Journal of Medical Entomology* 12(5):551–557.
[5] Heiling A (1999) Why do nocturnal orb-web spiders (Araneidae) search for light? *Behavioral Ecology and Sociobiology* 46(11):43–49.
[6] Heiling AM & Herberstein ME (1998) Activity patterns in different developmental stages and sexes of *Larinioides sclopetarius* (Clerck) (Araneae, Araneidae). In *Proceedings of the 17th European Colloquium of Arachnology, Edinburgh, 1997*, ed. PA Selden (British Arachnological Society, UK), 211–214.
[7] Frank KD (2009) Exploitation of artificial light by a jumping spider. *Peckhamia* 78(1):1–3.
[8] McCook, *American Spiders and Their Spinning Work*, vol. 2.
[9] Chuang C-Y, Yang E-C, & Tso I-M (2008) Deceptive color signaling in the night: A nocturnal predator attracts prey with visual lures. *Behavioral Ecology and Sociobiology* 19(2):237–244.
[10] Chuang C-Y, Yang E-C, & Tso I-M (2007) Diurnal and nocturnal prey luring of a colorful predator. *Journal of Experimental Biology* 210:3830–3837.
[11] McCook HC (1893) *American Spiders and Their Spinning Work*, vol. 3 (self-published, Philadelphia).
[12] Kleinteich A & Schneider JM (2011) Developmental strategies in an invasive spider: Constraints and plasticity. *Ecological Entomology* 36(1):82–93.

[13] Burgess JW & Uetz GW (1982) Social spacing strategies in spiders. In *Spider Communication, Mechanisms and Ecological Significance,* eds. PN Witt & JS Rovner (Princeton University Press, Princeton, NJ), 317–351.
[14] Kralj-Fišer S & Schneider JM (2012) Individual behavioural consistency and plasticity in an urban spider. *Animal Behaviour* 84(1):197–204.
[15] Frank KD (1988) Impact of outdoor lighting on moths: An assessment. *Journal of the Lepidopterists' Society* 42(2):63–93.
[16] Kleinteich A (2009) *Life History of the Bridge Spider,* Larinioides sclopetarius *(Clerck, 1757)* (doctoral dissertation, University of Hamburg, Germany).
[17] Levi HW (1974) The orb-weaver genera *Araniella* and *Nuctenea* (Araneae: Araneidae). *Bulletin of the Museum of Comparative Zoology* 146(6):291–316.
[18] Hentz NM (1847) Descriptions and figures of the araneides of the United States. *Boston Journal of Natural History Containing Papers and Communications Read to the Boston Society of Natural History, 1845–1847* 5:443–478.
[19] Ibid.
[20] Uma DB (2010) *Behavioral Ecology of Wasp-Spider Interactions: The Role of Webs, Chemicals, and Deception* (PhD dissertation, Georgetown University, Washington, DC).
[21] Rau P (1935) The spider prey of the mud wasp, *Sceliphron caementarium* (Araneae, Hymen: Sphegidae). *Ent News* 46(10):268–270; Rau P & Rau NL (1916) The sleep of insects: An ecological study. *Annals of the Entomological Society of America* 9(3):227–274.

Chapter 10: Black and Yellow Mud Dauber

[1] Bartram J & Collinson P (1745) An account of some very curious wasp nests made of clay in Pennsylvania; by Mr. John Bartram: Communicated by Mr. Peter Collinson, F.R.S. *Philosophical Transactions* 43:363–366.
[2] Redlich F (1945) The Philadelphia Water Works in relation to the Industrial Revolution in the United States. *The Pennsylvania Magazine of History and Biography* 69(3):243–256.
[3] Carter EC II (1971/1972) Benjamin Henry Latrobe and the growth and development of Washington, 1798–1818. In *Records of the Columbia Historical Society, Washington, DC,* vol. 71/72, 128–149.
[4] Latrobe BH (1806) On two species of the sphex or wasp, found in Virginia and Pennsylvania, and probably existing through all the United States. *The Philosophical Magazine* 25(99):236–241.
[5] Ibid., 240.
[6] Ibid., 240–241.
[7] Ibid., 241.
[8] McCook HC (1890) *American Spiders and Their Spinning Work,* vol. 2 (self-published, Philadelphia), 381–382.
[9] Hentz NM (1847) Descriptions and figures of the araneides of the United States. *Boston Journal of Natural History Containing Papers and Communications Read to the Boston Society of Natural History* 5:443–478.
[10] Eberhard W (1970) The predatory behavior of two wasps, *Agenoideus humilis* (Pompilidae) and *Sceliphron caementarium* (Sphecidae), on the orb weaving spider, *Araneus cornutus* (Araneidae). *Psyche* 77:243–251.
[11] Ibid.
[12] Obin MS (1982) Spiders living at wasp nesting sites: What constrains predation by mud-daubers? *Psyche* 89:321–335.
[13] Bohart RM & Menke AS (1976) *Sphecid Wasps of the World: A Generic Revision* (University of California Press, Berkeley).
[14] Fowler HG (1983) Human effects on nest survivorship of urban synanthropic wasps. *Urban Ecology* 7(2):137–143.
[15] Krakker JJ (2012) Mud wasp nests as markers of Middle Holocene house structures in the Central Mississippi Valley. *Journal of American Antiquity* 77(4):800–807.

Chapter 11: Yellowjackets

[1] Cresson ET (1887) Supplementary volume: Synopsis of the families and genera of the Hymenoptera of America, north of Mexico together with a catalogue of the described species and bibliography. *Transactions of the American Entomological Society and Proceedings of the Entomological Section of the Academy of Natural Sciences* (American Entomological Society, Philadelphia), 123.
[2] Morse RA, Eikwort GC, & Jacobson RS (1977) The economic status of an immigrant yellowjacket, *Vespula germanica* (Hymenoptera: Vespidae), in northeastern United States. *Environmental Entomology* 6(1):109–110.
[3] Lord WD (1977) The occurrence of pestiferous *Vespula* spp. in northern Delaware (Hymenoptera: Vespidae). *Entomological News* 88:193–196.
[4] Parrish HD & Roberts RB (1982) Successful establishment of the German yellowjacket in urban New Jersey indicated by relative abundance of *Vespula germanica* and *V. maculifrons* (Hymenoptera: Vespidae). *Journal of the Kansas Entomological Society* 55(2):272–276.

[5] MacDonald JF & Akre RD (1984) Range extension and emergence of subterranean nesting by the German yellowjacket, *Vespula germanica*, in North America (Hymenoptera:Vespidae). *Entomological News* 95(1):5–8.
[6] Stein KJ & Wrensch DL (1988) The pest status of yellowjackets in Ohio (Hymenoptera:Vespidae). *The Great Lakes Entomologist* 21(2):83–90.
[7] Duncan CD (1939) *A Contribution to the Biology of North American Vespine Wasps* (Stanford University Press, Redwood City, CA).
[8] Wood G, Hopkins D, & Schellhorn N (2006) Preference by *Vespula germanica* (Hymenoptera:Vespidae) for processed meats: Implications for toxic baiting. *Journal of Economic Entomology* 99(2):263–267.
[9] Reid BL, MacDonald JF, & Ross DR (1995) Foraging and spatial dispersion in protein-scavenging workers of *Vespula germanica* and *V. maculifrons* (Hymenoptera:Vespidae). *Journal of Insect Behavior* 8(3):315–330; Richter MR & Tisch VL (1999) Resource choice of social wasps: Influence of presence, size and species of resident wasps. *Insectes Sociaux* 46(2):131–136.
[10] Taylor B, Schalk D, & Jeanne R (2010) Yellowjackets use nest-based cues to differentially exploit higher-quality resources. *Naturwissenschaften* 97(12):1041–1046.
[11] Cameron SA, Lozier JD, Strange JP, Koch JB, et al. (2011) Patterns of widespread decline in North American bumble bees. *Proceedings of the National Academy of Sciences* 108(2):662–667.
[12] Bischof RO (1996) Seasonal incidence of insect stings: Autumn "yellow jacket delirium." *Journal of Family Practice* 43:271–273.
[13] Lynch JP & Rothstein RD (1997) A gastric "bee-zoar." *New England Journal of Medicine* 336:1763–1764; Sobonya R & Schmidt J (1997) Not a "bee-zoar," but a wasp. *New England Journal of Medicine* 337:575.
[14] Dibs SD & Baker MD (1997) Anaphylaxis in children: A 5-year experience. *Pediatrics* 99(1):E7.
[15] Waddell LS & Drobatz KJ (1999) Massive envenomation by *Vespula* spp. in two dogs. *Journal of Veterinary Emergency and Critical Care* 9(2):67–71.
[16] Davis HG (1978) Yellowjacket wasps in urban environments. In *Perspectives in Urban Entomology*, eds. GW Frankie & CS Koehler (Academic Press, New York), 163–185.
[17] West PL, McKeown NJ, & Hendrickson RG (2011) Massive hymenoptera envenomation in a 3-year-old. *Pediatric Emergency Care* 27(1):46–48.
[18] Landolt PJ, Heath RR, Reed HC, & Manning K (1995) Pheromonal mediation of alarm in the eastern yellowjacket (Hymenoptera:Vespidae). *The Florida Entomologist* 78(1):101–108.
[19] Simon RP & Benton AW (1968) Winter activities of *Vespula maculifrons*. *Annals of the Entomological Society of America* 61(2):542–543.
[20] Davis, Yellowjacket wasps in urban environments.
[21] Langley RL & Morrow WE (1997) Deaths resulting from animal attacks in the United States. *Wilderness and Environmental Medicine* 8:8–16.
[22] Greene A, Breisch NL, Golden DBK, Kelly D, et al. (2012) Sting embedment and avulsion in yellowjackets (Hymenoptera:Vespidae): A functional equivalent to autotomy. *American Entomologist* 58(1):50–57.
[23] Shorter JR & Rueppell O (2012) A review on self-destructive defense behaviors in social insects. *Insectes Sociaux* 59(1):1–10.
[24] Friedman LS, Modi P, Liang S, & Hryhorczuk D (2010) Analysis of Hymenoptera stings reported to the Illinois Poison Center. *Journal of Medical Entomology* 47(5):907–912.
[25] Cervo R, Zacchi F, & Turillazzi S (2000) *Polistes dominulus* (Hymenoptera, Vespidae) invading North America: Some hypotheses for its rapid spread. *Insectes Sociaux* 47(2):155–157.
[26] MacDonald JF, Akre RD, & Keyel RE (1980) The German yellowjacket (*Vespula germanica*) problem in the United States (Hymenoptera:Vespidae). *Bulletin of the Entomological Society of America* 26(4):436–444.
[27] Hopwood JM, Vaughan M, Shepherd M, Biddinger D, et al. (2012) *Are Neonicotinoids Killing Bees? A Review of Research into the Effects of Neonicotinoid Insecticides on Bees, with Recommendations for Action* (The Xerces Society for Invertebrate Conservation, Portland, OR).
[28] Jacobson RS (1991) *Polistes dominulus* spreading in USA. *Sphecos* 21:14.
[29] Wilson-Rich N (2010) The *Polistes* war: Weak immune function in the invasive *P. dominulus* relative to the native *P. fuscatus*. *Insectes Sociaux* 57:47–52.
[30] Fox-Wilson G (1946) Factors affecting populations of social wasps, *Vespula* species, in England (Hymenoptera). *Proceedings of the Royal Entomological Society of London. Series A, General Entomology* 21(4-6):17–27; Archer ME (2001) Changes in abundance of *Vespula germanica* and *V. vulgaris* in England. *Ecological Entomology* 26(1):1–7; Akre RD & Reed HC (1981) Population cycles of yellowjackets (Hymenoptera:Vespinae) in the northwest Pacific. *Environmental Entomology* 10(3):267–274.
[31] Archer ME (2012) *Vespine Wasps of the World: Behavior, Ecology & Taxonomy of the Vespinae* (Siri Scientific Press, Manchester, UK).

Chapter 12: Common Eastern Firefly

[1] De Geer C (1774) Second Memoire: Des lampyres, des thelephores, et des colliures. In *Mémoires pour servir à l'histoire des insectes, Tome IV* (Pierre Hesselberg, Stockholm), 52–53; pl. 17, fig. 28; Jones FM (1930) *Dynastes tityus* in Pennsylvania and Delaware (Coleop: Scarabaeidae). *Entomological News* 41:305–306.

[2] Leconte JL (1851) Synopsis of the Lampyrides of temperate North America. *Proceedings of the Academy of Natural Sciences of Philadelphia* 5:331–356.

[3] General Assembly of the Commonwealth of Pennsylvania. *Firefly—Official Insect of Pennsylvania.* Act of April 10, 1974, Public Law 247, No. 59, CI 71; as amended December 5, 1988, Public Law 1101, No. 130.

[4] McDermott FA (1911) Some further observations on the light-emission of American Lampyridae: The photogenic function as a mating adaptation in the Photinini. *Canadian Entomologist* 43:399–406.

[5] Ibid., 401.

[6] Ibid.

[7] Ibid., 403.

[8] Williams FX (1917) Notes on the life-history of some North American Lampyridæ. *Journal of the New York Entomological Society* 25(1):11–33.

[9] Ibid., 24.

[10] Lloyd JE (1965) Aggressive mimicry in *Photuris*: Firefly femmes fatales. *Science* 149(3684):653–654.

[11] Ibid.

[12] Lloyd JE (1975) Aggressive mimicry in *Photuris* fireflies: Signal repertoires by femmes fatales. *Science* 187(4175):452–453.

[13] Lloyd JE (1980) Male *Photuris* fireflies mimic sexual signals of their females' prey. *Science* 210(4470):669–671.

[14] Lloyd JE & Wing SR (1983) Nocturnal aerial predation of fireflies by light-seeking fireflies. *Science* 222(4624):634–635.

[15] El-Hani CN, Queiroz J, & Stjernfelt F (2010) Firefly femmes fatales: A case study in the semiotics of deception. *Biosemiotics* 3(1):33–55.

[16] Eisner T, Goetz MA, Hill DE, Smedley SR, et al. (1997) Firefly "femmes fatales" acquire defensive steroids (Lucibufagins) from their firefly prey. *Proceedings of the National Academy of Sciences of the United States of America* 94(18):9723–9728.

[17] González A, Hare JF, & Eisner T (1999) Chemical egg defense in *Photuris* firefly "femmes fatales." *Chemoecology* 9(4):177–185.

[18] Gronquist M, Schroeder FC, Ghiradella H, Hill D, et al. (2006) Shunning the night to elude the hunter: Diurnal fireflies and the "femmes fatales." *Chemoecology* 16:39–43.

[19] Nelson S, Carlson AD, & Copeland J (1975) Mating-induced behavioural switch in female fireflies. *Nature* 255:628–629.

[20] Barber HS (1951) *North American fireflies of the genus Photuris. Smithsonian Miscellaneous Collections vol. 117, no. 1 (Smithsonian publication number 4051). With preface and notes by Frank A. McDermott.* (Smithsonian Institution, Washington, DC); McDermott FA (1967) The North American fireflies of the genus *Photuris* Dejean: A modification of Barber's key (Coleoptera; Lampyridae). *The Coleopterists Bulletin* 21(4):106–116.

[21] McDermott FA (1958) *The Fireflies of Delaware,* 2nd ed. (Society of Natural History of Delaware, Wilmington).

[22] Heckscher CM (2010) Delaware *Photuris* fireflies (Coleoptera: Lampyridae): New state records, conservation status, and habitat associations. *Entomological News* 121(5):498–505.

[23] De Geer, Second memoire, 53.

[24] McDermott, *The Fireflies of Delaware.*

[25] Glassman S; Bolger B, ed. (2002) *National Historic Landmark Nomination. The College of Physicians of Philadelphia Building* (National Register of Historic Places Registration Form of United States Department of Interior, National Park Service), 21, http://www.nps.gov/nhl/designations/samples/pa/CollegeofPhysicians.pdf.

[26] Lloyd JE (1973) Firefly parasites and predators. *The Coleopterists Bulletin* 27(2):91–106.

[27] Walsh BD & Riley CV (1868) Fire-flies. *American Entomologist* 1(1):19.

[28] McDermott, *The Fireflies of Delaware.*

[29] Steinberg DA, Pouyat RV, Parmelee RW, & Groffman PM (1997) Earthworm abundance and nitrogen mineralization rates along an urban-rural land use gradient. *Soil Biology and Biochemistry* 29(3–4):427–430.

[30] Smetak KM, Johnson-Maynard JL, & Lloyd JE (2007) Earthworm population density and diversity in different-aged urban systems. *Applied Soil Ecology* 37(1–2):161–168.

[31] Reid BJ & Watson R (2005) Lead tolerance in *Aporrectodea rosea* earthworms from a clay pigeon shooting site. *Soil Biology and Biochemistry* 37(3):609–612.

[32] Spurgeon DJ & Hopkin SP (1996) The effects of metal contamination on earthworm populations around a smelting works: Quantifying species effects. *Applied Soil Ecology* 4(2):147–160.

[33] Buschman LL (1984) Larval biology and ecology of *Photuris* fireflies (Lampyridae: Coleoptera) in northcentral Florida. *Journal of the Kansas Entomological Society* 57(1):7–16.

[34] Lewis SM, Personal communication, March 24, 2010.

[35] Lloyd JE (2006) Stray light, fireflies, and fireflyers. In *Ecoloagical Consequences of Artificial Night Lighting,* eds. C Rich & T Longcore (Island Press, Washington, DC), 345–364.

[36] Museum of Science, Boston (2012) Firefly Watch, https://www.mos.org/fireflywatch/.
[37] Kipling R (1911) Philadelphia. In *Rewards and Fairies* (Doubleday, Page & Company, New York), 344.
[38] Reynolds JW (2008) The earthworms (Oligochaeta: Acanthodrilidae, Lumbricidae, Megascolecidae, and Sparganophilidae) of Pennsylvania, USA. *Megadrilogica* 11(March):131–146.

Chapter 13: Land Planarian

[1] Ogren RE & Sheldon JK (1991) Ecological observations on the land planarian *Bipalium pennsylvanicum* Ogren, with references to phenology, reproduction, growth tale and food niche. *Journal of the Pennsylvania Academy of Science* 65(3–9); Ogren RE (1987) Description of a new three-lined land planarian of the genus *Bipalium* (Turbellaria: Tricladida) from Pennsylvania, USA. *Transactions of the American Microscopical Society* 106(1):21–30.
[2] Ogren, Description of a new three-lined land planarian.
[3] Ibid.
[4] Ducey PK, West L-J, Shaw G, & Lisle JD (2005) Reproductive ecology and evolution in the invasive terrestrial planarian *Bipalium adventitium* across North America. *Pedobiologia* 49(4):367–377.
[5] Ducey PK & Noce S (1998) Successful invasion of New York State by the terrestrial flatworm, *Bipalium adventitium*. *Northeastern Naturalist* 5(3):199–206.
[6] Klots AB (1960) A terrestrial flatworm well established outdoors in the northeastern United States. *Systematic Zoology* 9(1):33–34.
[7] Fiore C, Tull JL, Zehner S, & Ducey PK (2004) Tracking and predation on earthworms by the invasive terrestrial planarian *Bipalium adventitium* (Tricladida, Platyhelminthes). *Behavioural Processes* 67(3):327–334.
[8] Dindal DL (1970) Feeding behavior of a terrestrial turbellarian *Bipalium adventitium*. *American Midland Naturalist* 83(2):635–637.
[9] Watermolen DJ & Fojut P (2008) *An Introduced Flatworm,* Bipalium adventitium *(Tricladida: Terricola), in Wisconsin and Its Potential Impacts.* Miscellaneous Publication PUB-SS-1041 2008 (Bureau of Science Services, Department of Natural Resources, Madison, WI).
[10] Zaborski ER (2002) Observations on feeding behavior by the terrestrial flatworm *Bipalium adventitium* (Platyhelminthes: Tricladida: Terricola) from Illinois. *American Midland Naturalist* 148(2):401–408.
[11] Ogren & Sheldon, Ecological observations on the land planarian *Bipalium pennsylvanicum* Ogren.
[12] Christensen OM & Mather JG (1998) The "New Zealand flatworm," *Artioposthia triangulata*, in Europe: The Faroese situation. *Pedobiologia* 42(5/6):532–540; Blackshaw RP & Stewart VI (1992) *Artioposthia triangulata* (Dendy, 1894), a predatory terrestrial planarian and its potential impact on lumbricid earthworms. *Agricultural Zoology Reviews* 5:201–219.
[13] Boag B & Yeates GW (2001) The potential impact of the New Zealand flatworm, a predator of earthworms, in western Europe. *Ecological Applications* 11(5):1276-1286; Watermolen DJ & Fojut P (2008) *An Introduced Flatworm, Bipalium adventitium*.
[14] Ogren RE (1985) The human factor in the spread of an exotic land planarian in Pennsylvania. *Proceedings of the Pennsylvania Academy of Science* 59:117–118.
[15] Ducey PK, Email correspondence, April 29, 2012.
[16] Leidy J (1862) *Report to the Councils of Philadelphia on Some of the Insects Injurious to Our Shade Trees* (s.n., Philadelphia), 1–11. (Leidy's home address is recorded on the last page of this document.)
[17] Leidy J (1851 [1852]) Helminthological contributions Nr. 3. *Proceedings of the Academy of Natural Sciences of Philadelphia* 5:239–244; Leidy J (1851 [1852]) Contributions to helminthology. *Proceedings of the Academy of Natural Sciences of Philadelphia* 5:349-351.
[18] Ogren RE (1981) Land planarians in Pennsylvania. *Proceedings of the Pennsylvania Academy of Science* 55:52–56.
[19] Ogren RE (1989) Identification features of the two-lined land planarian *Rhynchodemus sylvaticus*, with evidence that *Rhynchodemus americanus* is conspecific. *Transactions of the American Microscopical Society* 108(1):40–44.
[20] Ogren RE (1995) Predation behaviour of land planarians. *Hydrobiologia* 305(1–3):105–111.
[21] Ogren R (1955) Ecological observations on the occurrence of *Rhynchodemus*, a terrestrial turbellarian. *Transactions of the American Microscopical Society* 74(54–60).
[22] Ogren & Sheldon, Ecological observations on the land planarian *Bipalium pennsylvanicum* Ogren.

Chapter 14: American Robin

[1] Catesby M (1731) *The Natural History of Carolina, Florida and the Bahama Islands: Containing the Figures of Birds, Beasts, Fishes, Serpents, Insects and Plants: Particularly, the Forest—Trees, Shrubs, and Other Plants, not Hitherto Described, or Very Incorrectly Figured by Authors. Together with Their Descriptions in English and French. To Which Are Added Observations on the Air, Soil, and Waters: With Remarks upon Agriculture, Grain, Pulse Roots &C. To the Whole Is Prefixed a New and Correct Map of the Countries Treated of.* vol. I (printed at the expence of the author, and sold by W. Innys and R. Manby, at the West End of St. Paul's, by Mr. Hauksbee, at the Royal Society House, and by the author, at Mr. Bacon's in Hoxton, London).

[2] Wilson A (1808) Robin. *Turdus migratorius.* In *American Ornithology; Or the Natural History of the Birds of the United States,* vol. I (Bradford and Inskeep, Philadelphia), 35–39.

[3] Ibid., 36–37.

[4] Morris GS (1907) William Bartram. *Cassinia* 10:1–9.

[5] Wilson A, Robin. *Turdus migratorius*, 36–37.

[6] Shultz B (1795) *An Inaugural Botanico-Medical Dissertation, on the* Phytolacca decandra *of Linnaeus* (submitted to the examination of the Rev. John Ewing, S.T.P. provost; the trustees and medical faculty of the University of Pennsylvania, for the degree of doctor of medicine, on the twenty-first day of May 1795, University of Pennsylvania, Philadelphia), 22.

[7] Audubon JJ (1834) *Ornithological Biography, Or an Account of the Habits of the Birds of the United States of America: Accompanied by Descriptions of the Objects Represented in the Work Entitled* The Birds of America, *and Interspersed with Delineations of American Scenery and Manners,* vol. 3 (Adam and Charles Black, Edinburgh), 191.

[8] Kalm P (1771) *Travels into North America; Containing Its Natural History, and a Circumstantial Account of Its Plantations and Agriculture in General, with the Civil, Ecclesiastical and Commercial State of the Country, the Manners of the Inhabitants, and Several Curious and Important Remarks on Various Subjects,* vol. 2. trans. JR Forster (T. Lowndes, London), 90.

[9] Stone W (1937) *Bird Studies at Old Cape May,* vol. II (Delaware Valley Ornithological Club and the Academy of Natural Sciences of Philadelphia, Philadelphia), 768.

[10] Bent AC (1964) Life histories of North American thrushes, kinglets and their allies. In reprint of *Smithsonian Institution, United States National Museum Bulletin 196,* 1949, U.S. Government Printing Office (Dover, New York).

[11] Sauer JR, Hines JE, Fallon JE, Pardieck KL, et al. (2012) *The North American Breeding Bird Survey Results and Analysis, 1966–2011. Version 7.03.2013* (USGS Patuxent Wildlife Research Center, Laurel, MD), http://www.mbr-pwrc.usgs.gov/bbs/.

[12] McWilliam GM & Brauning DW (2000) *The Birds of Pennsylvania* (Cornell University Press, Ithaca, NY).

[13] Sauer et al., *The North American Breeding Bird Survey Results and Analysis, 1966–2011.*

[14] Anderson TR (2006) *The Biology of the Ubiquitous House Sparrow: From Genes to Populations* (Oxford University Press, Oxford, UK).

[15] Johnston RF & Janiga M (1995) *Feral Pigeons* (Oxford University Press, New York).

[16] Eiserer LA (1980) Effects of grass length and mowing on foraging behavior of the American robin (*Turdus migratorius*). *The Auk* 97:576–580.

[17] Todd TC, James SW, & Seastedt TR (1992) Soil invertebrate and plant responses to mowing and carbofuran application in a North American tallgrass prairie. *Plant and Soil* 144(1):117–124; Baker SW, Firth SJ, & Binns DJ (2000) The effect of mowing regime and the use of acidifying fertiliser on rates of earthworm casting on golf fairways. *Journal of Turfgrass Science* 76:2–11; Byrne LB, Bruns MA, & Kim KC (2008) Ecosystem properties of urban land covers at the aboveground–belowground interface. *Ecosystems* 11:1065–1077.

[18] Edwards C & Lofty J (1972) *Biology of Earthworms* (Chapman and Hall, London).

[19] Lutz FE (1941) *A Lot of Insects: Entomology in a Suburban Garden* (G. P. Putnam and Sons, New York); Owen J (2010) *Wildlife of a Garden: A Thirty-Year Study* (Royal Horticultural Society, Peterborough, UK); Smith R, Warren P, Thompson K, & Gaston K (2006) Urban domestic gardens (VI): Environmental correlates of invertebrate species richness. *Biodiversity and Conservation* 15(8):2415–2438; Smith RM, Gaston KJ, Warren PH, & Thompson K (2006) Urban domestic gardens (VIII): Environmental correlates of invertebrate abundance. *Biodiversity and Conservation* 15(8):2515–2545.

[20] Wheelwright NT (1986) The diet of American robins: An analysis of U.S. Biological Survey records. *The Auk* 103(4):710–725.

[21] Peters HS (1936) A list of external parasites from birds of the eastern part of the United States. *Bird-Banding* 7(1):9–27.

[22] Stafford KC, Bladen VC, & Magnarelli LA (1995) Ticks (Acari: Ixodidae) infesting wild birds (Aves) and white-footed mice in Lyme, CT. *Journal of Medical Entomology* 32(4):453–466.

[23] Ginsberg HS, Buckley PA, Balmforth MG, Zhioua E., et al. (2005) Reservoir competence of native North American birds for the Lyme disease spirochete, *Borrelia burgdorferi*. *Journal of Medical Entomology* 42(3):445–449.

[24] Peters R (2009) *Avian Tick Burdens across an Urban to Forest Land-Use Gradient* (master of science thesis, George Mason University, Fairfax, VA).

[25] Emslie SD (1981) Birds and prehistoric agriculture: The New Mexican pueblos. *Human Ecology* 9(3):305–329.

[26] McConaughy MA (2008) Current issues in paleoethnobotanical research from Pennsylvania and vicinity. In *Current Northeast Paleoethnobotany II*. New York State Museum Bulletin Series 512, ed. JP Hart (The New York State Education Department, Albany, NY), 9–27.

[27] Delcourt PA & Delcourt HR (2004) *Prehistoric Native Americans and Ecological Change: Human Ecosystems in Eastern North America since the Pleistocene* (Cambridge University Press, Cambridge, UK).

[28] Emslie, Birds and prehistoric agriculture; Hargrave LL (1939) Bird bones from abandoned Indian dwellings in Arizona and Utah. *Condor* 41:206–210.

[29] Voelker G, Rohwer S, Bowie R, & Outlaw D (2007) Molecular systematics of a speciose, cosmopolitan songbird genus: Defining the limits of, and relationships among, the *Turdus* thrushes. *Molecular Phylogenetics and Evolution* 42(2):422–434; Nylander JAA, Olsson U, Alstrom P, & Sanmartin I (2008) Accounting for phylogenetic uncertainty in biogeography: A Bayesian approach to dispersal-vicariance analysis of the thrushes (Aves: *Turdus*). *Systematic Biology* 57(2):257–268.

[30] Gleditsch JM & Carlo TA (2011) Fruit quantity of invasive shrubs predicts the abundance of common native avian frugivores in central Pennsylvania. *Diversity and Distributions* 17(2):244–253; Greenberg CH & Walter ST (2010) Fleshy fruit removal and nutritional composition of winter-fruiting plants: A comparison of non-native invasive and native species. *Natural Areas Journal* 30(3):312–321; Suthers HB, Bickal JM, & Rodewald PG (2000) Use of successional habitat and fruit resources by songbirds during autumn migration in central New Jersey. *The Wilson Bulletin* 112(2):249–260.

[31] Wheelwright, The diet of American robins.

[32] Morneau F, Lépine C, Décarie R, Villard M-A, et al. (1995) Reproduction of American robin (*Turdus migratorius*) in a suburban environment. *Landscape and Urban Planning* 32(1):55–62.

[33] Barton WPC (1818) *Compendium Florae Philadelphicae Containing a Description of the Indigenous and Naturalized Plants Found within a Circuit of Ten Miles around Philadelphia,* vol. 1 (M. Carey and Son, Philadelphia), 128.

[34] Gleditsch & Carlo, Fruit quantity of invasive shrubs predicts the abundance of common native avian frugivores in central Pennsylvania.

[35] Barton, *Compendium Florae Philadelphicae*, 154.

[36] Sauer et al., *The North American Breeding Bird Survey Results and Analysis, 1966–2011.*

[37] Wauer RH (1999) *The American Robin* (University of Texas Press, Austin).

[38] Bird J, Alcock J, & Erckmann WJ (1973) Starlings stealing worms from robins. *The Wilson Bulletin* 85(4):480–482.

[39] Sauer et al., *The North American Breeding Bird Survey Results and Analysis, 1966–2011.*

[40] LaDeau SL, Kilpatrick AM, & Marra PP (2007) West Nile virus emergence and large-scale declines of North American bird populations. *Nature* 447:710–713; Rahbek C (2007) The silence of the robins. *Nature* 447:652–653.

[41] Philadelphia Department of Public Health, Division of Disease Control (2010) Annual Report, https://hip.phila.gov/xv/Portals/0/HIP/Annual_Reports/DDC_Annual_Report_%202010_revised.pdf

[42] Wertheimer AM (2012) West Nile virus: An update on recent developments. *Clinical Microbiology Newsletter* 34(9):67–71.

[43] Loss SR, Will T, & Marra PP (2013) The impact of free-ranging domestic cats on wildlife of the United States. *Nature Communications* 4:1396. http://dx.doi.org/10.1038/ncomms2380.

[44] Wauer, *The American Robin.*

[45] Wilson A (1810) Cow bunting. *Emberiza pecoris.* In *American Ornithology; Or the Natural History of the Birds of the United States: Illustrated with Plates Engraved and Colored from Original Drawings Taken from Nature*, vol. 2 (Bradford and Inskeep, Philadelphia), 145–146.

[46] Rothstein SI (1975) An experimental and teleonomic investigation of avian brood parasitism. *Condor* 77:250–271.

[47] Rothstein SI (1982) Mechanisms of avian egg recognition: Which egg parameters elicit responses by rejecter species? *Behavioral Ecology and Sociobiology* 11(4):229–239.

[48] Smith JNM, Taitt MJ, & Zanette L (2002) Removing brown-headed cowbirds increases seasonal fecundity and population growth in song sparrows. *Ecology* 83:3037–3047.

[49] Wauer, *The American Robin*; Friedmann H (1929) *The Cowbirds. A Study of the Biology of Social Parasitism* (Charles C. Thomas, Baltimore, MD).

[50] Rothstein SI & Peer BD (2005) Conservation solutions for threatened and endangered cowbird (*Molothrus* spp.) hosts: Separating fact from fiction. *Ornithological Monographs* 57:98–114.

[51] Barton BS (1779) *Fragments of a Natural History of Pennsylvania* (Way & Groff, Philadelphia) [Willughby Society reprint, edited by Osbert Salvin and published in 1883 by Taylor and Francis, London], 16.

[52] Burhans DE & Thompson FR (2006) Songbird abundance and parasitism differ between urban and rural shrublands. *Ecological Applications* 16:394–405.

[53] Briskie JV, Sealy SG, & Hobson KA (1992) Behavioral defenses against avian brood parasitism in sympatric and allopatric host populations. *Evolutionary Biology* 46(2):334–340.

[54] Avilés JM, Stokke BG, & Parejo D (2006) Relationship between nest predation suffered by hosts and brown-headed cowbird parasitism: A comparative study. *Evolutionary Ecology* 20(2):97–111.

[55] Kilner RM (2006) The evolution of egg colour and patterning in birds. *Biological Reviews* 81(3):383–406.
[56] Brittingham MC & Temple SA (1983) Have cowbirds caused forest songbirds to decline? *BioScience* 33(1):31–35; Rodewald AD (2009) Urban-associated habitat alteration promotes brood parasitism of Acadian Flycatchers. *Journal of Field Ornithology* 80(3):234–241.
[57] Stone W (1913) Bird migration records of William Bartram 1802–22. *The Auk* 30:325–358.
[58] Sallabanks R & Courtney SP (1992) Frugivory, seed predation, and insect-vertebrate interactions. *Annual Review of Entomology* 37:377–400.
[59] Wilson A, Robin. *Turdus migratorius*, 35–36.
[60] Environment Canada (2012) Canadian Atlas of Bird Banding. Encounter Map Data Table. American Robin (*Turdus migratorius*) 761.0, http://www.ec.gc.ca/aobc-cabb/index.aspx?lang=en&nav=encounter-Map&aou=761&map=3.
[61] Nese J & Schwartz G (2002) *The Philadelphia Area Weather Book* (Temple University Press, Philadelphia).
[62] Zhang X, Friedl MA, Schaaf CB, Strahler AH, et al. (2004) The footprint of urban climates on vegetation phenology. *Geophysical Research Letters* 31:L12209. doi:10.1029/2004GL020137.
[63] Oka M (2011) The Influence of urban street characteristics on pedestrian heat comfort levels in Philadelphia. *Transactions in GIS* 15(1):109–123.
[64] Fan H & Sailor DJ (2005) Modeling the impacts of anthropogenic heating on the urban climate of Philadelphia: A comparison of implementations in two PBL schemes. *Atmospheric Environment* 39(1):73–84.
[65] Shustack D (2008) *Reproductive Timing of Passerines in Urbanizing Landscapes* (dissertation presented in partial fulfillment of the requirements for the degree of doctor of philosophy, Ohio State University, Columbus).
[66] Inouye DW, Barr B, Armitage KB, & Inouye BD (2000) Climate change is affecting altitudinal migrants and hibernating species. *Proceedings of the National Academy of Science* 97(4):1630–1633.
[67] Wilson WH Jr. (2007) Spring arrival dates of migratory breeding birds in Maine: Sensitivity to climate change. *The Wilson Journal of Ornithology* 119(4):665–677.
[68] Miller MW (2006) Apparent effects of light pollution on singing behavior of American robins. *Condor* 108(1):130–139.
[69] Kempenaers B, Borgström P, Loës P, Schlicht E, et al. (2010) Artificial night lighting affects dawn song, extra-pair siring success, and lay date in songbirds. *Current Biology* 20(19):1735–1739.
[70] Fuller RA, Warren PH, & Gaston KJ (2007) Daytime noise predicts nocturnal singing in urban robins. *Biology Letters* 3(4):368–370.
[71] Seger KD (2007) *Avian Bioacoustics in Urbanizing Landscapes: Relationships between Urban Noise and Avian Singing Behavior* (honors thesis presented in partial fulfillment of the requirements for graduation with research distinction in the undergraduate colleges of Ohio State University, Columbus); Seger-Fullam KD, Rodewald AD, & Soha JA (2011) Urban noise predicts song frequency in northern cardinals and American robins. *Bioacoustics* 20(3):267–276.
[72] Dowling JL, Luther DA, & Marra PP (2012) Comparative effects of urban development and anthropogenic noise on bird songs. *Behavioral Ecology* 23(1):201–209.
[73] Wilson A, Robin. *Turdus migratorius*, 37–38.
[74] Eason PK, Sherman PT, Rankin O, & Coleman B (2006) Factors affecting flight initiation distance in American robins. *The Journal of Wildlife Management* 70(6):1796–1800.
[75] Clucas B & Marzluff JM (2012) Attitudes and actions toward birds in urban areas: Human cultural differences influence bird behavior. *The Auk* 129(1):8–16.

Chapter 15: Chinese Mantid

[1] Laurent P (1898) A species of Orthoptera. *Entomological News* 9:144–145.
[2] Ibid., 144.
[3] Jacobs E (1898) *Tenodera sinensis*. *Entomological News* 9:170.
[4] Meehan T (1858) *How to Plant Trees, Including a Catalogue of the Ornamental and Fruit Trees, Shrubs, etc. Cultivated and for Sale at the Germantown Nurseries near Philadelphia, PA* (A. Ketterlinus, Philadelphia).
[5] Thierolf WR (1928) The economic importance of *Paratenodera sinensis*. *Entomological News* 39:112–116, 140–145.
[6] Gurney AB (1950) Praying mantids of the United States, native and introduced. In *Annual Report of the Board of Regents of the Smithsonian Institution* (Smithsonian Institution, Washington, DC), 339–362.
[7] BugGuide (2012) Subspecies *Tenodera sinensis sinensis* —Chinese Mantid, http://bugguide.net/node/view/10098/data.
[8] Bartley JA (1982) Movement patterns in adult male and female mantids, *Tenodera aridifolia sinensis* Saussure (Orthoptera: Mantodea). *Environmental Entomology* 11(5):1108–1111.
[9] Hurd LE (1999) Ecology of praying mantids. In *The Praying Mantids*, eds. FR Prete, H Wells, PH Wells, & LE Hurd, vol. 3 (Johns Hopkins University Press, Baltimore, MD), 43–60.
[10] Jensen D, Svenson GJ, Song H, & Whiting MF (2010) Phylogeny and evolution of male genitalia within the praying mantis genus *Tenodera* (Mantodea: Mantidae). *Invertebrate Systematics* 23(5):409–421.
[11] Gurney, Praying mantids of the United States, native and introduced.

[12] Rehn JAG (1935) The Orthoptera of Costa Rica, part I: Mantidae. *Proceedings of the Academy of Natural Sciences of Philadelphia* 87:167–272.
[13] Hebard M (1937) Where and when to find the Orthoptera of Pennsylvania, with notes on the species which in distribution reach nearest this state. *Entomological News* 48(8):219–226.
[14] Eisenberg RM, Hurd LE, Fagan WF, Tilmon KJ, et al. (1992) Adult dispersal of *Tenodera aridifolia sinensis* (Mantodea: Mantidae). *Environmental Entomology* 21(2):350–353.
[15] Grimaldi DA (2003) A revision of Cretaceous mantises and their relationships, including new taxa (Insecta: Dictyoptera: Mantodea). In *American Museum novitates*; no. 3412 (American Museum of Natural History, New York).
[16] Gurney, Praying mantids of the United States, native and introduced.
[17] Ibid.
[18] Rau P & Rau N (1913) Biology of *Stagmomantis carolina*. *Transactions of the Academy of Science of St. Louis* 22(1):1–58; pl. 1–18.
[19] Skinner H (1899) Minutes of the meeting of the American Entomological Society, Academy of Natural Sciences of Philadelphia. Entomological Section. *Entomological News* 10(March):79–80.
[20] Hurd, Ecology of praying mantids.
[21] Breed GA, Stichter S, & Crone EE (2012) Climate-driven changes in northeastern US butterfly communities. *Nature Climate Change* 3:142–145.
[22] Colinvaux P (1978) *Why Big Fierce Animals Are Rare: An Ecologist's Perspective* (Princeton University Press, Princeton, NJ).

Chapter 16: Pillbugs

[1] Smigel JT & Gibbs AG (2008) Conglobation in the pill bug, *Armadillidium vulgare*, as a water conservation mechanism. *Journal of Insect Science* 44:1–9.
[2] Say T (1818) An account of the Crustacea of the United States (concluded). *Journal of the Academy of Natural Sciences of Philadelphia* 1(2):423–458.
[3] Garthwaite R, Lawson R, & Sassaman C (1995) Population genetics of *Armadillidium vulgare* in Europe and North America. *Crustacean Issues* 9:145–199.
[4] Jass J & Klausmeier B (2000) Endemics and immigrants: North American terrestrial isopods (Isopoda, Oniscidea) north of Mexico. *Crustaceana* 73(7):771–799.
[5] French H, Demitroff M, & Newell WL (2009) Past permafrost on the mid-Atlantic Coastal Plain, eastern United States. *Permafrost and Periglacial Processes* 20(3):285–294.
[6] Fowler HW (1911) Crustacea of New Jersey. In *Annual Report of the New Jersey State Museum* (New Jersey State Museum, Trenton), 29–650; pl. 1–150.
[7] Ganter PF & Hanton WK (1984) A note on the cause of skewed sex ratios in populations of terrestrial isopods in North Carolina. *Crustaceana* 46(2):154–159.
[8] Rigaud T, Soutygrosset C, Raimond R, Mocquard JP, & Juchault P (1991) Feminizing endocytobiosis in the terrestrial crustacean *Armadillidium vulgare* Latr. (Isopoda) - recent acquisitions. *Endocytobiosis and Cell Research* 7(3):259 - 273.
[9] Hertig M & Wolbach SB (1924) Studies on Rickettsia-like micro-organisms in insects. *The Journal of Medical Research* 44:329–374, pl. XXVII–XXX.
[10] Stouthamer R, Breeuwer J, & Hurst G (1999) *Wolbachia pipientis*: Microbial manipulator of arthropod reproduction. *Annual Review of Microbiology* 53:71–102.
[11] Bouchon D, Rigaud T, & Juchault P (1998) Evidence for widespread *Wolbachia* infection in isopod crustaceans: Molecular identification and host feminization. *Proceedings of the Royal Society of London B* 265(1401):1081–1090.
[12] Jiggins FM & Hurst GDD (2011) Rapid insect evolution by symbiont transfer. *Science* 332(6026):185–186.
[13] Hilgenboecker K, Hammerstein P, Schlattmann P, Telschow A, et al. (2008) How many species are infected with *Wolbachia*? A statistical analysis of current data. *FEMS Microbiology Letters* 281:215–220.
[14] Werren JH, Baldo L, & Clark ME (2008) *Wolbachia*: Master manipulators of invertebrate biology. *Nature Reviews Microbiology* 6(10):741–751.
[15] Ibid.
[16] Stouthamer R, Luck R, & Hamilton W (1990) Antibiotics cause parthenogenetic *Trichogramma* (Hymenoptera/Trichogrammatidae) to revert to sex. *Proceedings of the National Academy of Sciences of the United States of America* 87(7):2424–2427.
[17] Himler AG, Adachi-Hagimori T, Bergen JE, Kozuch A, et al. (2011) Rapid spread of a bacterial symbiont in an invasive whitefly is driven by fitness benefits and female bias. *Science* 332(6026):254–256.
[18] Hosokawa T, Koga R, Kikuchi Y, Meng X-Y, et al. (2010) *Wolbachia* as a bacteriocyte-associated nutritional mutualist. *Proceedings of the National Academy of Science* 107(2):769–774.
[19] Paris OH (1965) The vagility of P32-labeled isopods in grassland. *Ecology* 46:635–648.
[20] Kight SL (2009) Reproductive ecology of terrestrial isopods (Crustacea: Oniscidea). *Terrestrial Arthropod Reviews* 1(2):95–110.

[21] Moreau J, Bertin A, Caubet Y, & Rigaud T (2001) Sexual selection in an isopod with *Wolbachia*-induced sex reversal: Males prefer real females. *Journal of Evolutionary Biology* 14(3):388–394.

[22] Rigaud T & Moreau J (2004) A cost of *Wolbachia*-induced sex reversal and female-biased sex ratios: Decrease in female fertility after sperm depletion in a terrestrial isopod. *Proceedings of the Royal Society of London B* 271:1941–1946.

[23] Gavotte L, Mercer DR, Stoeckle JJ, & Dobson SL (2010) Costs and benefits of *Wolbachia* infection in immature *Aedes albopictus* depend upon sex and competition level. *Journal of Invertebrate Pathology* 105:341–346.

[24] Rigaud T, Mocquard J, & Juchault P (1992) The spread of parasitic sex factors in populations of *Armadillidium vulgare* Latr (Crustacea, Oniscidea): Effects on sex ratio. *Genetics Selection Evolution* 24(1):3–18.

[25] Koehncke A, Telschow A, Werren JH, & Hammerstein P (2009) Life and death of an influential passenger: *Wolbachia* and the evolution of CI-modifiers by their hosts. *PLoS ONE* 4(2):e4425. doi:10.1371/journal.pone.0004425.

[26] Cooper CL & Crites JL (1976) Community ecology of helminth parasitism in an insular passerine avifauna. *The Journal of Parasitology* 62(1):105–110.

[27] Van Cleave HJ (1918) The Acanthocephala of North American birds. *Transactions of the American Microscopical Society* 37(1):19–48.

[28] Van Cleave HJ (1924) A critical study of the Acanthocephala described and identified by Joseph Leidy. *Proceedings of the Academy of Natural Sciences of Philadelphia* 76:279–334.

[29] Van Cleave HJ (1942) A reconsideration of *Plagiorhynchus formosus* and observations on Acanthocephala with atypical lemnisci. *Transactions of the American Microscopical Society* 61(2):206–210.

[30] Bush AO (2001) *Parasitism: The Diversity and Ecology of Animal Parasites* (Cambridge University Press, Cambridge, UK).

[31] Sinitsin DT (1929) *Journal of Parasitology* 15(4):287. This is a report recorded in the minutes of the *Proceedings of the Helminthological Society of Washington.* It designates *Plagiorhynchus cylindraceus* as *Plagiorhynchus formosus,* a synonym. See Schmidt GD (1981) *Plagiorhynchus formosus* Van Cleave, 1918, a synonym of *Plagiorhynchus cylindraceus* (Goeze, 1782) Schmidt and Kuntz, 1966. *The Journal of Parasitology* 67(4):597–598.

[32] Schmidt GD & Olsen OW (1964) Life cycle and development of *Prosthorhynchus formosus* (Van Cleave, 1918) Travassos, 1926, an Acanthocephalan parasite of birds. *The Journal of Parasitology* 50(6):721–730.

[33] Connors VA & Nickol BB (1991) Effects of *Plagiorhynchus cylindraceus* (Acanthocephala) on the energy metabolism of adult starlings, *Sturnus vulgaris*. *Parasitology Research* 103:395–402.

[34] Schmidt & Olsen, Life cycle and development of *Prosthorhynchus formosus*.

[35] Nickol BB & Dappen GE (1982) *Armadillidium vulgare* (Isopoda) as an intermediate host of *Plagiorhynchus cylindraceus* (Acanthocephala) and isopod response to infection. *The Journal of Parasitology* 68(4):570–575.

[36] Moore J (1983) Responses of an avian predator and its isopod prey to an Acanthocephalan parasite. *Ecology* 64(5):1000–1015.

[37] Bouchon, Rigaud, & Juchault, Evidence for widespread *Wolbachia* infection in isopod crustaceans.

[38] Cooper & Crites, Community ecology of helminth parasitism in an insular passerine avifauna.

[39] Williams T (2008) Natural invertebrate hosts of iridoviruses (Iridoviridae). *Neotropical Entomology* 37(6):615–632.

[40] Schultz GA (1961) Distribution and establishment of a land isopod in North America. *Systematic Zoology* 10(4):193–196.

[41] Adamkewicz S (1969) Colour polymorphism in the land isopod *Armadillidium nasatum*. *Heredity (Edinb)* 24(2):249–264.

[42] Pollard SD, Jackson RR, Olphen AV, & Robertson MW (1995) Does *Dysdera crocata* (Araneae, Dysderidae) prefer woodlice as prey? *Ethology Ecology & Evolution* 7(3):271–275; Rezác M & Pekár S (2007) Evidence for woodlice-specialization in *Dysdera* spiders: Behavioural versus developmental approaches. *Physiological Entomology* 32(4):367–371; Rezác M, Král J, & Pekár S (2007) The spider genus *Dysdera* (Araneae, Dysderidae) in central Europe: Revision and natural history. *Journal of Arachnology* 35:432–462.

[43] Pollard SD (1986) Prey capture in *Dysdera crocata* (Araneae: Dysderidae), a long fanged spider. *New Zealand Journal of Zoology* 13(149–150); Rezác M, Pekár S, & Lubin Y (2008) How oniscophagous spiders overcome woodlouse armour. *Journal of Zoology* 275(1):64–71.

[44] Sutton SL (1972) *Invertebrate Types: Woodlice* (Ginn & Company Limited, London).

[45] Kaston BJ (1981) Spiders of Connecticut, rev. ed. *State Geological and Natural History Survey of Connecticut* (Department of Environmental Protection. Bulletin 70, Hartford).

[46] Allee W (1926) Studies in animal aggregations: Causes and effects of bunching in land isopods. *Journal of Experimental Zoology* 45(1):255–277.

[47] Allee WC (1931) *Animal Aggregations: A Study in General Sociology* (University of Chicago Press, Chicago), 43–44.

[48] Takeda N (1983) The aggregation phenomenon in terrestrial isopods. *Symposia of the Zoological Society of London* 53: 381–404.

[49] Hassall M, Edwards DP, Carmenta R, Derhé MA, et al. (2010) Predicting the effect of climate change on aggregation behaviour in four species of terrestrial isopods. *Behaviour* 147(2):151–164.

[50] Devigne C, Broly P, & J-L D (2011) Individual preferences and social interactions determine the aggregation of woodlice. *PLoS ONE* 6(2):e17389. doi:10.1371/journal. pone.0017389.

[51] Hamilton WD (1971) Geometry of the selfish herd. *Journal of Theoretical Biology* 31:295–311.

[52] Wise DH (1995) *Spiders in Ecological Webs* (Cambridge University Press, Cambridge, UK).

[53] Cook A (1981) Huddling and the control of water loss by the slug *Limax pseudoflavus* Evans. *Animal Behaviour* 29(1):289–298.

[54] Welsford IG, Banta PA, & Prior DJ (1990) Size-dependent responses to dehydration in the terrestrial slug, *Limax maximus* L.: Locomotor activity and huddling behavior. *Journal of Experimental Zoology* 253(2):229–234; South A (1992) *Terrestrial Slugs: Biology, Ecology and Control* (Chapman & Hall, London); Waite TA (1988) Huddling and postural adjustments in response to desiccating conditions in *Deroceras reticulatum*. *Journal of Molluscan Studies* 54(2):249–250.

[55] *Lehmannia valentiana* is also known as *Ambigolimax valentianus*. Common names include threeband gardenslug, greenhouse slug, and Valencia slug. See ME Paustian, *Ambigolimax valentianus,* at Terrestrial Slugs Web, http://terrslugs.lifedesks.org/pages/31328, accessed November 24, 2013.

[56] Simberloff D & Holle BV (1999) Positive interactions of nonindigenous species: Invasional meltdown? *Biological Invasions* 1(1):21–32.

Chapter 17: Common Milkweed

[1] Barton BS (1804) Memorandum concerning a new vegetable muscipula. *Transactions of the American Philosophical Society* 6(1):79–82. Quote is from pp. 79–80.

[2] Morse DH (1981) Modification of bumblebee foraging: The effect of milkweed pollinia. *Ecology* 62(1):89–97.

[3] Corry TH (1883) On the structure and development of the gynostegium and the mode of fertilization of *Asclepias Cornuti*, Decaisne (*A. syriaca*, L.). *Transactions of the Linnean Society of London*, 2nd Series: Botany. 2(8):173–207.

[4] Barton WPC (1818) *Compendium Florae Philadelphicae Containing a Description of the Indigenous and Naturalized Plants Found within a Circuit of Ten Miles around Philadelphia,* vol. 1 (M. Carey and Son, Philadelphia), 131.

[5] Bhowmik P & Bandeen J (1976) The biology of Canadian weeds: 19. *Asclepias syriaca* L. *Canadian Journal of Plant Science* 56(3):579–589.

[6] Baker HG (1955) Self-compatibility and establishment after "long-distance" dispersal. *Evolution* 9(3):347–349.

[7] Stebbins GL (1957) Self-fertilization and population variability in the higher plants. *American Naturalist* 91(337–354).

[8] Sparrow FK & Pearson NL (1948) Pollen compatibility in *Asclepias syriaca*. *Journal of Agricultural Research* 77:187–199.

[9] Evetts LL & Burnside OC (1974) Root distribution and vegetative propagation of *Asclepias syriaca* L. *Weed Research* 14(5):283–288.

[10] Kephart SR (1981) Breeding systems in *Asclepias incarnata* L., *A. syriaca* L., and *A. verticillata* L. *American Journal of Botany* 68(2):226–232.

[11] Gascoigne J, Berec L, Gregory S, & Courchamp F (2009) Dangerously few liaisons: A review of mate-finding Allee effects. *Population Ecology* 51(3):355–372.

[12] Livshultz T, Mead JV, Goyder DJ, & Brannin M (2011) Climate niches of milkweeds with plesiomorphic traits (Secamonoideae; Apocynaceae) and the milkweed sister group link ancient African climates and floral evolution. *American Journal of Botany* 98(12):1966–1977.

[13] Harder LD & Johnson SD (2008) Function and evolution of aggregated pollen in angiosperms. *International Journal of Plant Sciences* 169:59–78.

[14] Wyatt R, Broyles SB, & Lipow SR (2000) Pollen-ovule ratios in milkweeds (Asclepiadaceae): An exception that probes the rule. *Systematic Botany* 25(2):171–180.

[15] Frost SW (1958) *Halysidota tessellaris* S & A and pollenia. *Entomological News* 69(5):137–138.

[16] Frost SW (1965) Insects and pollinia. *Ecology* 46(4):556–558.

[17] The Academy of Natural Sciences, Philadelphia (2004) The Titian Peale Butterfly and Moth Collection, http://clade.ansp.org/entomology/collections/peale/index.html.

[18] Tietz HM (1952) *The Lepidoptera of Pennsylvania: A Manual* (Pennsylvania State College, School of Agriculture, Agricultural Experiment Station, State College, PA).

[19] Morse DH & Fritz RS (1983) Contributions of diurnal and nocturnal insects to the pollination of common milkweed (*Asclepias syriaca* L.) in a pollen-limited system. *Oecologia* 60(2):190–197.

[20] Jennersten O & Morse DH (1991) The quality of pollination by diurnal and nocturnal insects visiting common milkweed, *Asclepias syriaca*. *American Midland Naturalist* 125(1):18–28.

[21] Morse & Fritz, Contributions of diurnal and nocturnal insects to the pollination of common milkweed.

[22] Jennersten & Morse, The quality of pollination by diurnal and nocturnal insects visiting common milkweed.

[23] McFrederick QS, Kathilankal JC, & Fuentes JD (2008) Air pollution modifies floral scent trails. *Atmospheric Environment* 42(10):2336–2348.

[24] Duchelle SF & Skelly JM (1981) Response of common milkweed to oxidant air pollution in the Shenandoah National Park in Virginia. *Plant Diseases* 65(8):661–663; City of Philadelphia Department of Public Health Air Management Services (2010) Philadelphia's Air Quality Report, http://www.phila.gov/health/pdfs/airmanagement/AQR_210_Final.pdf.

[25] Gregg JW, Jones CG, & Dawson TE (2003) Urbanization effects on tree growth in the vicinity of New York City. *Nature* 424:183–187.
[26] Mulligan GA & Findlay JN (1970) Reproductive systems and colonization in Canadian weeds. *Canadian Journal of Botany* 48:859–860.
[27] Lyman JC & Ellstrand NC (1984) Clonal diversity in *Taraxacum officinale* (Compositae), an apomict. *Heredity* 53:1–10.
[28] Stephenson AG & Thomas WW (1977) Diurnal and nocturnal pollination of *Catalpa speciosa* (Bignoniaceae). *Systematic Botany* 2(3):191-198.
[29] Aldrich PR, Brusa A, Heinz CA, Greer GK, et al. (2008) Floral visitation of the invasive stinking ash in western suburban Chicago. *Transactions of the Illinois State Academy of Science* 101(1&2):1–12.
[30] Kowarik I & Säumel I (2007) Biological flora of Central Europe: *Ailanthus altissima* (Mill.) Swingle. *Perspectives in Plant Ecology, Evolution and Systematics* 8(4):207–237.
[31] Aldrich et al. Floral visitation of the invasive stinking ash in western suburban Chicago
[32] Thompson JS (2008) *Pollination Biology of* Ailanthus altissima *(Mill.) Swingle (Tree-of-Heaven) in the Mid-Atlantic United States* (master's thesis, Department of Entomology, Virginia Polytechnic Institute and State University, Blacksburg, VA).
[33] Turkington R & Burdon JJ (1983) The biology of Canadian weeds: 57. *Trifolium repens* L. *Canadian Journal of Plant Science* 63(1):243–266.
[34] Bhowmik & Bandeen, The biology of Canadian weeds: 19. *Asclepias syriaca* L.
[35] Ibid.
[36] Hartzler RG (2010) Reduction in common milkweed (*Asclepias syriaca*) occurrence in Iowa cropland from 1999 to 2009. *Crop Protection* 29(12):1542–1544.
[37] Pleasants JM & Oberhauser KS (2013) Milkweed loss in agricultural fields because of herbicide use: Effect on the monarch butterfly population. *Insect Conservation and Diversity* 6(2):135–144. doi: 10.1111/j.1752-4598.2012.00196.x.
[38] Fry JT (1996) An international catalogue of North American trees and shrubs: The Bartram Broadside, 1783. Bartram's garden catalogue of North American plants. *Journal of Garden History* 16(1):1–66.

Chapter 18: Purple-stemmed cliffbrake

[1] Gardner ATE (1955) A Philadelphia masterpiece: Haviland's prison. *The Metropolitan Museum of Art Bulletin*, New Series, 14(4):103–108; Dolan FX (2007) *Eastern State Penitentiary* (Arcadia Publishing, Mount Pleasant, SC); Eastern State Penitentiary, http://www.easternstate.org/learn/timeline.
[2] Tryon AF (1957) A revision of the fern genus *Pellaea* section *Pellaea*. *Annals of the Missouri Botanical Garden* 44(2):125–193.
[3] Muhlenberg H (1793) Index florae Lancastriensis. *Transactions of the American Philosophical Society of Philadelphia* 3:157–184.
[4] Darlington W (1837) *Flora Cestrica: An Attempt to Enumerate and Describe the Flowering and Filicoid Plants of Chester County in the State of Pennsylvania. With Brief Notices of Their Properties, and Uses, in Medicine, Domestic and Rural Economy, and the Arts* (self-published, Westchester, PA).
[5] Barton WPC (1818) *Compendium Florae Philadelphicae Containing a Description of the Indigenous and Naturalized Plants Found within a Circuit of Ten Miles around Philadelphia,* 2 vols. (M. Carey and Son, Philadelphia).
[6] Keller I & Brown S (1905) *Handbook of the Flora of Philadelphia and Vicinity* (Philadelphia Botanical Club, Philadelphia).
[7] Steil WN (1911) Apogamy in *Pellaea atropurpurea*. *Botanical Gazette* 52(5):400–401.
[8] Wagner WH Jr., Farrar DR, & Chen KL (1965) A new sexual form of *Pellaea glabella* var. *glabella* from Missouri. *American Fern Journal* 55(4):171–178.
[9] Steil WN. Apogamy in *Pellaea atropurpurea*.
[10] Pickett F (1931) Notes on xerophytic ferns. *American Fern Journal* 21(2):49–57.
[11] Jenks MA & Wood AJ (2007) Plant desiccation tolerance: Diversity, distribution, and real world applications. In *Plant Desiccation Tolerance*, eds. MA Jenks & AJ Wood (Blackwell, Ames, Iowa), 3–10.
[12] Pickett, Notes on xerophytic ferns.
[13] Wherry ET (1920) The soil reactions of certain rock ferns: I. *American Fern Journal* 10(1):15–22.
[14] Wild M & Gagnon D (2005) Does lack of available suitable habitat explain the patchy distributions of rare calcicole fern species? *Ecography* 28(2):191–196.
[15] Moran RC (2008) Diversity, biogeography, and floristics. In *Biology and Evolution of Ferns and Lycophytes*, eds. TA Ranker & CH Haufler (Cambridge University Press, NY), 367–394.
[16] McVaugh R (1935) Studies on the spores of some northeastern ferns. *American Fern Journal* 25(3):73–85.
[17] LaTorre C & Bhushan B (2005) Nanotribological characterization of human hair and skin using atomic force microscopy. *Ultramicroscopy* 105:155–175.
[18] Moran, Diversity, biogeography, and floristics.

[19] Thorsteinson A (1988) Urban airflow dynamics and mosquito infestations. *Bulletin of the Society of Vector Ecologists* 13(1):97–101.
[20] Hendrix SD (1980) An evolutionary and ecological perspective of the insect fauna of ferns. *American Naturalist* 115:171–196.
[21] Wherry ET (1969) Growing ferns in rocks. In *Handbook on Ferns: A Special Printing of Plants & Gardens,* vol.25, no.1, eds. HS Hull & MJ Dietz (Brooklyn Botanic Garden, Brooklyn, NY), 34.
[22] Foster FG (1999) *Ferns to Know and Grow* (Timber Press, Portland, OR).
[23] Peattie DC (1925) The flora of an historic wall. *American Midland Naturalist* 9(9):381–383.
[24] Benedict R (1952) Ferns on a prison wall. *American Fern Journal* 42(1):19–20.
[25] Strong MT, Strong JL, & Kelloff CL (2009) *Pellaea atropururea* (L.) Link growing in the mortar of the historical brick kiln here (Smithsonian Institution, Washington, DC), http://collections.si.edu/search/results.htm?q=record_ID:nmnhbotany_2866430.
[26] Hill EJ (1899) The habitats of the *Pellaeas*. *Bulletin of the Torrey Botanical Club* 26(11):596–598.
[27] Gorbushina AA & Broughton WJ (2009) Microbiology of the atmosphere-rock interface: How biological interactions and physical stresses modulate a sophisticated microbial ecosystem. *Annual Review of Microbiology* 63:431–450.
[28] Dyer AF & Lindsay S (1992) Soil spore banks of temperate ferns. *American Fern Journal* 82(3):89–123.
[29] Gilbert OL (1992) *Rooted in Stone: The Natural Flora of Urban Walls* (English Nature, Peterborough, UK).
[30] McKee HJ (1973) *Introduction to Early American Masonry: Stone, Brick, Mortar and Plaster.* (National Trust for Historic Preservation, and Columbia University, Washington, DC).
[31] Pickett FL & Manuel ME (1926) An ecological study of certain ferns: *Pellaea atropurpurea* (L.) Link and *Pellaea glabella* Mettenius. *Bulletin of the Torrey Botanical Club* 53(1):1–5.

Chapter 19: Mugwort

[1] Barton WPC (1818) *Compendium Florae Philadelphicae Containing a Description of the Indigenous and Naturalized Plants Found within a Circuit of Ten Miles around Philadelphia*, 2 vols. (M. Carey and Son, Philadelphia).
[2] Tatnall RR (1938–1939) Nuttall's plant collections in southern Delaware. *Bartonia* 20:1–6; Barney JN (2006) North American history of two invasive plant species: Phytogeographic distribution, dispersal vectors, and multiple introductions. *Biological Invasions* 8(4):703–717.
[3] Cutler M (1785 [1903]) An account of some of the vegetable productions, naturally growing in this part of America, botanically arranged by the Rev. Manasseh Cutler, F.A.A. and M.S., and Member of the Philosophical Society of Philadelphia [Facsimile edition published in 1903]. *Bulletin of the Lloyd Library of Botany, Pharmacy and Materia Medica. Bulletin #7.* Reproduction Series No. 4 (JU Lloyd & CG Lloyd, Cincinnati, OH).
[4] Fernald ML (1900) Some Jesuit influences upon our northeastern flora. *Rhodora* 2(19):133–142.
[5] Salmon W (1710) *Botanologia, the English Herbal, or, History of Plants* (Dawks, Rhodes, and Taylor, London), 740.
[6] Miller J (1722) *Botanicum Officinale; Or a Compendious Herbal: Giving an Account of All Such Plants as Are Now Used in the Practice of Physick. With Their Descriptions and Virtues* (E. Bell, J. Senex, W. Taylor, and J. Osborn, London), 52.
[7] Barney, North American history of two invasive plant species.
[8] Barney JN, Whitlow TH, & DiTommaso A (2009) Evolution of an invasive phenotype: Shift to belowground dominance and enhanced competitive ability in the introduced range. *Plant Ecology* 202(2):275–284.
[9] Stebbins G (1970) Adaptive radiation of reproductive characteristics in angiosperms, I: Pollination mechanisms. *Annual Review of Ecology and Systematics* 1:307–326.
[10] Torrey J & Gray A (1841) *A Flora of North America: Containing Abridged Descriptions of All Known Indigenous and Naturalized Plants Growing North of Mexico*, vol. 1 (Wiley and Putnam, New York), 421.
[11] Barney, North American history of two invasive plant species.
[12] Burk I (1877) List of plants recently collected on ships' ballast in the neighborhood of Philadelphia. *Proceedings of the Academy of Natural Sciences of Philadelphia* 29(1877):105–109.
[13] Barney JN & DiTommaso A (2003) The biology of Canadian weeds. 118. *Artemisia vulgaris* L. *Canadian Journal of Plant Science* 83:205–215.
[14] Barney, North American history of two invasive plant species.
[15] Darlington W (1837) *Flora Cestrica: An Attempt to Enumerate and Describe the Flowering and Filicoid Plants of Chester County in the State of Pennsylvania. With Brief Notices of Their Properties, and Uses, in Medicine, Domestic and Rural Economy, and the Arts* (self-published, Westchester, PA), 491.
[16] Stone HE (1945) *A Flora of Chester County, Pennsylvania: With Especial Reference to the* Flora Cestrica *of Dr. William Darlington*, 2 vols. (Academy of Natural Sciences of Pennsylvania, Philadelphia); Stone HE (1929) A centennial survey of the Chester County flora. *Bartonia* 11:36–48.
[17] Holm L, Doll J, Holm E, Pancho J, et al. (1997) *World Weeds: Natural History and Distribution* (John Wiley and Sons, New York).
[18] Rhoads AF & Klein WM, Jr. (1993) *The Vascular Flora of Pennsylvania: Annotated Checklist and Atlas* (American Philosophical Society, Philadelphia).

[19] Beck LC (1833) *Botany of the Northern and Middle States; or a Description of the Plants Found in the United States, North of Virginia, Arranged According to the Natural System. With a Synopsis of the Genera According to the Linnaean System—a Sketch of the Rudiments of Botany, and a Glossary of Terms* (Webster and Skinners, Albany, NY).

[20] Kosisky SE, Marks MS, & Nelson MR (2010) Pollen aeroallergens in the Washington, DC, metropolitan area: A 10-year volumetric survey (1998–2007). *Annals of Allergy, Asthma and Immunology* 104:223–235.

[21] Holm L et al., *World Weeds*.

[22] Barney et al., The biology of Canadian weeds. 118. Artemisia vulgaris L.

Chapter 20: Freshwater sponge

[1] Leidy J (1851 [1852]) *Spongilla fragilis. Proceedings of the Academy of Natural Sciences of Philadelphia* 5:277–278.

[2] Leidy J (1870) Further observations on *Urnatella. Proceedings of the Academy of Natural Sciences of Philadelphia* 22:100.

[3] Potts E (1887) *Fresh Water Sponges: A Monograph.* Including "Diagnosis of European Spongillidae" by Franz Vejdovsky (Academy of Natural Sciences of Philadelphia, Philadelphia).

[4] Ibid., 200.

[5] Leidy J (1876) Bituminous sediment of the Schuylkill River. In *Proceedings of the Academy of Natural Sciences of Philadelphia*, ed. EJ Nolan (Academy of Natural Sciences, Philadelphia), 193.

[6] Roberts WM, et al. (1875) *Report on the Water Supply for the City of Philadelphia: Made by the Commission of Engineers Appointed by the Mayor under the Ordinance of Councils, Approved June 5th* (Commission on Water-Supply, Philadelphia), 7.

[7] Ibid., 66.

[8] McCaffery TF III (1980–1982) Hazardous waste regulation: An evaluation from an historical perspective. *Columbia Journal of Environmental Law* 7:251–288.

[9] Smith RA (1852) *Philadelphia as It Is in 1852: Being a Correct Guide to All the Public Buildings; Literary, Scientific, and Benevolent Institutions; and Places of Amusement; Remarkable Objects; Manufacturies; Commercial Warehouses; and Wholesale and Retail Stores in Philadelphia and Its Vicinity* (Lindsay and Blakiston, Philadelphia).

[10] Romig CL (1980) Schuylkill River desilting project. In *The Schuylkill River Symposium*, ed. LF Berseth (The Academy of Natural Sciences of Philadelphia, Philadelphia), 18–21.

[11] Strickland W, Gill EH, & Campbell HR, eds. (1841) *The Philadelphia Gas Works* (J. Weale, London).

[12] Patrick R (1980) The Schuylkill River in changing times. In *The Schuylkill River Symposium*, ed. LF Berseth (The Academy of Natural Sciences of Philadelphia, Philadelphia), 9–12.

[13] Thorp JH & Rogers DC (2011) *Field Guide to Freshwater Invertebrates of North America* (Academic Press, Burlington, MA).

[14] Schuylkill River Project Engineers, Pennsylvania Water and Power Resources Board (1951) *The Schuylkill River Desilting Project: Final Report of the Schuylkill River Project Engineers 1 July 1951* (Commonwealth of Pennsylvania, Dept. of Forests and Waters, Harrisburg, PA).

[15] Romig, Schuylkill River desilting project.

[16] Patrick, The Schuylkill River in changing times, 10.

[17] Schuylkill River Project Engineers, *The Schuylkill River Desilting Project,* pl. 12.

[18] Kramek N & Loh L (2007) *The History of Philadelphia's Water Supply and Sanitation System: Lessons in Sustainability for Developing Urban Water Systems.* (Philadelphia Global Water Initiative, Philadelphia), http://esa.un.org/iys/docs/san_lib_docs/Philalessonsinsustainability%5B1%5D.pdf.

[19] Baker A (2002) *The Schuylkill River Park Public Art Process: An Ethnographic Focus on a Philadelphia Urban Park's Development* (dissertation submitted to the Temple University Graduate Board in partial fulfillment of the requirements for the degree of doctor of philosophy, Temple University, Philadelphia, PA).

[20] Perillo JA & Butler LH (2009) Evaluating the use of Fairmount Dam fish passage facility with application to anadromous fish restoration in the Schuylkill River, Pennsylvania. *Journal of the Pennsylvania Academy of Science* 83(1):24–33.

[21] Romig, Schuylkill River desilting project.

[22] Raymond PA & Oh N-H (2009) Long term changes of chemical weathering products in rivers heavily impacted from acid mine drainage: Insights on the impact of coal mining on regional and global carbon and sulfur budgets. *Earth and Planetary Science Letters* 284:50–55.

[23] Holst A (2007) The Philadelphia Water Department and the burden of history. *Public Works Management Policy* 11(3):233–238.

[24] Delaware Riverkeeper Network (2007) Protecting streams in Pennsylvania: A resource for municipal officials, http://www.delawareriverkeeper.org/pdf/Protecting%20Streams%20in%20PA.pdf.

[25] Kauffman GJ, Homsey AR, Belden AC, & Sanchez JR (2011) Water quality trends in the Delaware River basin (USA) from 1980 to 2005. *Environmental Monitoring and Assessment* 177(1–4):193–225.

[26] Interlandi SJ & Crockett CS (2003) Recent water quality trends in the Schuylkill River, Pennsylvania, USA: A preliminary assessment of the relative influences of climate, river discharge and suburban development. *Water Research* 37(8):1737–1748.

[27] Raymond & Oh, Long term changes of chemical weathering products in rivers heavily impacted from acid mine drainage.

[28] Perillo & Butler, Evaluating the use of Fairmount Dam fish passage facility.

[29] Ettinger WS (1982) Macrobenthos of the freshwater tidal Schuylkill River at Philadelphia, Pennsylvania. *Journal of Freshwater Ecology* 1(6):599–606.

[30] Leidy, *Spongilla fragilis*.

[31] Potts E (1884 [1885]) On the wide distribution of some American sponges. *Proceedings of the Academy of Natural Sciences of Philadelphia* 36:215–217.

[32] Penney JT & Racek AA (1968) Comprehensive revision of a worldwide collection of freshwater sponges (Porifera: Spongillidae). In *United States National Museum Bulletin 272* (Smithsonian Institution Press, Washington, DC).

[33] Smith DG (2001) *Pennak's Freshwater Invertebrates of the United States: Porifera to Crustacea*, 4th ed. (John Wiley and Sons, New York).

[34] Fell PE & Fell AE (1987) Cold hardiness of the gemmules of *Eunapius fragilis* (Porifera: Spongillidae). *Transactions of the American Microscopical Society* 106(2):187–189; Fell PE (1990) Environmental factors affecting dormancy in the freshwater sponge *Eunapius fragilis* (Leidy). *Invertebrate Reproduction & Development* 18(3):213–219; Fell PE (1992) Salinity tolerance of the gemmules of *Eunapius fragilis* (Leidy) and the inhibition of germination by various salts. *Hydrobiologia* 242(1):33–39; Loomis SH, Hand SC, & Fell PE (1996) Metabolism of gemmules from the freshwater sponge *Eunapius fragilis* during diapause and post-diapause states. *Biological Bulletin* 191(3):385–392.

[35] Fell, Environmental factors affecting dormancy in the freshwater sponge *Eunapius fragilis* (Leidy).

[36] Kipp R, Bailey SA, MacIsaac HJ, & Ricciardi A (2010) Transoceanic ships as vectors for nonindigenous freshwater bryozoans. *Diversity and Distributions* 16(1):77–83.

[37] Smith, *Pennak's Freshwater Invertebrates of the United States.*

[38] Leidy J (1851 [1852]) *Cristatella (Pectinatella) magnifica. Proceedings of the Academy of Natural Sciences of Philadelphia* 5:265–266.

[39] Wood TS (2010) Bryozoans. In *Ecology and Classification of North American Freshwater Invertebrates,* 3rd ed., eds. JH Thorp & AP Covich (Elsevier, New York), 437–454.

[40] Brown HP (1952) The life history of *Climacia areolaris* (Hagen), a neuropterous "parasite" of fresh water sponges. *American Midland Naturalist* 47(1):130–160.

[41] Ibid.

[42] Bilger MD, Riva-Murray K, & Wall GL (2005) *A Checklist of the Aquatic Invertebrates of the Delaware River Basin, 1990–2000.* U.S. Geological Survey Data Series 116. (U.S. Department of the Interior, U.S. Geological Survey, Reston, VA).

[43] Roberts et al., *Report on the Water Supply for the City of Philadelphia.*

[44] Carson R (1962) *Silent Spring* (Houghton Mifflin, Boston).

[45] Potts, *Fresh Water Sponges: A Monograph.*

[46] Leidy, Bituminous sediment of the Schuylkill River.

[47] Roberts et al., *Report on the Water Supply for the City of Philadelphia.*

[48] Potts E (1884 [1885]) Freshwater sponges as improbable causes of the pollution of river-water. *Proceedings of the Academy of Natural Sciences of Philadelphia* 36:28–30.

[49] Harrison E (1974) Sponges (Porifera: Spongillidae). In *Pollution Ecology of Freshwater Invertebrates,* eds. CW Hart Jr & SLH Fuller (Academic Press, New York), 29–66.

[50] Roback SS (1974) Insects (Arthropoda: Insecta). In *Pollution Ecology of Freshwater Invertebrates*, eds. CW Hart Jr & SLH Fuller (Academic Press, New York), 314–376.

[51] Reiswig HM, Frost TM, & Ricciardi A (2010) Porifera. In *Ecology and Classification of North American Freshwater Invertebrates,* 3rd ed., eds. JH Thorp & AP Covich (Elsevier, New York), 91–123.

Chapter 21: Brown bullhead

[1] Lesueur CA (1819) Notice de quelques poissons découverts dans les lacs du Haut-Canada, durant l'été de 1816. *Mémoires du Muséum d'Histoire Naturelle* 5:148–161.

[2] Norris T (1808) *American Fish-Culture: Embracing All the Details of Artificial Breeding and Rearing of Trout; the Culture of Salmon, Shad, and Other Fishes* (Porter & Coates, Philadelphia).

[3] Ibid., 214.

[4] Fowler HW (1914) Fishes in polluted waters. *Copeia* 5:4.

[5] Ibid.

[6] Jordan DS & Evermann BW (1905) *American Food and Game Fishes: A Popular Account of All the Species Found in America North of the Equator, with Keys for Ready Identification, Life Histories and Methods of Capture* (Doubleday, Page & Co, New York).

[7] Ibid., 27.

[8] DeWalt RE, Rash VH, & Hilsenhoff WL (2010) Diversity and classification of insects and Collembola. In *Ecology and Classification of North American Freshwater Invertebrates*, 3rd ed., eds. JH Thorp & AP Covich (Academic Press, New York), 587-657.

[9] Carr J & Hiltunen J (1965) Changes in the bottom fauna of western Lake Erie from 1930 to 1961. *Limnology and Oceanography* 10(4):551–569.

[10] Doherty MSE, Hudson PL, Ciborowski JJH, & Schloesser DW (1999) Morphological deformities in larval Chironomidae (Diptera) from the western basin of Lake Erie: A historical comparison. In *Proceedings of the 25th Annual Aquatic Toxicity Workshop: October 18–21, 1998, Quebec City,* eds. RV Collie, R Chasse, L Hare, C Julien, et al. Canadian Technical Report of Fisheries and Aquatic Sciences, No. 2269:134.

[11] Lesko LT, Smith SB, & Blouin MA (1996) The effect of contaminated sediments on fecundity of the brown bullhead in three Lake Erie tributaries. *Journal of Great Lakes Research* 22(4):830–837.

[12] Johnson MS & Munger F (1930) Observations on excessive abundance of the midge *Chironomus plumosus* at Lake Pepin. *Ecology* 11(1):110–126.

[13] Ali A (1995) A concise review of chironomid midges (Diptera: Chironomidae) as pests and their management. *Journal of Vector Ecology* 21(2):105–121.

[14] Bilger MD, Riva-Murray K, & Wall GL (2005) *A Checklist of the Aquatic Invertebrates of the Delaware River Basin, 1990–2000, U.S. Geological Survey Data Series 116* (U.S. Department of the Interior, USGS, Reston, VA).

[15] Carr & Hiltunen, Changes in the bottom fauna of western Lake Erie from 1930 to 1961.

[16] Govedich FR, Bain BA, Moser WE, Gelder SR, et al. (2010) Annelida (Clitellata): Oligochaeta, Branchiobdellida, Hirundinida, Acanthobdellida. In *Ecology and Classification of North American Freshwater Invertebrates*, 3rd ed., eds. JH Thorp & AP Covich (Academic Press, New York), 385–436.

[17] Klarberg DP & Benson A (1975) Food habits of *Ictalurus nebulosus* in acid polluted water of northern West Virginia. *Transactions of the American Fisheries Society* 104(3):541–547.

[18] Ettinger WS (1982) Macrobenthos of the freshwater tidal Schuylkill River at Philadelphia, Pennsylvania. *Journal of Freshwater Ecology* 1(6):599–606.

[19] Leidy J (1876) Bituminous sediment of the Schuylkill River. In *Proceedings of the Academy of Natural Sciences of Philadelphia*, ed. EJ Nolan (Academy of Natural Sciences, Philadelphia), 193.

[20] Beddard FE (1895) *A Monograph of the Order of Oligochaeta* (Clarendon Press, Oxford).

[21] Ettinger, Macrobenthos of the freshwater tidal Schuylkill River.

[22] Klarberg & Benson, Food habits of *Ictalurus nebulosus* in acid polluted water of northern West Virginia.

[23] Ibid.

[24] Marvin DE & Heath AG (1968) Cardiac and respiratory responses to gradual hypoxia in three ecologically distinct species of fresh-water fish. *Comparative Biochemistry and Physiology* 27(1):349–355.

[25] Wismer DA & Christie AE (1987) *Relationships of Great Lakes Fishes: A Data Compilation. Great Lakes Fisheries Commission Special Publication 87-3* (Great Lakes Fisheries Commission, Ann Arbor, MI).

[26] Reisinger HJ II (1981) Trace metals in a conceptualized Schuylkill River food chain. In *Proceedings of the Schuylkill River Symposium*, ed. LF Berseth (Samuel S. Fels Fund and the Academy of Natural Sciences of Philadelphia, Philadelphia), 53–65.

[27] Hightower JE, Fleming WJ, & Hayman MA (1999) Effects of contaminated sediments on brown bullhead age structure, growth, and condition in a North Carolina tidal creek. In *American Fisheries Society Symposium 24. Catfish 2000: Proceedings of the International Ictalurid Symposium, held at Davenport, Iowa, June 23–25, 1998*, eds. E Irwin, W Hubert, C Rabeni, H Schramm Jr, et al. (American Fisheries Society, Bethesda, MD), 161–172.

[28] Blumer LS (1986) The function of parental care in the brown bullhead *Ictalurus nebulosus. American Midland Naturalist* 115(2):234–238.

[29] Grabarkiewicz JD & Davis WS (2008) *An Introduction to Freshwater Fishes as Biological Indicators. EPA-260-R-08-016* (U.S. Environmental Protection Agency, Office of Environmental Information, Washington, DC).

[30] Fowler, Fishes in polluted waters.

[31] Steyermark AC, Spotila JR, Gillette D, & Isseroff H (1999) Biomarkers indicate health problems in brown bullheads from the industrialized Schuylkill River, Philadelphia. *Transactions of the American Fisheries Society* 128(2):328–338.

[32] Pennsylvania Department of Environmental Protection (2012) *Commonwealth of Pennsylvania Fish Consumption Advisory Listing for 2012, by River Basin*. Revised September 23, 2011, http://files.dep.state.pa.us/Water/Drinking%20Water%20and%20Facility%20Regulation/WaterQualityPortalFiles/FishConsumption/FishAdvisory/FishConsAdvTables2012-Rev092311-Final.pdf.

[33] Weintraub M & Birnbaum LS (2008) Catfish consumption as a contributor to elevated PCB levels in a non-Hispanic black subpopulation. *Environmental Research* 107(3):412–417.

[34] Grasman KA, Scanlon PF, & Fox GA (1998) Reproductive and physiological effects of environmental contaminants in fish-eating birds of the Great Lakes: A review of historical trends. *Environmental Monitoring and Assessment* 53(1):117–145.

[35] Matz AC & Parsons KC (2004) Organochlorines in black-crowned night heron (*Nycticorax nycticorax*) eggs reflect persistent contamination in Northeastern U.S. estuaries. *Archives of Environmental Contamination and Toxicology* 46(2):270–274.

[36] Larson JM, Karasov WH, Sileo L, Stromborg KL, et al. (1996) Reproductive success, developmental anomalies, and environmental contaminants in double-crested cormorants (*Phalacrocorax auritus*). *Environmental Toxicology and Chemistry* 15(4):553–559.

[37] Toschik PC, Rattner BA, McGowan PC, Christman MC, et al. (2005) Effects of contaminant exposure on reproductive success of ospreys (*Pandion haliaetus*) nesting in Delaware River and Bay, USA. *Environmental Toxicology and Chemistry* 24(3):617–628.

[38] Barker JL (1984) Organochlorine and polychlorinated biphenyl residues at four trophic levels in the Schuylkill River, Pennsylvania. Paper 2262. In *Selected Papers in the Hydrologic Sciences 1984: U.S. Geological Survey Water-Supply*, ed. EL Meyer (Superintendent of Documents, U.S. Government Printing Office, Washington, DC), 25–31.

[39] Zaranko DT, Griffiths RW, & Kaushik NK (1997) Biomagnification of polychlorinated biphenyls through a riverine food web. *Environmental Toxicology and Chemistry* 16(7):1463–1471.

[40] Lucké B & Schlumberger HG (1941) Transplantable epitheliomas of the lip and mouth of catfish: I. Pathology. Transplantation to anterior chamber of eye and into cornea. *Journal of Experimental Medicine* 74(5):397–408.

[41] Pinkney AE, Harshbarger JC, Karouna-Renier NK, Jenko K, et al. (2011) Tumor prevalence and biomarkers of genotoxicity in brown bullhead (*Ameiurus nebulosus*) in Chesapeake Bay tributaries. *Science of the Total Environment* 410–411:248–257.

[42] Pinkney A, Harshbarger J, & Roberts M (2004) *Tumor prevalence in brown bullheads (*Ameiurus nebulosus*) from Darby Creek, John Heinz National Wildlife Refuge at Tinicum, Philadelphia, PA* (U.S. Fish and Wildlife Service, Annapolis, MD, and State College, PA).

[43] Delaware River Basin Commission (1988) *Fish Health and Contamination Study* (Superintendent of Documents, U.S. Government Printing Office, Washington, DC), http://www.gpo.gov/fdsys/pkg/CZIC-sh174-f57-1988/html/CZIC-sh174-f57-1988.htm.

[44] Sindermann CJ (1979) Pollution-associated diseases and abnormalities of fish and shellfish: A review. *Fishery Bulletin* 76(4):717–749.

[45] Ibid.

[46] Baumann PC (1998) Epizootics of cancer in fish associated with genotoxins in sediment and water. *Mutation Research/Reviews in Mutation Research* 411(3):227–233.

[47] Baumann PC & Harshbarger JC (1995) Decline in liver neoplasms in wild brown bullhead catfish after coking plant closes and environmental PAHs plummet. *Environmental Health Perspectives* 103(2):168–170.

[48] Spitsbergen JM & Wolf MJ (1995) The riddle of hepatic neoplasia in brown bullheads from relatively unpolluted waters in New York State. *Toxicologic Pathology* 23(6):716–725.

[49] Pinkney et al., Tumor prevalence and biomarkers of genotoxicity in brown bullhead (*Ameiurus nebulosus*) in Chesapeake Bay tributaries.

[50] Boyer MR (1995) Catfish. In *Living Resources of the Delaware Estuary*, eds. LE Dove & RM Nyman (The Delaware Estuary Program, Wilmington), 157–166.

[51] Brown J, Perillo J, Kwak T, & Horwitz R (2005) Implications of *Pylodictis olivaris* (flathead catfish) introduction into the Delaware and Susquehanna drainages. *Northeastern Naturalist* 12(4):473–484; Perillo JA & Butler LH (2009) Evaluating the use of Fairmount Dam fish passage facility with application to anadromous fish restoration in the Schuylkill River, Pennsylvania. *Journal of the Pennsylvania Academy of Science* 83(1):24–33.

[52] Harmon PL (1980) Abundance of distribution of fishes in the Schuylkill River. In *Proceedings of the Schuylkill River Symposium, September 24–25, 1980,* ed. LF Berseth (Samuel S. Fels Fund and the Academy of Natural Sciences of Philadelphia), 85–100.

[53] Mulfinger RM & Kaufman M (1980) Fish passage at the Fairmount Fishway in 1979 and 1980 with implications for the Schuylkill River fisheries through future fishway construction. In *Proceedings of the Schuylkill River Symposium, September 24–25, 1980,* ed. LF Berseth (Samuel S. Fels Fund and the Academy of Natural Sciences of Philadelphia), 101–124.

[54] Leo Sheng, Conversation with the author, September 8, 2012.

[55] Boyer, Catfish.

[56] U.S. Environmental Protection Agency (2012) Polychlorinated Biphenyls (PCBs), http://www.epa.gov/epawaste/hazard/tsd/pcbs/index.htm.

[57] Brown et al., Implications of *Pylodictis olivaris* (flathead catfish) introduction into the Delaware and Susquehanna drainages.

[58] Boyer, Catfish.

[59] Perillo J, Conversation with the author, September 7, 2013.

[60] Saylor RK, Lapointe NWR, & Angermeier PL (2012) Diet of non-native northern snakehead (*Channa argus*) compared to three co-occurring predators in the lower Potomac River, USA. *Ecology of Freshwater Fish* 21(3):443–452.

[61] Ettinger, Macrobenthos of the freshwater tidal Schuylkill River.

[62] Exelon Generation Company (2008) *Biological Source Water Characterization for Schuylkill Generating Station*. Prepared for Exelon Generation Company, LLC, 300 Exelon Way, Kennett Square, PA 19348 by GB Waterfield, BW Lees, & RW Blye Jr. of Normandeau Associates, Inc., 400 Old Reading Pike, Building A, Suite 101, Stowe, PA 19464 (Exelon Generation Company, Kennett Square, PA).

[63] Delaware River Basin Commission (2012) Notice of Commission Meeting and Public Hearing [FR Doc No: 2012-15398]. *Federal Register* 77(122):37887–37890, http://www.gpo.gov/fdsys/pkg/FR-2012-06-25/html/2012-15398.htm; Exelon Corporation (2012) *Schuylkill Generating Station,* http://www.exeloncorp.com/PowerPlants/schuylkill/Pages/profile.aspx;Veolia Energy (2011) Veolia Energy, Philadelphia, PA, http://www.veoliaenergyna.com/veolia-energy-north-america/locations/philadelphia.htm.

[64] U.S. Energy Information Administration (2012) *Pennsylvania Nuclear Power Plants 2010,* http://www.eia.gov/nuclear/state/pennsylvania/.

[65] Exelon Generation Company (2008) *Design and Construction Technology Plan, Schuylkill Generating Station, Philadelphia, December 2008.* Technical Consultants: ARCADIS; Veritas Economic Consulting; Normandeau Associates, Inc. (Exelon Generation Company, Kennett Square, PA); Exelon Generation Company (2008) *Impingement Mortality and Entrainment Characterization Study, Schuylkill Generating Station, Philadelphia, December 2008.* Technical Consultants: Normandeau Associates, Inc; Veritas Economic Consulting; ARCADIS (Exelon Generation Company, Kennett Square, PA).

[66] Exelon Generation Company, *Design and Construction Technology Plan, Schuylkill Generating Station, Philadelphia, December 2008.*

[67] Exelon Generation Company (2008) *Historical Impingement and Entrainment Comparisons for Schuylkill Generating Station.* Prepared for Exelon Generation Company, LLC, 300 Exelon Way, Kennett Square, PA 19348 by GB Waterfield, BW Lees, & RW Blye Jr. of Normandeau Associates, Inc., 400 Old Reading Pike, Building A, Suite 101, Stowe, PA 19464. # 20645.005 (Exelon Generation Company, Kennett Square, PA).

[68] Lane JA, Portt CB, & Minns CK (1996) *Spawning Habitat Characteristics of Great Lakes Fishes. Canadian Manuscript Report of Fisheries and Aquatic Sciences No. 2368* (Department of Fisheries and Oceans, ONT, Canada).

[69] Massengill RR (1973) Change in feeding and body condition of brown bullheads overwintering in the heated effluent of a power plant. *Chesapeake Science* 14(2):138–141.

[70] U.S. Geological Survey (2012) *USGS 01474500, Schuylkill River at Philadelphia, PA,* http://waterdata.usgs.gov/pa/nwis/uv?cb_00060=on&cb_00065=on&cb_00010=on&format=gif_stats&period=&begin_date=2012-08-01&end_date=2012-08-01&site_no=01474500.

[71] Exelon Generation Company, *Design and Construction Technology Plan, Schuylkill Generating Station, Philadelphia, December 2008.*

[72] U.S. Environmental Protection Agency (2012) *Cooling Water Intake Structures—CWA §316(b),* http://water.epa.gov/lawsregs/lawsguidance/cwa/316b/index.cfm.

[73] Exelon Generation Company, *Design and Construction Technology Plan, Schuylkill Generating Station, Philadelphia, December 2008.*

[74] Ibid.

[75] Foster JR & Wheaton TJ (1981) Losses of juvenile and adult fishes at the Nanticoke Thermal Generating Station due to entrapment, impingement, and entrainment. *Journal of Great Lakes Research* 7(2):162–170.

[76] Bauers S (2012) Single parenthood at the Franklin hawk nest? (blog), http://www.philly.com/philly/blogs/greenliving/Single-parenthood-at-the-Franklin-hawk-nest.html.

[77] Norris, *American Fish-Culture.*

[78] Peck GW (1880) A plea for the bullhead. In *Peck's Fun: Being Extracts from the "La Crosse Sun" and "Peck's Sun," Milwaukee, Carefully Selected with the Object of Affording the Public in One Volume the Cream of Mr. Peck's Writings of the Past Ten Years,* ed. VW Richardson (Belford, Clarke & Co, Chicago), 134–135.

[79] Bauers S (2012) Bald eagle chick hatches in Tinicum nest (blog), http://www.philly.com/philly/blogs/greenliving/Chick-.html.

[80] Viverette CB, Garman GC, McIninch SP, Markham AC, et al. (2007) Finfish-waterbird trophic interactions in tidal freshwater tributaries of the Chesapeake Bay. *Waterbirds* 30(sp1):50–62.

[81] Ibid.

Chapter 22: Red Back Salamander

[1] Green J (1818) Descriptions of several species of North American Amphibia, accompanied with observations. *Journal of the Academy of Natural Sciences of Philadelphia* 1(1):348–359.

[2] Kurtz DW (1991) Jacob Green's chemical philosophy: Insights on the teaching of American college chemistry 160 years ago. *Journal of Chemical Education* 68(3):186–189.

[3] Reed CF (1960) *Plethodon erythronotus* (Raf.), the red-backed salamander. *Herpetologica* 16(3):207–213.

[4] Stuckey RL (1986) Opinions of Rafinesque expressed by his American botanical contemporaries. *Bartonia* 52:26–41. (Rafinesque's disreputable status may explain why people chose to ignore his naming of the red back salamander.)

[5] Rafinesque CS (1818 [March]) Description of a new American salamander—the red back salamander from the Highlands. *The Scientific Journal of New York* 1(2):25–26. (I accessed a photoreproduction of this publication in: Goodwin GH Jr. (1960) Unrecorded papers of Rafinesque and Jacob Green. *Systematic Biology* 9(1):35–36.)

[6] Green, Descriptions of several species of North American Amphibia.

[7] Rafinesque, Description of a new American salamander.

[8] Holbrook JE (1842) *North American Herpetology Or, a Description of the Reptiles Inhabiting the United States*, vol. 5 (J. Dobson, Philadelphia).

[9] Hallowell E (1856 [1857]) Urodeles most abundant in the neighborhood of Philadelphia. *Proceedings of the Academy of Natural Sciences of Philadelphia* 8:101; Stone W (1906) Notes on reptiles and batrachians of Pennsylvania, New Jersey and Delaware. *The American Naturalist* 40(471):159–170; Hulse AC, McCoy CJ, & Censky EJ (2001) *Amphibians and Reptiles of Pennsylvania and the Northeast* (Cornell University Press, Ithaca, NY).

[10] Burger JW (1935) *Plethodon cinereus* (Green) in eastern Pennsylvania and New Jersey. *The American Naturalist* 69(725):578–586; Reams RD, Searcy R, Wyatt JE III, & Gehrmann WH (2008) Habitat utilization by reptiles and amphibians at an urban state park in Indiana. *Bulletin of the Chicago Herpetological Society* 43(2):17–20; Noël S & Lapointe F-J (2010) Urban conservation genetics: Study of a terrestrial salamander in the city. *Biological Conservation* 143(11):2823–2831; Walton BM & Canterbury R (1999) Amphibian and avian fauna of an urban riparian woodland in northeast Ohio. *International Association for Great Lakes Research: Great Lakes, Great Science, Great Cities.* Program and Abstracts, A-114.

[11] Fowler HW & Dunn ER (1917) Notes on salamanders. *Proceedings of the Academy of Natural Sciences of Philadelphia* 69:7–28.

[12] Walton BM, Tsatiris D, & Rivera-Sostre M (2006) Salamanders in forest-floor food webs: Invertebrate species composition influences top-down effects. *Pedobiologia* 50(4):313–321.

[13] Maerz J, Karuzas J, Madison D, & Blossey B (2005) Introduced invertebrates are important prey for a generalist predator. *Diversity and Distributions* 11:83–90.

[14] Ransom TS (2012) Behavioral responses of a native salamander to native and invasive earthworms. *Biological Invasions* 14(12):2601–2616.

[15] Hulse, McCoy, & Censky, *Amphibians and Reptiles of Pennsylvania and the Northeast.*

[16] Wake DB & Hanken J (1996) Direct development in the lungless salamanders: What are the consequences for developmental biology, evolution and phylogenesis? *International Journal of Developmental Biology* 40:859–869.

[17] Cochran ME (1911) The biology of the red-backed salamander (*Plethodon cinereus erythronotus* Green). *Biological Bulletin* 20(6):332–349.

[18] Taub FB (1961) The distribution of the red-backed salamander, *Plethodon c. cinereus*, within the soil. *Ecology* 42(4):681–698.

[19] Heatwole H (1960) Burrowing ability and behavioral responses to desiccation of the salamander, *Plethodon cinereus. Ecology* 41(4):661–668.

[20] Test FH (1946) Relations of the red-backed salamander (*Plethodon cinereus*) to light and contact. *Ecology* 27(3):246–254.

[21] Hood HH (1934) A note on the red-backed salamander at Rochester, New York. *Copeia* 1934(3):141–142.

[22] Jaeger R, Kalvarsky D, & Shimizu N (1982) Territorial behaviour of the red-backed salamander: Expulsion of intruders. *Animal Behaviour* 30:490–496.

[23] Marsh DM, Thakur KA, Bulka KC, & Clarke LB (2004) Dispersal and colonization through open fields by a terrestrial, woodland salamander. *Ecology* 85:3396–3405.

[24] Wind E (2000) Effects of habitat fragmentation on amphibians: What do we know and where do we go from here? In *Proceedings of a Conference on the Biology and Management of Species and Habitats at Risk, Kamloops, BC, February 15–19, 1999,* vol. 2, ed. LM Darling (Ministry of Environment, Lands and Parks, Victoria, BC, and University College of the Cariboo, Kamloops, BC), 885–890.

[25] Bishop SC (1941) *The Salamanders of New York. New York State Museum Bulletin No. 324* (University of the State of New York, Albany).

[26] Gibbs JP (1998) Amphibian movements in response to forest edges, roads, and streambeds in southern New England. *The Journal of Wildlife Management* 62(2):584–589.

[27] Kleeberger SR & Werner JK (1982) Home range and homing behavior of *Plethodon cinereus* in northern Michigan. *Copeia* 1982(2):409–415.

[28] Ousterhout BH & Liebgold EB (2010) Dispersal versus site tenacity of adult and juvenile red-backed salamanders (*Plethodon cinereus*). *Herpetologica* 66(3):269–275.

[29] Cheptou P-O, Carrue O, Rouifed S, & Cantarel A (2008) Rapid evolution of seed dispersal in an urban environment in the weed *Crepis sancta*. *Proceedings of the National Academy of Sciences* 105(10):3796–3799.

[30] Hanson HC (1997) *The Giant Canada Goose,* rev. ed. (Southern Illinois University Press, Carbondale and Edwardsville).

[31] Conover MR (2011) Population growth and movements of Canada geese in New Haven County, Connecticut, during a 25-year period. *Waterbirds 30 (Special Publication 1)* 34(4):412–421.

[32] Gosser AL, Conover MR, & Messmer TA (1997) *Managing Problems Caused by Urban Canada Geese. Berryman Institute Publication 13* (Utah State University, Logan).

[33] Stephenson T & Van Ballenberghe V (1995) Wolf, *Canis lupus,* predation on dusky Canada geese, *Branta canadensis occidentalis. Canadian Field-Naturalist* 109(2):253–255.

[34] Roff DA (1990) The evolution of flightlessness in insects. *Ecological Monographs* 60(4):389–421.

[35] Usinger RL (1966) *Monograph of the Cimicidae (Hemiptera–Heteroptera)* (Entomological Society of America, Lanham, MD).

[36] Piersol WH (1908–9) The habits and larval state of *Plethodon cinereus erythronotus. Transactions of the Canadian Institute* 8:469–493.

[37] Highton R & Savage T (1961) Functions of the brooding behavior in the female red-backed salamander, *Plethodon cinereus. Copeia* 1961(1):95–98.

[38] Sayler A (1966) The reproductive ecology of the red-backed salamander, *Plethodon cinereus,* in Maryland. *Copeia* 1966(2):183–193.

[39] Wake & Hanken, Direct development in the lungless salamanders.

[40] Vieites DR, Min M-S, & Wake DB (2007) Rapid diversification and dispersal during periods of global warming by plethodontid salamanders. *Proceedings of the National Academy of Sciences* 104(50):19903–19907; Wake DB & Marks SB (1993) Development and evolution of plethodontid salamanders: A review of prior studies and a prospectus for future research. *Herpetologica* 49(2):194–203.

[41] Saylor, The reproductive ecology of the red-backed salamander; Hood, A note on the red-backed salamander at Rochester, New York.

[42] Jameson EW Jr. (1944) Food of the red-backed salamander. *Copeia* 1944(3):145–147.

[43] Ibid; Cochran, The biology of the red-backed salamander.

[44] Highton & Savage, Functions of the brooding behavior in the female red-backed salamander.

[45] Piersol, The habits and larval state of *Plethodon cinereus erythronotus.*

[46] Edwards CA & Lofty JR (1972) *Biology of Earthworms* (Chapman and Hall, London).

[47] Saylor, The reproductive ecology of the red-backed salamander.

Chapter 23: Liverwort

[1] Muhlenberg H (1799) Supplementum indicis florae Lancastriensis. *Transactions of the American Philosophical Society* 4:235–242.

[2] Vanderpoorten A & Goffinet B (2009) *Introduction to Bryophytes* (Cambridge University Press, Cambridge, UK).

[3] Schuster RM (1992) *The Hepaticae and Anthocerotae of North America East of the Hundredth Meridian*, vol. VI (Field Museum of Natural History, Chicago).

[4] Konrat MV, Söderström L, Renner MAM, Hagborg A, et al. (2010) Early land plants today (ELPT): How many liverwort species are there? *Phytotaxa* 9:22–40.

[5] Evert RF & Eichhorn SE (2012) *Raven Biology of Plants* (W. H. Freeman, New York).

[6] Hicks ML (2004) Aytoniacea – Reboulia. *Bryophyte Flora of North America, Provisional Publication*, (Missouri Botanical Garden), http://www.mobot.org/plantscience/bfna/V3/AytoReboulia.htm.

[7] Manville GC (1994) *Hepatics of Pennsylvania. Final Report: 1994* (self-published); Porter TC (1904) *Catalogue of the Bryophyta (Hepatics, Anthocerotes and Mosses) and Pteridophyta (Ferns and Fern-Allies) Found in Pennsylvania* (Gin & Company, Boston).

[8] Boisselier-DuBayle MC, Lambourdiere J, & Bischler H (1998) Taxa delimitation in *Reboulia* investigated with morphological, cytological, and isozyme markers. *Bryologist* 101(1):61–69.

[9] Schuster, *The Hepaticae and Anthocerotae of North America East of the Hundredth Meridian,* 159.

[10] Olmsted FL, Kelsey HP, & Officers of the American Civic Association (1908) *The Smoke Nuisance* (American Civic Association, Department of Nuisances, Philadelphia), 4.

[11] Gilbert OL (1968) Bryophytes as indicators of air pollution in the Tyne Valley. *New Phytologist* 67(1):15–30.

[12] Ibid.

[13] Bates JW (2000) Mineral nutrition, substratum ecology and pollution. In *Bryophyte Biology*, eds. AJ Shaw & B Goffinet (Cambridge University Press, Cambridge, UK), 248–311.
[14] Gilbert, Bryophytes as indicators of air pollution in the Tyne Valley.
[15] Zvereva EL & Kozlov MV (2011) Impacts of industrial polluters on bryophytes: A meta-analysis of observational studies. *Water, Air, Soil Pollution* 218:573–586.
[16] City of Philadelphia Department of Public Health Air Management Services. History of Air Pollution Control in Philadelphia, http://www.phila.gov/health/pdfs/air/History.pdf.
[17] Bates, Mineral nutrition, substratum ecology and pollution; City of Philadelphia Department of Public Health Air Management Services. 2012–2013 Air Monitoring Network Plan, http://www.phila.gov/health/pdfs/airmanagement/2012-13%20AMNP%20-%20FINAL.pdf.
[18] Ibid.
[19] Tutt JW (1896) *British Moths* (George Routledge and Sons, Ltd, Manchester), 305.
[20] Laxen DPH & Thompson MA (1987) Sulphur dioxide in Greater London, 1931–1985. *Environmental Pollution* 43(2):103–114.
[21] Rose CI & Hawksworth DL (1981) Lichen recolonization in London's cleaner air. *Nature* 289:289–292.
[22] Majerus MEN (1998) *Melanism: Evolution in Action* (Oxford University Press, Oxford, UK).
[23] Sargent T, Millar C, & Lambert D (1998) The "classical" explanation of industrial melanism: Assessing the evidence. *Evolutionary Biology,* vol. 30, eds. MK Hecht, RJ MacIntyre, & MT Clegg (Plenum Press, New York), 299–322; Cook LM (2000) Changing views on melanic moths. *Biological Journal of the Linnean Society* 69(3):431–441; Rudge D (2006) Myths about moths: A study in contrasts. *Endeavour* 30(1):19–23.
[24] Van't Hof A & Saccheri I (2010) Industrial melanism in the peppered moth is not associated with genetic variation in canonical melanisation gene candidates. *PLoS ONE* 5(5):e10889; Jiggins CD (2011) A Peppered icon enters the genomic era. *Bioscience* 61(9):655–656; Cook LM, Grant BS, Saccheri IJ, & Mallet J (2012) Selective bird predation on the peppered moth: The last experiment of Michael Majerus. *Biology Letters.* doi:10.1098/rsbl.2011.1136.
[25] Majerus MEN (2009) Industrial melanism in the peppered moth, *Biston betularia*: An excellent teaching example of Darwinian evolution in action. *Evolution Education Outreach* 2:63–74.
[26] Owen DF (1961) Industrial melanism in North American moths. *American Naturalist* 95:227–233.
[27] U.S. Environmental Protection Agency (2012). Clean Air Act Requirements and History, http://epa.gov/air/caa/caa_history.html.
[28] Grant BS & Wiseman LL (2002) Recent history of melanism in American peppered moths. *The Journal of Heredity* 93(2):86–90.
[29] Rao DN (1982) Responses of bryophytes to air pollution. In *Bryophyte Ecology*, ed. AJE Smith (Chapman and Hall, London), 445–471.
[30] Grout AJ (1916) *The Moss Flora of New York City and Vicinity* (self-published, New Dorp, NY), 5.
[31] Tees GM (1933) *An Annotated Check List of the Mosses of Philadelphia and Vicinity* (master of science thesis, Graduate School of Arts and Sciences, University of Pennsylvania, Philadelphia).
[32] Atwood JJ, Allen B, & Pursell RA (2009) *Checklist of Pennsylvania Mosses. Final Report to the Pennsylvania Department of Conservation and Natural Resources/Wild Resource Conservation Program: 22 September 2009* (published by the authors, St. Louis, MO); Porter, *Catalogue of the Bryophyta . . . and Pteridophyta . . . Found in Pennsylvania.*
[33] Sayre G (1984) Thomas Potts James: a bio-bibliography. *Cryptogamie: Bryologie, Lichenologie.* 5(1-2):51-62.
[34] Larsen RS, Bell JN, James PW, Chimonides PG, et al. (2007) Lichen and bryophyte distribution on oak in London in relation to air pollution and bark acidity. *Environmental Pollution* 146(2):332–340.
[35] Stevenson R & Hill M (2008) Urban myths exploded: Results of a bryological survey of King's Lynn (Norfolk, UK). *Journal of Bryology* 30(1):12–22; Sabovljevic M & Sabovljevic A (2009) Biodiversity within urban areas: A case study on bryophytes of the city of Cologne (NRW, Germany). *Plant Biosystems* 143(3):473–481; Sabovljevic M & Grdovic Z (2009) Bryophyte diversity within urban areas: Case study of the city of Belgrade (Serbia). *International Journal of Botany* 5(1):85–92.
[36] Miller NG & McDaniel SF (2004) Bryophyte dispersal inferred from colonization of an introduced substratum on Whiteface Mountain, New York. *American Journal of Botany* 91(8):1173–1182.
[37] Davies L (2007) Diversity and sensitivity of epiphytes to oxides of nitrogen in London. *Environmental Pollution* 146(2):299–310.
[38] Academy of Natural Sciences of Philadelphia (1952) Specimen of *Reboulia hemisphaerica* collected May 15, 1952, by E.W. Johnson. Specimen (barcode 45311) in Hepatics of Pennsylvania in the general collection of the herbarium of the Academy of Natural Sciences of Drexel University.
[39] Gilbert, Bryophytes as indicators of air pollution in the Tyne Valley.
[40] Manville, Hepatics of Pennsylvania. Final Report: 1994.
[41] Uva RH, Neal JC, & Ditomaso JM (1997) *Weeds of the Northeast* (Comstock Publishing Company/Cornell University Press, Ithaca, NY), 397.
[42] During HJ (1979) Life strategies of bryophytes: A preliminary review. *Lindbergia* 5(1):2–18.
[43] Clare D & Terry T (1960) Dispersal of *Bryum argenteum*. *Transactions of the British Bryological Society* 3:748.
[44] Greenslade P & Convey P (2012) Exotic Collembola on subantarctic islands: Pathways, origins and biology. *Biological Invasions* 14(2):405–417.

[45] Marino P, Raguso R, & Goffinet B (2009) The ecology and evolution of fly dispersed dung mosses (Family Splachnaceae): Manipulating insect behaviour through odour and visual cues. *Symbiosis* 47:61–76.
[46] Rudolphi J (2009) Ant-mediated dispersal of asexual moss propagules. *The Bryologist* 112(1):73–79.
[47] Johnson SD & Jürgens A (2010) Convergent evolution of carrion and faecal scent mimicry in fly-pollinated angiosperm flowers and a stinkhorn fungus. *South African Journal of Botany* 76(4):796–807.
[48] Boisselier-DuBayle, Lambourdiere, & Bischler, Taxa delimitation in *Reboulia* investigated with morphological, cytological, and isozyme markers.
[49] Vanderpoorten A, Gradstein SR, Carine MA, & Devos N (2010) The ghosts of Gondwana and Laurasia in modern liverwort distributions. *Biological Reviews* 85(3):471–487.
[50] Boisselier-DuBayle, Lambourdiere, & Bischler, Taxa delimitation in *Reboulia* investigated with morphological, cytological, and isozyme markers.
[51] O'Hanlon ME (1930) Gametophyte development in *Reboulia hemisphaerica. American Journal of Botany* 17(8):765–769.
[52] Akers TG, Edmonds RL, Kramer CL, Lighthart B, et al. (1979) Sources and characteristics of airborne materials. In *Aerobiology: The Ecological Systems Approach*, ed. RL Edmonds (Dowden, Hutchinson and Ross, Inc., Stroudsburg, PA), 11–84.
[53] Lewis WH, Vinay P, & Zenger VE (1983) *Airborne and Allergenic Pollen of North America* (Johns Hopkins University Press, Baltimore).
[54] Akers et al., Sources and characteristics of airborne materials.
[55] O'Hanlon, Gametophyte development in *Reboulia hemisphaerica.*
[56] Vanderpoorten et al., The ghosts of Gondwana and Laurasia in modern liverwort distributions.
[57] Studlar SM, Eddy C, & Spencer J (2007) Survival of four mosses from West Virginia after two hours in the stratosphere. *Evansia* 24(1):17–21.

Chapter 24: Silver-Haired Bat

1. Palisot de Beauvois A-M-F-J (Undated) *Catalogue raisonné du museum, de Mr. C. W. Peale, membre de la société philosophique de Pensylvanie. Rédigé par A. M. F. J. Beauvois, membre de la société des sciences et arts du Cap français, île et côte Saint-Domingue* (De L'Imprimerie de Parent, Philadelphia) Gale Reprint: ECCO ed.
2. Peale CW & Palisot de Beauvois A-M-F-J (1796) *A Scientific and Descriptive Catalogue of Peale's Museum by C. W. Peale, Member of the American Philosophical Society, and A. M. F. J. Beauvois, Member of the Society of Arts and Sciences of St. Domingo, of the American Philosophical Society, and Correspondent to the Museum of Natural History at Paris* (Samuel H. Smith, Philadelphia) Reprint: ECCO print ed.
3. Allen H (1864) *Monograph of the Bats of North America.* Smithsonian Miscellaneous Collections vol. 165 (Smithsonian Institution, Washington, DC).
4. Merritt JF (1987) *Guide to the Mammals of Pennsylvania* (University of Pittsburgh Press, Pittsburgh, PA).
5. Ibid.
6. As quoted in: Godman JD (1826) *American Natural History,* vol. I, pt. I. *Mastology,* (H. C. Carey and I. Lea, Philadelphia), 56–57.
7. Merritt, *Guide to the Mammals of Pennsylvania;* Harlan R (1825) *A Fauna Americana; Being a Description of the Mammiferous Animals Inhabiting North America* (Anthony Finley, Philadelphia).
8. Allen, *Monograph of the Bats of North America;* Kirkland J, Gordon L. & Hart JA (1999) Recent distributional records for ten species of small mammals in Pennsylvania. *Northeastern Naturalist* 6(1):1–18.
9. Allen, *Monograph of the Bats of North America.*
10. Kunz TH (1982) Lasionycteris noctivagans. *Mammalian Species* 172:1–5.
11. Parsons HJ, Smith DA, & Whittam RF (1986) Maternity colonies of silver-haired bats, *Lasionycteris noctivagans*, in Ontario and Saskatchewan. *Journal of Mammalogy* 67(3):598–600.
12. Perry RW, Saugey DA, & Crump BG (2010) Winter roosting ecology of silver-haired bats in an Arkansas forest. *Southeastern Naturalist* 9(3):563–572; Barclay RMR, Faure PA, & Farr DR (1988) Roosting behavior and roost selection by migrating silver-haired bats (*Lasionycteris noctivagans*). *Journal of Mammalogy* 69(4):821–825.
13. Parsons, Smith, & Whittam, Maternity colonies of silver-haired bats; Kurta A (2010) Reproductive timing, distribution, and sex ratios of tree bats in Lower Michigan. *Journal of Mammalogy* 91(3):586–592; McGuire LP, Guglielmo CG, Mackenzie SA, & Taylor PD (2011) Migratory stopover in the long distance migrant silver haired bat, *Lasionycteris noctivagans. Journal of Animal Ecology* 81(2):377–385.
14. Hamilton WJ & Whitaker JO Jr. (1979) *Mammals of the Eastern United States* (Comstock Publishing Associates, Ithaca, NY).
15. Nese J & Schwartz G (2002) *The Philadelphia Area Weather Book* (Temple University Press, Philadelphia).
16. Hamilton & Whitaker, *Mammals of the Eastern United States.*
17. Krutzsch PH (1966) Remarks on silver-haired and Leib's bats in eastern United States. *Journal of Mammalogy* 47(1):121.
18. Perry, Saugey, & Crump, Winter roosting ecology of silver-haired bats in an Arkansas forest.

19 Speakman JR & Thomas DW (2005) Physiological ecology and energetics of bats. In *Bat Ecology*, eds. TH Kunz & MB Fenton (University of Chicago Press, Chicago), 430–490.

20 Barclay, Faure & Farr, Roosting behavior and roost selection by migrating silver-haired bats (*Lasionycteris noctivagans*).

21 Dunbar MB (2007) Thermal energetics of torpid silver-haired bats, *Lasionycteris noctivagans. Acta Theriologica* 52(1):65–68.

22 Seibert HC (1949) Differences between migrant and non-migrant birds in food and water intake at various temperatures and photoperiods. *The Auk* 66(2):128–153.

23 Barbour RW & Davis WH (1969) *Bats of America* (University of Kentucky Press, Lexington).

24 Barclay, Faure, & Farr, Roosting behavior and roost selection by migrating silver-haired bats (*Lasionycteris noctivagans*).

25 Ibid.

26 Perry, Saugey, & Crump, Winter roosting ecology of silver-haired bats in an Arkansas forest.

27 Messenger SL, Rupprecht CE, & Smith JS (2003) Bats, emerging virus infections, and the rabies paradigm. In *Bat Ecology*, eds. TH Kunz & MB Fenton (University of Chicago Press, Chicago), 622–679.

28 Frick WF, Pollock JF, Hicks AC, Langwig KE, et al. (2010) An emerging disease causes regional population collapse of a common North American bat species. *Science* 329(5992):679–682; Reeder D, Frank CL, Turner GG, Meteyer CU, et al. (2012) Frequent arousal from hibernation linked to severity of infection and mortality in bats with white-nose syndrome. *PLoS ONE* 7(6):e38920. doi:38910.31371/journal.pone.0038920.

29 Merritt, *Guide to the Mammals of Pennsylvania.*

30 Frick et al., An emerging disease causes regional population collapse of a common North American bat species.

31 Warnecke L, Turner JM, Bollinger TK, Lorch JM, et al. (2012) Inoculation of bats with European *Geomyces destructans* supports the novel pathogen hypothesis for the origin of white-nose syndrome. *Proceedings of the National Academy of Sciences* 109:6999–7003.

32 Reeder et al., Frequent arousal from hibernation linked to severity of infection and mortality in bats with white-nose syndrome.

33 Verant M, Boyles J, Waldrep W Jr, Wibbelt G, et al. (2012) Temperature-dependent growth of *Geomyces destructans*, the fungus that causes bat white-nose syndrome. *PLoS ONE* 7(9):e46280. doi:46210.41371/journal.pone.0046280; Hallam TG & Federico P (2012) The panzootic white-nose syndrome: An environmentally constrained disease? *Transboundary and Emerging Diseases* 59(3):269–278.

34 Messenger SL, Smith JS, Orciari LA, Yager PA, et al. (2003) Emerging pattern of rabies deaths and increased viral infectivity. *Emerging Infectious Diseases* 9(2):151–154.

35 Messenger SL, Smith JS, & Rupprecht CE (2002) Emerging epidemiology of bat-associated cryptic cases of rabies in humans in the United States. *Clinical Infectious Disease* 35(6):738–747.

36 Feder HM Jr., Petersen BW, Robertson KL, & Rupprecht CE (2012) Rabies: Still a uniformly fatal disease? Historical occurrence, epidemiological trends, and paradigm shifts. *Current Infectious Disease Reports* 14(4):408–422.

37 Messenger, Smith, & Rupprecht, Emerging epidemiology of bat-associated cryptic cases of rabies in humans in the United States.

38 Blanton JD, Dyer J, McBrayer J, & Rupprecht CE (2012) Rabies surveillance in the United States during 2011. *Journal of the American Veterinary Medical Association* 241(6):712–722.

39 Klug BJ, Turmelle AS, Ellison JA, Baerwald EF, et al. (2011) Rabies prevalence in migratory tree-bats in Alberta and the influence of roosting ecology and sampling method on reported prevalence of rabies in bats. *Journal of Wildlife Diseases* 47(1):64–77.

40 Serres GD, Dallaire F, Côte M, & Skowronski DM (2008) Bat rabies in the United States and Canada from 1950 through 2007: Human cases with and without bat contact. *Clinical Infectious Diseases* 46:1329–1337.

41 Pennsylvania Department of Agriculture Veterinary Lab (2012) Rabies Map. Pennsylvania positive Animal Rabies for 2011. http://www.agriculture.state.pa.us/portal/server.pt/gateway/PTARGS_0_2_24476_10297_0_43/agwebsite/Files/Publications/RABIES%20MAP%202011.pdf.

42 Serres GD, Dallaire F, Côte M, & Skowronski DM (2008)

43 Curran EB, Holle RL, & López RE (2000) Lightning casualties and damages in the United States from 1959 to 1994. *Journal of Climate* 13(19):3448–3464.

44 Serres GD, Dallaire F, Côte M, & Skowronski DM (2008)

45 Mader EC Jr, Maury JS, Santana-Gould L, Craver RD, et al. (2012) Human rabies with initial manifestations that mimic acute brachial neuritis and Guillain-Barré syndrome. *Clinical Medicine Insights: Case Reports* 5:49–55; Petersen BW & Rupprecht CE (2011) Human rabies epidemiology and diagnosis. In *Non-Flavivirus Encephalitis*, ed. S Tkachev (InTech, Rijeka, Croatia), 247–278.

46 Ford WM, Britzke ER, Dobony CA, Rodrigue JL, et al. (2011) Patterns of acoustical activity of bats prior to and following white-nose syndrome occurrence. *Journal of Fish and Wildlife Management* 2(2):125–134.

47 Feder et al., Rabies: Still a uniformly fatal disease?

48 Ibid.; Foley J, Clifford D, Castle K, Cryan P, et al. (2011) Investigating and managing the rapid emergence of white-nose syndrome, a novel, fatal, infectious disease of hibernating bats. *Conservation Biology* 25:223–231.

[49] Langwig KE, Frick WF, Bried JT, Hicks AC, et al. (2012) Sociality, density-dependence and microclimates determine the persistence of populations suffering from a novel fungal disease, white-nose syndrome. *Ecology Letters* 15:1050–1057.
[50] Feder et al., Rabies: Still a uniformly fatal disease?
[51] Racey PA & Entwistle AC (2003) Conservation ecology of bats. In *Bat Ecology*, eds. TH Kunz & MB Fenton (University of Chicago Press, Chicago), 680–743.
[52] Mackiewicz J & Backus RH (1956) Oceanic records of *Lasionycteris noctivagans* and *Lasiurus borealis*. *Journal of Mammalogy* 37(3):442–443; Miller GS (1897) Migration of bats on Cape Cod, Massachusetts. *Science* 5(118):541–543; Johnson JB, Gates JE, & Zegre NP (2011) Monitoring seasonal bat activity on a coastal barrier island in Maryland, USA. *Environmental Monitoring and Assessment* 173(1-4):685–699.
[53] Dvorak MJ, Corcoran BA, Ten Hoeve JE, McIntyre NG, et al. (2012) US East Coast offshore wind energy resources and their relationship to peak-time electricity demand. *Wind Energy* 16(7):977–997. doi:10.1002/we.1524.
[54] Arnett EB, Brown WK, Erickson WP, Fiedler JK, et al. (2008) Patterns of bat fatalities at wind energy facilities in North America. *Journal of Wildlife Management* 72(1):61–78.
[55] Rollins KE, Meyerholz DK, Johnson GD, Capparella AP, et al. (2012) A forensic investigation into the etiology of bat mortality at a wind farm: Barotrauma or traumatic injury? *Veterinary Pathology* 49(2):362–371.
[56] Kunz TH, Arnett EB, Erickson WP, Hoar AR, et al. (2007) Ecological impacts of wind energy development on bats: Questions, research needs, and hypotheses. *Frontiers in Ecology and the Environment* 5(6):315–324.
[57] Dunbar MB & Brigham RM (2010) Thermoregulatory variation among populations of bats along a latitudinal gradient. *Journal of Comparative Physiology B* 180:885–893.
[58] Speakman JR & Thomas DW (2005) Physiological ecology and energetics of bats. In *Bat Ecology*, eds. TH Kunz & MB Fenton (University of Chicago Press, Chicago), 430–490.
[59] Dunbar MB, Whitaker JO Jr., & Robbins LW (2007) Winter feeding by bats in Missouri. *Acta Chiropterologica* 9(1):305–310.
[60] Cryan PM & Veilleux JP (2007) Migration and use of autumn, winter and spring roosts by tree bats. In *Bats in Forests: Conservation and Management*, eds. MJ Lacki, JP Hayes, & A Kurta (Johns Hopkins University Press, Baltimore, MD), 153–175.

Chapter 25: Canada Goose

[1] Barton BS (1799) *Fragments of a Natural History of Pennsylvania* (Willughby Society reprint, ed. Osbert Salvin, Taylor and Francis, London, 1883).
[2] Wilson A (1814) Canada goose. In *American Ornithology, or, The Natural History of the Birds of the United States, Illustrated with Plates, Engraved and Colored from Original Drawings Taken from Nature,* vol. 8 (Bradford and Inskeep, Philadelphia), 53–59.
[3] Ibid., 57.
[4] Ibid., 56–57.
[5] Ibid., 55.
[6] Bryant HC, et al. (1936) Report of the Committee on Bird Protection, American Ornithologists' Union. *The Auk* 53(1):70–73.
[7] Nelson HK (1963) Restoration of breeding Canada goose flocks in the North Central States. In *Transactions of the Twenty-eighth North American Wildlife and Natural Resources Conference*, ed. JB Trefethen (Wildlife Management Institute, Detroit, MI), 133–150.
[8] Trimble R (1940) Changes in the bird life at Pymatuning Lake, Pennsylvania. *Annals of the Carnegie Museum* 28:83–132.
[9] Hartman FE (1992) Canada goose (*Branta canadensis*). In *Atlas of Breeding Birds in Pennsylvania*, ed. DW Brauning (University of Pittsburgh Press, Pittsburgh, PA), 66–67.
[10] Kallman H, Agee CP, Goforth WR, & Linduska JP, eds. (1987) *Restoring America's Wildlife 1937–1987. The First 50 Years of the Federal Aid in Wildlife Restoration (Pittman-Robertson) Act* (U.S. Department of the Interior, Fish and Wildlife Service, Washington, DC).
[11] Ibid.
[12] U.S. Fish and Wildlife Service (2012) Migratory bird program—Geese management, http://www.fws.gov/migratorybirds/RegulationsPolicies/geese.html; Kosack J (1995) *The Pennsylvania Game Commission 1895–1995* (Pennsylvania Game Commission, Harrisburg).
[13] Hartman, Canada goose (*Branta canadensis*); Jacobs KJ (2002) *Project Annual Job Report: Canada Goose Nest Ecology and Gosling Survival at Pymatuning Wildlife Management Area* (Pennsylvania Game Commission, Bureau of Wildlife Management, Research Division, Harrisburg, PA).
[14] Kosack, *The Pennsylvania Game Commission 1895–1995.*
[15] Canada Goose Committee of the Atlantic Flyway Technical Section (1999) Atlantic Flyway Resident Canada Goose Management Plan, http://www.mdwfa.org/flyway/CAGO_RPManagementPlan.pdf.
[16] Hartman, Canada goose (*Branta canadensis*).

[17] Sheaffer SE & Malecki RA (1998) Status of Atlantic Flyway resident nesting Canada geese. In *Biology and Management of Canada Geese: Proceedings of the International Canada Goose Symposium*, eds. DH Rusch et al. (Milwaukee, WI), 29–34.

[18] Surrendi DC (1970) The mortality, behavior, and homing of transplanted juvenile Canada geese. *The Journal of Wildlife Management* 34(4):719–733.

[19] Lee FB, Sherwood GA, & Schoonover LJ (1970) The free-flying flock. In *Home Grown Honkers*, eds. HH Dill & FB Lee (reprint, International Wildfowl Association, Jamestown, ND), 52–54.

[20] McWilliams GM & Brauning DW (2000) *The Birds of Pennsylvania* (Cornell University Press, Ithaca, NY).

[21] Hanson HC (1997) *The Giant Canada Goose,* rev. ed. (Southern Illinois University Press, Carbondale and Edwardsville).

[22] Nelson HK & Oetting R (1998) Giant Canada goose flocks in the United States. In *Biology and Management of Canada Geese: Proceedings of the International Canada Goose Symposium*, eds. DH Rusch et al. (Milwaukee, WI), 483–495.

[23] Hanson, *The Giant Canada Goose.*

[24] Lefebvre EA & Raveling DG (1967) Distribution of Canada geese in winter as related to heat loss at varying environmental temperatures. *Journal of Wildlife Management* 31(3):538–546.

[25] Delacour J (1954) *Waterfowl of the World,* vol. 1 (Country Life Ltd., London).

[26] Nelson, Restoration of breeding Canada goose flocks in the North Central States; Trimble, Changes in the bird life at Pymatuning Lake, Pennsylvania.

[27] Hanson, *The Giant Canada Goose.*

[28] Wilson, Canada goose.

[29] Conover MR (1992) Ecological approach to managing problems caused by urban Canada geese. *Proceedings of the Fifteenth Vertebrate Pest Conference 1992,* http://digitalcommons.unl.edu/vpc15/19.

[30] Sheaffer & Malecki, Status of Atlantic Flyway resident nesting Canada geese; Lyngs P (2003) *Migration and Winter Ranges of Birds in Greenland: An Analysis of Ringing Recoveries* (Danish Ornithological Society, Copenhagen).

[31] Hestbeck J & Nichols J (1991) Estimates of movement and site fidelity using mark-resight data of wintering Canada geese. *Ecology* 72(2):523–533; Hartman FE & Dunn JP (1998) The Canada goose in Pennsylvania: From none to too many (abstract). In *Biology and Management of Canada Geese: Proceedings of the International Canada Goose Symposium*, eds. DH Rusch et al. (Milwaukee, WI), 477.

[32] Thompson JE, Hill MRJ, Merendino MT, & Ankney CD (1999) Improving use of morphometric discrimination to identify Canada goose subspecies. *Wildlife Society Bulletin* 27(2):274–280; Wagner CEV & Baker AJ (1990) Association between mitochondrial DNA and morphological evolution in Canada geese. *Journal of Molecular Evolution* 31(5):373–382; Scribner KT, et al. (2003) Genetic methods for determining racial composition of Canada goose harvests. *The Journal of Wildlife Management* 67(1):122–135.

[33] Malecki RA & Trost RE (1998) Status of breeding Canada geese in North America. In *Biology and Management of Canada Geese: Proceedings of the International Canada Goose Symposium*, eds. DH Rusch et al. (Milwaukee, WI), 3–8.

[34] Kallman et al., *Restoring America's Wildlife 1937–1987.*

[35] Hartman & Dunn, The Canada goose in Pennsylvania: From none to too many (abstract); Frye B (2006) *Deer Wars: Science, Tradition, and the Battle over Managing Whitetails in Pennsylvania* (Penn State University Press, University Park).

Chapter 26: Stink Bug Hunter

[1] Hoebeke ER & Carter ME (2003) *Halyomorpha halys* (Stål) (Heteroptera: Pentatomidae): A polyphagous plant pest from Asia newly detected in North America. *Proceedings of the Entomological Society of Washington* 105(1):225–237.

[2] Ibid., 225–226.

[3] Ibid.

[4] Seise A (October 27, 2007) One thing's for sure: You don't want to squash 'em. Invasion of the stink bugs. *Philadelphia Inquirer*, http://articles.philly.com/2007-10-27/news/25232839_1_stink-bugs-marmorated-gypsy-moth.

[5] Jacobs S (2012) Brown marmorated stink bug, *Halyomorpha halys*. In *Entomological Notes* (Department of Entomology, College of Agricultural Sciences, Cooperative Extension, Penn State, University Park), http://ento.psu.edu/extension/factsheets/brown-marmorated-stink-bug.

[6] Wermelinger B, Wyniger D, & Forster B (2008) First records of an invasive bug in Europe: *Halyomorpha halys* Stål (Heteroptera: Pentatomidae), a new pest on woody ornamentals and fruit trees? *Mitteilungen der Schweizerischen Entomologischen Gesellschaft (Bulletin de la Société Entomologique Suisse)* 81:1–8.

[7] Harris AC (2010) *Halyomorpha halys* (Hemiptera: Pentatomidae) and *Protaetia brevitarsis* (Coleoptera: Scarabaeidae: Cetoniinae) intercepted in Dunedin. *The Weta* 40:42–44.

[8] Leskey TC, Short BD, Butler BR, & Wright SE (2012) Impact of the invasive brown marmorated stink bug, *Halyomorpha halys* (Stål), in mid-Atlantic tree fruit orchards in the United States: Case studies of commercial management. *Psyche* 2012: Article ID 535062, http://dx.doi.org/10.1155/2012/535062.

[9] Nielsen AL & Hamilton GC (2009) Life history of the invasive species *Halyomorpha halys* (Hemiptera: Pentatomidae) in northeastern United States. *Annals of the Entomological Society of America* 102(4):608–616.

[10] Leskey et al., Impact of the invasive brown marmorated stink bug, *Halyomorpha halys* (Stål), in mid-Atlantic tree fruit orchards in the United States.

[11] Funayama K (2012) Control effect on the brown-marmorated stink bug, *Halyomorpha halys* (Hemiptera: Pentatomidae), by combined spraying of pyrethroid and neonicotinoid insecticides in apple orchards in northern Japan. *Applied Entomology and Zoology* 47(1):75–78.

[12] Hopwood J, Vaughan M, Shepherd M, Biddinger D, et al. (2012) *Are Neonicotinoids Killing Bees? A Review of Research into the Effects of Neonicotinoid Insecticides on Bees, with Recommendations for Action* (The Xerces Society for Invertebrate Conservation, Portland, OR).

[13] Gouli V, Gouli S, Skinner M, Hamilton G, et al. (2012) Virulence of select entomopathogenic fungi to the brown marmorated stink bug, *Halyomorpha halys* (Stål) (Heteroptera: Pentatomidae). *Pest Management Science* 68(2):155–157.

[14] Arakawa R & Namura Y (2002) Effects of temperature on development of three *Trissolcus* spp. (Hymenoptera: Scelionidae), egg parasitoids of the brown marmorated stink bug, *Halyomorpha halys* (Hemiptera: Pentatomidae). *Entomological Science* 5(2):215–218; Yang Z-Q, Yao Y-X, Qiu L-F, & Li Z-X (2009) A new species of *Trissolcus* (Hymenoptera: Scelionidae) parasitizing eggs of *Halyomorpha halys* (Heteroptera: Pentatomidae) in China with comments on its biology. *Annals of the Entomological Society of America* 102(1):39–47.

[15] Biddinger DJ, Hull L, & Krawczyk G (2011) Effects of pesticides controlling brown marmorated stink bug on apple IPM in Pennsylvania and the potential of biological control. In *Great Lakes Fruit, Vegetable & Farm Market EXPO* (DeVos Place Convention Center, Grand Rapids, MI), http://www.glexpo.com/summaries/2011summaries/webApple_II.pdf.

[16] Aldrich JR, Khrimian A, Zhang A, & Shearer PW (2006) Bug pheromones (Hemiptera, Heteroptera) and tachinid fly host-finding. *Denisia* 19:1015–1031, http://naldc.nal.usda.gov/download/17381/PDF.

[17] Pickett CH *Trichopoda pennipes* (Diptera: Tachinidae). Biological Control: A Guide to Natural Enemies in North America, ed. A Shelton (Cornell University College of Agriculture and Life Sciences, Department of Entomology, Ithaca, NY), accessed September 21, 2013, at http://www.biocontrol.entomology.cornell.edu/parasitoids/trichopoda.html.

[18] Biddinger, Hull, & Krawczyk, Effects of pesticides controlling brown marmorated stink bug on apple IPM in Pennsylvania and the potential of biological control.

[19] Gentry TG (1878) *The House Sparrow at Home and Abroad, with Some Concluding Remarks upon Its Usefulness, and Copious References to the Literature of the Subject* (Claxton, Remsen, and Haffelfinger, Philadelphia).

[20] Sanderson C & Cathcart RLW (1919) Analyses of materials sold as insecticides and fungicides during 1919. *Bulletin 339 of the New Jersey Agricultural Experiment Station* (New Jersey Agricultural Experiment Station, New Brunswick); Riley CV (1887) Our shade trees and their insect defoliators, being a consideration of the four most injurious species which affect the trees of the capital; with means of destroying them. *U.S. Department of Agriculture, Division of Entomology, Bulletin Number 10* (U.S. Government Printing Office, Washington, DC).

[21] Biddinger, Hull, & Krawczyk, Effects of pesticides controlling brown marmorated stink bug on apple IPM in Pennsylvania and the potential of biological control.

[22] Harris VE & Todd JW (1982) Longevity and reproduction of the southern green stink bug, *Nezara viridula*, as affected by parasitization by *Trichopoda pennipes*. *Entomologia Experimentalis et Applicata* 31(4):409–412.

[23] Aldrich et al., Bug pheromones (Hemiptera, Heteroptera) and tachinid fly host-finding.

[24] Say T (1824) Appendix, pt. I. Natural History, Zoology, pt. E, Class Insecta, Order Hymenoptera, Monedula. In *Narrative of an Expedition to the Source of St. Peter's River, Lake Winnepeck, Lake of the Woods, &c. &c: Performed in the Year 1823, by Order of the Hon. J. C. Calhoun, Secretary of War, under the Command of Stephen H. Long, Major U.S.T.E.*, vol. 2, ed. WH Keating (H. C. Carey & I. Lea, Philadelphia), 336.

[25] Evans HE (1966) *The Comparative Ethology and Evolution of the Sand Wasps* (Harvard University Press, Cambridge, MA), 144–175.

[26] Ibid.

[27] Ibid.

[28] Biddinger, Hull, & Krawczyk, Effects of pesticides controlling brown marmorated stink bug on apple IPM in Pennsylvania and the potential of biological control.

[29] Gall WK (1992) Further eastern range extension and host records for *Leptoglossus occidentalis* (Heteroptera: Coreidae): Well-documented dispersal of a household nuisance. *Great Lakes Entomologist* 25(3):159–171.

[30] Biddinger, Hull, & Krawczyk, Effects of pesticides controlling brown marmorated stink bug on apple IPM in Pennsylvania and the potential of biological control.

[31] Evans, *The Comparative Ethology and Evolution of the Sand Wasps,* 144–175.

[32] Ibid.

1 Melsheimer FV (1806) *A Catalogue of Insects of Pennsylvania Part First* (W.D. Lepper, Hanover, York County, PA) [Note: This publication is limited to beetles. Planned supplements never appeared.]

2 Anon. (1975) Entomological Society of Pennsylvania rededication of the F.V. Melsheimer Memorial, New Holland, PA Sept 26, 1974. *Bulletin of the Entomological Society of Pennsylvania* 21(1):17.

3 Schwarz EA (1995) Some notes on Melsheimer's catalogue of the Coleoptera of Pennsylvania. *Proceedings of the Entomological Society of Washington* 3(3):134-138.

4 Mallis A (1971) *American Entomologists* (Rutgers University Press, New Brunswick, NJ); Prowell GR (1903) Frederick Valentine Melsheimer. A pioneer entomologist and a noted clergyman and author. A paper read before the Historical Society of York County, April 8, 1897 by George R. Prowell. *Proceedings of the Historical Society of York County* 1(2):17-26.

5 Say T (1823 (1824)) Descriptions of coleopterous insects collected in the late Expedition to the Rocky Mountains, performed by order of Mr. Calhoun, Secretary of War, under the command of Major Long. *Journal of the Academy of Natural Sciences of Philadelphia* 4(1):83-99.

6 Calvert PP (1942) Entomology, scientific and human aspects. *Proceedings of the American Philosophical Society* 86(1, Symposium on the Early History of Science and Learning in America):123-129.

7 Hagen HA (1884) The Melsheimer family and Melscheimer collection. *Canadian Entomologist* 16:191-197.

8 Riley CV (1893) *Parasitic and Predaceous Insects in Applied Entomology* — Read at the fifth annual meeting of the Association of Economic Entomologists, Madison, Wisconsin, August 15, 1893, and reprinted from *Insect Life* 6(2):130-141. (United States Government Printing Office, Washington, D.C.).

9 Hoebeke ER & Wheeler Jr. AG (1980) New distribution records of *Coccinella septempunctata* L. in the eastern United States (Coleoptera: Coccinellidae). *The Coleopterists Bulletin* 34(2):209-212.

10 Schaefer PW, Dysart RJ, & Specht HB (1987) North American distribution of *Coccinella septempunctata* (Coleoptera: Coccinellidae) and its mass appearance in coastal Delaware. *Environmental Entomology* 16:368–373.

11 Wheeler AG, Jr. & Hoebeke ER (1995) *Coccinella novemnotata* in northeastern North America: historical occurrence and current status (Coleoptera: Coccinellidae). *Proceedings of the Entomolological Society of Washington* 97:701-716.

12 Reitz SR & Trumble JT (2002) Competitive displacement among insects and arachnids. *Annual Review of Entomology* 47:435-465.

13 Harmon JP, Stephens E, & Losey J (2007) The decline of native coccinellids (Coleoptera: Coccinellidae) in the United States and Canada. *Journal of Insect Conservation* 11(1):85-94.

14 Hodek I & Mkchaud JP (2008) Why is *Coccinella septempunctata* so successful? (A point-of-view) *European Journal of Entomology* 105:1-12.

15 A. G. Wheeler J & Stoops CA (1996) Status and spread of the Palearctic lady beetles *Hippodamia Variegata* and *Propylea quatuordecimpunctata* (Coleoptera: Coccinellidae) in Pennsylvania, 1993-1995. *Entomological News* 107(5):291-298.

16 Brown PM, *et al.* (2011) The global spread of *Harmonia axyridis* (Coleoptera: Coccinellidae): distribution, dispersal and routes of invasion. *BioControl* 56(4):623-641.

17 Hodek & Mkchaud, Why is *Coccinella septempunctata* so successful? (A point-of-view).

18 Pell JK, Baverstock J, Roy HE, Ware RL, & Majerus MEN (2008) Intraguild predation involving *Harmonia axyridis:* a review of current knowledge and future perspectives *BioControl* 53(1).

19 Sato S, Kushibuchi K, & Yasuda H (2009) Effect of reflex bleeding of a predatory ladybird beetle, *Harmonia axyridis* (Pallas) (Coleoptera: Coccinellidae), as a means of avoiding intraguild predation and its cost. *Applied Entomology and Zoology* Vol. 44(2):203-206.

20 Snyder WE, Joseph SB, Preziosi RF, & Moore AJ (2000) Nutritional benefits of cannibalism for the lady beetle *Harmonia axyridis* (Coleoptera: Coccinellidae) when prey quality is poor. *Environmental Entomology* 29(6):1173-1179.

21 Joseph SB, Snyder WE, & Moore AJ (1999) Cannibalizing *Harmonia axyridis* (Coleoptera: Coccinellidae) larvae use endogenous cues to avoid eating relatives. *Journal of Evolutionary Biology* 12(4):792-797.

22 Koch RL (2003) The multicolored Asian lady beetle, *Harmonia axyridis*: A review of its biology, uses in biological control, and non-target impacts. *Journal of Insect Science* 3(32): http://www.ncbi.nlm.nih.gov/pmc/articles/PMC524671/.

23 Vilcinskas A, Stoecker K, Schmidtberg H, Röhrich CR, & Vogel H (2013) Invasive harlequin ladybird carries biological weapons against native competitors. *Science* 340(6134):862-863.

24 Vilcinskas A, Mukherjee K, & Vogel H (2013) Expansion of the antimicrobial peptide repertoire in the invasive ladybird *Harmonia axyridis*. . *Proceedings of the Royal Society B: Biological Sciences* 280(1750):2012-2113.

25 Vilcinskas A, *et al.* (2014) Evolutionary ecology of microsporidia associated with the invasive ladybird *Harmonia axyridis*. *Insect Science* 21 (4): doi: 10.1111/1744-7917.12159.

26 Sloggett JJ, *et al.* (2011) The chemical ecology of *Harmonia axyridis*. *BioControl* 56(4):643-661.

27 Sloggett JJ, Haynes KF, & Obrycki JJ (2009) Hidden costs to an invasive intraguild predator from chemically defended native prey. *Oikos* 118::1396-1404.

[28] Roy HE, *et al.* (2011) Living with the enemy: parasites and pathogens of the ladybird *Harmonia axyridis. BioControl, (4), 663-679.* 56(4):663-679.

[29] Comont RF, *et al.* (2014) Escape from parasitism by the invasive alien ladybird, *Harmonia axyridis. Insect Conservation and Diversity* 7(4):334-342.

[30] Riddick EW & Schaefer PW (2005) Occurrence, Density, and Distribution of Parasitic Fungus *Hesperomyces virescens* (Laboulbeniales: Laboulbeniaceae) on Multicolored Asian Lady Beetle (Coleoptera: Coccinellidae) *Annals of the Entomological Society of America* 98(4):615-624.

[31] Riddick EW (2010) Ectoparasitic mite and fungus on an invasive lady beetle: parasite coexistence and influence on host survival. *Bulletin of Insectology* 63(1):13-20.

[32] Cottrell TE & Riddick EW (2012) Limited transmission of the ectoparasitic fungus *Hesperomyces virescens* between lady beetles. *Psyche* 2012:Article ID 814378. doi:10.1155/2012/814378.

[33] Sutherland AM & Parrella MP (2009) Biology and co-occurrence of *Psyllobora vigintimaculata taedata* (Coleoptera: Coccinellidae) and powdery mildews in an urban landscape of California. *Annals of the Entomological Society of America* 102(3):484-491.

[34] Day WH, Prokrym DR, Ellis DR, & Chianese RJ (1994) The known distribution of the predator *Propylea quatuordecimpunctata* (Coleoptera: Coccinellidae) in the United States, and thoughts on the origin of this species and five other exotic lady beetles in eastern North America. *Entomological. News 105* 105:244–256.

[35] Brown PM, *et al.* (2011) The global spread of *Harmonia axyridis* (Coleoptera: Coccinellidae): distribution, dispersal and routes of invasion. *BioControl* 56(4):623-641.

[36] Miller GL & Cave RD (1987) Bionomics of *Micromus posticus* (Walker) (Neuroptera: Hemerobiidae) with descriptions of immature stages. *Proceedings of the Entomological Society of Washington* 89(4):776-789.

[37] Cutright CR (1923) Life history of *Micromus posticus* Walker. *Journal of Economic Entomology.* 16(5):448-456.

[38] Carpenter FM (1940) A Revision of the Nearctic Hemerobiidae, Berothidae, Sisyridae, Polystoechotidae and Dilaridae (Neuroptera). *Proceedings of the American Academy of Arts and Sciences* 74(7):193-280.

Chapter 28: Overview

[1] Levi HW (1957) The spider genera *Crustulina* and *Steatoda* in North America, Central America, and the West Indies (Araneae, Theridiidae). *Bulletin of the Museum of Comparative Zoology* 117(3):367–424; Chapin JB & Brou VA (1991) *Harmonia axyridis* (Pallas), the third species of the genus to be found in the United States (Coleoptera: Coccinellidae). *Proceedings of the Entomological Society of Washington* 93(3):630–635.)

[2] Crutzen PJ (2002) Geology of mankind. *Nature* 415:23.

[3] Gozzaldi MIJ (1903) Thomas Potts James. *Bryologist* 6(5):71–74.

[4] Atwood JJ, Allen B, & Pursell RA (2009) *Checklist of Pennsylvania Mosses. Final Report to the Pennsylvania Department of Conservation and Natural Resources/Wild Resource Conservation Program: 22 September 2009* (self-published, St. Louis, MO); Porter TC (1904) *Catalogue of the Bryophyta (Hepatics, Anthocerotes and Mosses) and Pteridophyta (Ferns and Fern-Allies) Found in Pennsylvania* (Gin & Company, Boston).

[5] Brimblecombe P, ed. (2003) *The Effects of Air Pollution on the Built Environment* (Imperial College Press, London).

[6] Fowler HW (1914) Fishes in polluted waters. *Copeia* 5:4; Hardy C III (1999) Fish or foul: A history of the Delaware River Basin through the perspective of the American shad, 1682 to the present. *Pennsylvania History* 66(4):506–534; Perillo JA & Butler LH (2009) Evaluating the use of Fairmount Dam fish passage facility with application to anadromous fish restoration in the Schuylkill River, Pennsylvania. *Journal of the Pennsylvania Academy of Science* 83(1):24–33.

[7] Nese J & Schwartz G (2002) *The Philadelphia Area Weather Book* (Temple University Press, Philadelphia).

[8] Fan H & Sailor DJ (2005) Modeling the impacts of anthropogenic heating on the urban climate of Philadelphia: A comparison of implementations in two PBL schemes. *Atmospheric Environment* 39(1):73–84.

[9] Zhang X, Friedl MA, Schaaf CB, Strahler AH, et al. (2004) The footprint of urban climates on vegetation phenology. *Geophysical Research Letters* 31(L12209). doi:10.1029/2004GL020137.

[10] Oka M (2011) The influence of urban street characteristics on pedestrian heat comfort levels in Philadelphia. *Transactions in GIS* 15(1):109–123.

[11] Meineke EK, Dunn RR, Sexton JO, & Frank SD (2013) Urban warming drives insect pest abundance on street trees. *PLoS ONE* 8(3):e59687. doi:59610.51371/journal.pone.0059687.

[12] Wood AJ (2007) The nature and distribution of vegetative desiccation-tolerance in hornworts, liverworts and mosses. *The Bryologist* 110(2):163–177.

[13] Arnold GE (1962) Thermal pollution of surface supplies. *Journal of the American Water Works Association* 54(11):1332–1346.

[14] Ibid., 1333–1334.

[15] Rupple F (2002) Steaming ahead: The Grays Ferry Cogeneration Project. *District Energy* 88(2):1–4.

[16] Council of the City of Philadelphia (2006) An ordinance on noise (Bill no. 050749), http://www.phila.gov/health/pdfs/air/Noise_Bill_050749.pdf.

[17] Francis CD, Ortega CP, & Cruz A (2009) Noise pollution changes avian communities and species interactions. *Current Biology* 19:1415–1419.

[18] Bermúdez-Cuamatzin E, Ríos-Chelén AA, Gil D, & Garcia CM (2011) Experimental evidence for real-time song frequency shift in response to urban noise in a passerine bird. *Biology Letters* 1:36–38.

[19] Bermúdez-Cuamatzin E, Ríos-Chelén AA, Gil D, & Garcia CM (2009) Strategies of song adaptation to urban noise in the house finch: Syllable pitch plasticity or differential syllable use? *Behaviour* 146(9):1269–1286.

[20] Seger-Fullam KD, Rodewald AD, & Soha JA (2011) Urban noise predicts song frequency in Northern cardinals and American robins. *Bioacoustics: The International Journal of Animal Sound and Its Recording* 20(3):267–276.

[21] Francis, Ortega, & Cruz, Noise pollution changes avian communities and species interactions.

[22] Lampe U, Schmoll T, Franzke A, & Reinhold K (2012) Staying tuned: Grasshoppers from noisy roadside habitats produce courtship signals with elevated frequency components. *Functional Ecology* 26(6):1348–1354.

[23] Schaub A, Ostwald J, & Siemers BM (2008) Foraging bats avoid noise. *Journal of Experimental Biology* 211:3174–3180.

[24] Foster KR, Jenkins MF, & Toogood AC (1998) The Philadelphia yellow fever epidemic of 1793. *Scientific American* 279:88–93.

[25] Holick J, Kyle A, Ferraro W, Delaney R, et al. (2002) Discovery of *Aedes albopictus* infected with West Nile virus in southeastern Pennsylvania. *Journal of the American Mosquito Control Association* 18(2):131.

[26] Lambrechts L, Scott T, & Gubler D (2010) Consequences of the expanding global distribution of *Aedes albopictus* for dengue virus transmission. *PLoS Negl Trop Dis* 4(5):e646; Bonizzoni M, Gasperi G, Chen X, & James AA (2013) The invasive mosquito species *Aedes albopictus*: Current knowledge and future perspectives. *Trends in Parasitology* 29(9):460–468.

[27] Holick et al., Discovery of *Aedes albopictus* infected with West Nile virus in southeastern Pennsylvania; Hawley W, Reiter P, Copeland R, Pumpuni C, et al. (1987) *Aedes albopictus* in North America: Probable introduction in used tires from northern Asia. *Science* 236(4805):1114–1116.

[28] Pennsylvania's West Nile Virus Control Program (2013) Surveillance Data and Maps, http://www.westnile.state.pa.us/surv.htm.

[29] ArboNET, Arboviral Diseases Branch, Centers for Disease Control and Prevention (2013) West Nile Virus Disease Cases Reported to CDC by State, 1999–2012, http://www.cdc.gov/westnile/resources/pdfs/cummulative/99_2012_cummulativeHumanCases.pdf

[30] ArboNET, Arboviral Diseases Branch, Centers for Disease Control and Prevention (2013) West Nile Virus Disease Cases and Deaths Reported to CDC by Year and Clinical Presentation, 1999–2012, http://www.cdc.gov/westnile/resources/pdfs/cummulative/99_2012_CasesAndDeathsClinicalPresentationHumanCases.pdf.

[31] Pennsylvania's West Nile Virus Control Program (2013) Mosquito Adulticiding Events in Pennsylvania in 2013, http://www.westnile.state.pa.us/events.htm#P.

[32] Holm LG, Plucknett DL, Pancho JV, & Herberger JP (1991) *The World's Worst Weeds: Distribution and Biology* (Krieger Publishing Company, Malabar, FL).

[33] Pennsylvania Department of Agriculture. Purple loosestrife. Accessed April 18, 2013, at http://www.agriculture.state.pa.us/portal/server.pt/gateway/PTARGS_0_2_75292_10297_0_43/AgWebsite/Files/Publications/Purple_Loosestrife.pdf.

[34] Taylor FH (1894) *The City of Philadelphia as It Appears in the Year 1894: A Compilation of Facts Supplied by Distinguished Citizens for the Information of Business Men, Travelers, and the World at Large* (G. S. Harris & Sons, Philadelphia).

[35] Wilson E (1984) *Biophilia: The Human Bond with Other Species* (Harvard University Press, Cambridge, MA).

Chapter 29: Conclusion

[1] McKibben B (1989) *The End of Nature* (Random House, New York).

[2] Philadelphia (PA) (2011) Title 14, Zoning and Planning. The Philadelphia Code, 10th ed. (American Legal Publishing Corporation, Cincinnati, OH), http://www.amlegal.com/nxt/gateway.dll/Pennsylvania/philadelphia_pa/thephiladelphiacode?f=templates$fn=default.htm$3.0$vid=amlegal:philadelphia_pa.

[3] McKinney ML (2006) Urbanization as a major cause of biotic homogenization. *Biological Conservation* 127(3):247–260.

[4] Hansen AA (1921) *Lawn Pennywort: A New Weed. Department Circular 165* (U.S. Department of Agriculture, Bureau of Plant Industry, U.S. Government Printing Office, Washington, DC); Anon. (1909) Abstract of the proceedings of the Philadelphia Botanical Club for 1909. *Bartonia* (2):28-31. [Refers to *H. sibthorpioides* as *H. rotundifolia*, a synonym.]

[5] Schumpeter JA (1950) *Capitalism, Socialism and Democracy, 3rd ed.* (reprint, Harper Perennial, New York, 1976).

[6] Holling CS (2001) Understanding the complexity of economic, ecological, and social systems. *Ecosystems* 4(5):390–405.

[7] Beatley T (2010) *Biophilic Cities: Integrating Nature into Urban Design and Planning* (Island Press, Washington, DC).

SPOTLIGHT ENDNOTES

Spotlight: Human body louse

1 Gerhard W (1837) On the Typhus Fever: which occurred at Philadelphia in the spring and summer of 1836; Illustrated by clinical observations at the Philadelphia Hospital; showing the distinction between this form of disease and dothinenteritis or the typhoid fever with alteration of the follicles of the small intestine. *The American Journal of the Medical Sciences* 38(February):289-322.

2 Li W, et al. (2010) Genotyping of human lice suggests multiple emergences of body lice from local head louse populations. *PLOS Neglected Tropical Diseases* 4(3):e641; Centers for Disease Control and Prevention, United States Department of Health and Human Services (2013) Parasites - Lice - Head Lice. http://www.cdc.gov/parasites/lice/head/gen_info/faqs.html.

Spotlight: Buckeye

1 Shapiro AM (1966) *Butterflies of the Delaware Valley. Special Publication of the American Entomological Society* (American Entomological Society, Philadelphia, PA).

Spotlight: Oriental cockroach

1 Rehn JAG (1945) Man's uninvited fellow traveler—The cockroach. *The Scientific Monthly* 61(4):265-276; Roth LM & Willis ER (1956) Parthenogenisis in cockroaches. *Annals of the Entomological Society of America* 49(3):195-204; Ibid (1957) The medical and veterinary importance of cockroaches. *Smithsonian Miscellaneous Collections* 134(10):1-147; Dobney K, Kenward H, Ottaway P, & Donel L (1998) Down, but not out: Biological evidence for complex economic organization in Lincoln in the late fourth century. *Antiquity* 72(276):417-424.

Spotlight: Common blue violet

1 Barton WPC (1818) *Compendium Florae Philadelphicae, Containing A Description Of The Indigenous And Naturalized Plants Found Within A Circuit Of Ten Miles Around Philadelphia* vol. 1 (M. Carey and Son, Philadelphia).

2 Brainerd E (1910) The evolution of new forms in Viola through hybridism. *The American Naturalist* 44(520):229-236. [*Viola papilionacea* discussed here is now regarded as a synonym for *V. sororia*.]

3 Gil-ad NL (1997) Systematics of *Viola* subsection Boreali-Americanae. Boissiera 53:1-130.

Spotlight: Common blue violet – continued

1 Solbrig OT, Newell SJ, & Kincaid DT (1980) The population biology of the genus *Viola:* I. The demography of *Viola sororia*. *Journal of Ecology* 68(2):521-546.

Spotlight: Common groundsel

1 Briggs D (1976) Genecological studies of lead tolerance in groundsel (*Senecio vulgaris* L.). *New Phytologist* 77,(1):173-186; Briggs D (1978) Genecological studies of salt tolerance in groundsel (*Senecio vulgaris* L.) with particular reference to roadside habitats. *New Phytologist* 81(2):381-389; Robinson DE, O'Donovan JT, Sharma MP, Doohan DJ, & Figueroa R (2003) The biology of Canadian weeds. 123. *Senecio vulgaris* L. *Canadian Journal of Plant Science* 83:629-644.

2 Barton WPC (1818) *Compendium Florae Philadelphicae, Containing A Description Of The Indigenous and Naturalized Plants Found Within A Circuit of Ten MIles Around Philadelphia* volume 2 (M. Carey and Son, Philadelphia), 125.

Spotlight: Eastern garter snake

1 Klepp SE & Smith BG eds (2005) *The Unfortunate. The Voyage and Adventures of William Moraley, an Indentured Servant* (The Pennsylvania State University Press, University Park) 2nd ed. (This is an edited version of William Moraley's autobiography originally published in 1743.); Linnaeus, C. (1758) *Systema Naturæ Per Regna Tria Naturæ, Secundum. Classes, Ordines, Genera, Species, Cum Characteribus, Differentiis, Synonymis, Locis. Tomus I. Editio Decima, Reformata.* (Impensis Laurentii Salvii, Stockholm), 222.

Spotlight: House finch

1 Frock Jr. RF (1969) House finch population studies in the Philadelphia area. *Cassinia* (51 [1968-69]):33-44.

2 Badyaev AV, Belloni V, & Hill GE (2012) House finch (*Haemorhous mexicanus*). In *The Birds of North America* online (Cornell Laboratory of Ornithology, New York).

3 Sauer JR, et al. (2014) *The North American Breeding Bird Survey, Results and Analysis 1966 - 2012.* Version 02.19.2014 USGS Patuxent Wildlife Research Center, Laurel, MD.

4 Dhondt A, Tessaglia D, & Slothower R (1998) Epidemic mycoplasmal conjunctivitis in house finches from eastern *North America. Journal of Wildlife Diseases* 34(2):265-280.

Spotlight: Lambsquarters

1 Barton WPC (1818) *Compendium Florae Philadelphicae Containing a Description of the Indigenous and Naturalized Plants Found within a Circuit of Ten Miles around Philadelphia,* vol 1 (M. Carey and Son, Philadelphia), 148.

2 de Schweinitz LD (1828) Remarks on the plants of Europe which have become naturalized in a more or less degree, in the United States. *Annals of The Lyceum of Natural History of New York* 3(1):148-155.

3 van Zeist W (1970) The Oriental Institute Excavations at Mureybi, Syria: Preliminary report on the 1965 campaign part III: The paleobotany. *Journal of Near Eastern Studies* 29(3):167-176.

4 Asingh P & Lynnerup N eds (2007) *Grauballe Man. An Iron Age Bog Body Revisited.* (Jutland Archeological Society and Moesgaard Museum, Moesgaard, Denmark).

5 Johnston A (1962) *Chenopodium album* as a food plant in Blackfoot Indian prehistory. *Ecology* 43(1):129-130.

Spotlight: Milkweed aphid

1 Harrison JS & Mondor EB (2011) Evidence for an invasive aphid "superclone": Extremely low genetic diversity in oleander aphid (*Aphis nerii*) populations in the southern United States. *PLOS ONE* 6(3): e17524. doi:17510.11371/journal.pone.0017524.

Spotlight: Pavement ant

1 McCook HC (1878) The mode of recognition among ants. *Proceedings of the Academy of Natural Sciences of Philadelphia* 30:15-20; Ibid (1879) Combats and nidification of the pavement ant, *Tetramorium cæspitum. Proceedings of the Academy of Natural Sciences of Philadelphia* 31(2):156-161; Weber NA (1965) Note on the European pavement ant, *Tetramorium caespitum,* in the Philadelphia area (Hymenoptera: Formicidae). *Entomological News* 76(5):137-139; Steiner FM, et al. (2008) Combined modelling of distribution and niche in invasion biology: a case study of two invasive *Tetramorium* ant species. *Diversity and Distributions* 14: 538–545.

Spotlight: Red-eared slider

1 Stone W (1906) Notes on reptiles and batrachians of Pennsylvania, New Jersey and Delaware. *The American Naturalist* 40(471):159-170; Ernst CH & Lovich JE (2009) *Turtles of the United States and Canada,* 2nd edition (Johns Hopkins University Press, Baltimore); Stone JE (2010) Distribution and abundance of non-native red-eared slider turtles (*Trachemys scripta elegans*) and native red-bellied turtles (*Pseudemys rubriventris*). Master's Thesis (Drexel University, Philadelphia).

Spotlight: Common ragweed

1 Wodehouse RP (1971) *Hayfever Plants,* second revised edition. (Hafner Publishing Company, New York).

2 Durham, OC (1935) The pollen content of air in North America. *Journal of Allergy* 6: 128-149; Solomon AM & Buell MF (1969) Effects of suburbanization upon airborne pollen. *Bulletin of the Torrey Botanical Club* 96(4):435-445; Port A, Hein J, Wolff A, & Bielory L (2006) Aeroallergen prevalence in the northern New Jersey-New York City metropolitan area: a 15-year summary. *Annals of Allergy, Asthma and Immunology* 96(5):687-691; Levetin E & Avery J (2008) Long term trends in airborne ragweed pollen in Tulsa, Oklahoma: 1987 to 2006. *Journal of Allergy and Clinical Immunology* 121(2, Supplement 1):S21.

Spotlight: Boxelder

1 Barton WPC (1818) *Compendium Florae Philadelphicae Containing A Description Of The Indigenous And Naturalized Plants Found Within A Circuit Of Ten Miles Around Philadelphia* v1 (M. Carey and Son, Philadelphia), 185.

2 Carr R (1831) *Catalogue Of American Trees, Plants And Seeds, Cultivated And For Sale At The Bartram Botanic Garden Near Philadelphia* (Russell & Martien, Philadelphia).

Spotlight: Japanese mazus

1 Pennell FW (1935) *The Scrophulariaceae of Eastern Temperate North America* (Academy of Natural Sciences of Philadelphia, Philadelphia, PA)

Index

A

B

C

D

E

F

G

H

I

J

K

L

M

N

O

P

T

U

V

W

X

Y

Z

www.ingramcontent.com/pod-product-compliance
Ingram Content Group UK Ltd.
Pitfield, Milton Keynes, MK11 3LW, UK
UKHW060017051026
15591UKWH00002B/241
* 9 7 8 0 6 9 2 5 0 6 4 8 6 *